REPRESENTATIONS OF THE ROTATION AND LORENTZ GROUPS AND THEIR APPLICATIONS

I. M. GEL'FAND, R. A. MINLOS
AND
Z. Ya. SHAPIRO

Representations of the rotation and Lorentz groups and their applications

TRANSLATED BY

G. CUMMINS AND T. BODDINGTON

ENGLISH TRANSLATION EDITOR

H. K. FARAHAT
Department of Mathematics
The University of Sheffield

A Pergamon Press Book

THE MACMILLAN COMPANY
NEW YORK
1963

THE MACMILLAN COMPANY
60 Fifth Avenue
New York 11, N.Y.

This book is distributed by
THE MACMILLAN COMPANY
pursuant to a special arrangement with
PERGAMON PRESS LIMITED
Oxford, England

Library of Congress Catalog Card Number 62-9193

This translation has been made from a book written by I. M. Gel'fand, R. A. Minlos, and Z. Ya. Shapiro, entitled "Predstavleniya gruppy vrashchenii i gruppy Lorentsa", published in Moscow, 1958 by Fizmatgiz.

Printed in Great Britain by Latimer, Trend & Co. Ltd., Whitstable

Page

CONTENTS

CONTENTS

Page

Section 6

CONTENTS

PREFATORY NOTE

This book is devoted to the description and detailed study of the representations of the rotation group of three-dimensional space and of the Lorentz group.

These groups are of fundamental importance in theoretical physics. Bearing in mind the requirements of physicists the authors have gathered into this book all the basic material of the theory of representations which is used in quantum mechanics.

The book is also designed for mathematicians studying the representations of Lie groups. For them the book can serve as an introduction to the general theory of representations.

PREFACE

The present book is devoted to a study of the rotation group of three-dimensional space and of the Lorentz group. The reader is assumed to be acquainted with the fundamentals of linear algebra, for instance to the extent of the first two chapters of Gelfand's book "Lectures on Linear Algebra".

The theory of representations, in particular of the three-dimensional rotation group and the Lorentz group, is used extensively in quantum mechanics. In this book we have gathered together all the fundamental material which, in our view, is necessary to quantum mechanical applications.

On the other hand, the study of the representations of the three-dimensional rotation group and the Lorentz group can serve as a useful introduction to the general representation theory of Lie groups; these two examples are all the more fortunate inasmuch as they clearly illustrate the difference between the representations of compact groups (the rotation group) and locally compact Lie groups (the Lorentz group). Moreover, from the material presented in the book the connexions between the theory of representations and other branches of mathematics (spherical functions, tensors, differential equations, etc.) emerge fairly clearly; these connexions have not always been well studied.

The first part of the book which is devoted to the rotation group is taken from a paper published by two of the authors, Gelfand and Shapiro, in "Uspekh-akh matematicheskikh nauk" (Advances in the mathematical sciences) for 1952 (vol. VII, No.1) under the title "Representations of the group of rotations of three-dimensional space and their applications".

We have added to this paper "paragraphs"* 6 and 7, section 9 and we have rewritten section 10 in which the Clebsch-Gordan coefficients are evaluated.

The second part of the book on the representations of the Lorentz group and relativistic-invarient equations was written by Minlos. The choice of the material and also the general plan and style of the exposition, however, were discussed in detail by all three authors. The last chapter (on relativistic-invariant equations) is based on the work by Gelfand and Yaglom: "General relativistic-invariant equations and infinite representations of the Lorentz group" (JETP, vol.18, No.8, 1948): this chapter can therefore be considered as a detailed and rather more complete presentation of the above paper.

* Translation Editors' Note: Each part of the book is divided into "sections" and each section is divided into subsections (§) which the authors call "paragraphs".

The inclusion in the book of relativistic-invariant equations is justified by their intrinsic interest and by the fact that the methods applied here are used extensively in the previous chapter for the study of the representations of the Lorentz group: thus each of the chapters of Part Two is complete in itself. In Part Two we lay the emphasis on finite representations since it is principally these which have been up to the present time essential for physical applications.

To the reader wishing to study the representations of the Lorentz group more thoroughly and at greater length we recommend the book by M.A.Naimark entitled "Representations of the Lorentz group"*.

The authors consider it their duty to mention the large amount of work which was done by the book's editor F.A.Berezin, which went far beyond the limits of ordinary editorial duties. His innumerable requests and his advice and comments have considerably raised the quality of the book. We thank him.

I.GELFAND
R.MINLOS
Z.SHAPIRO

---

* Translation Editor's Note: A translation of this book is in preparation by Pergamon Press.

PART I

REPRESENTATIONS OF THE GROUP OF ROTATIONS OF THREE-DIMENSIONAL SPACE

CHAPTER I

THE ROTATION GROUP AND ITS REPRESENTATIONS

Section 1. The Group of Rotations in Three-dimensional Space

§ 1. Definition of the Group of Rotations.

We consider the set of all rotations of three-dimensional space about a fixed point (the origin of coordinates). By the product of two rotations g_1 and g_2 we understand the rotation $g = g_1 g_2$ which consists in carrying out first the rotation g_2, followed by the rotation g_1.* It is not difficult to show that the set G^+ of all such rotations forms a group with this definition of multiplication; all the group axioms are satisfied. The unit element e of the group, the identity rotation, consists in a rotation through a zero angle.

The rotation g will transform each vector x into a vector x'. We shall express this by the equation

$$x' = gx. \tag{1}$$

We now proceed to an analytical description of rotations. We take a fixed orthogonal coordinate system in three-dimensional space. A rotation will then be described by equations

$$x'_i = \sum_{k=1}^{3} g_{ik} x_k, \tag{2}$$

where the x_k, x'_k are the components of the vectors x and x' in this coordinate system. The rotation is completely defined by the matrix $\|g_{ik}\|$ and we shall use the same letter g to denote this matrix. We shall now determine the conditions which must be satisfied by the elements g_{ik} of the matrix of a rotation. Since a rotation does not alter lengths or angles it must leave the scalar product of any two vectors invariant. Thus if $x' = gx$ and $y' = gy$, then

$$\sum_{i=1}^{3} x'_i y'_i = \sum_{k=1}^{3} x_k y_k. \tag{3}$$

Substituting on the left hand side of this equation for x'_i, y'_i as given by (2) we obtain

---

* This is the order usually followed in the multiplication of linear transformations.

$$\sum_{i,k,l} g_{ik} g_{il} x_k y_l = \sum_{k=1}^{3} x_k y_k.$$

Equating the coefficients of $x_k y_l$ we obtain:

$$\sum_{i=1}^{3} g_{ik} g_{il} = \delta_{kl}, \tag{4}$$

where δ_{kl} is equal to one if $k = l$ and is equal to zero if $k \neq l$. Equation (4) may be written in matrix form. On the right-hand side of the equation we have the $(k, l)^{\text{th}}$ element of the unit matrix e, and on the left-hand side the corresponding element of the product $g'g$ of the transpose g' of g with g itself. Hence

$$g'g = e \tag{5}$$

or

$$g' = g^{-1}. \tag{5'}$$

Matrices which satisfy equation (5) are called orthogonal matrices. If we take determinants on both sides of (5) we obtain $\operatorname{Det}(g') \cdot \operatorname{Det}(g) = 1$, i.e. $[\operatorname{Det}(g)]^2 = 1$, and hence

$$\operatorname{Det}(g) = \pm 1. \tag{6}$$

Every rotation is a rotation about some axis through a suitable angle φ. If we take this axis to be the axis of z, then the matrix of the rotation will have the form

$$\left\| \begin{matrix} \cos\varphi & -\sin\varphi & 0 \\ \sin\varphi & \cos\varphi & 0 \\ 0 & 0 & 1 \end{matrix} \right\|. \tag{7}$$

Since the matrix (7) has a determinant equal to unity, it follows that, for a rotation g, $\operatorname{Det}(g) = +1$. Orthogonal transformations g for which $\operatorname{Det}(g) = -1$, are called improper orthogonal transformations. An example of an improper orthogonal transformation is given by the matrix

$$g_- = \left\| \begin{matrix} -1 & 0 & 0 \\ 0 & -1 & 0 \\ 0 & 0 & -1 \end{matrix} \right\|,$$

which represents reflection in the origin of the coordinate system. If g is an improper orthogonal transformation, $\operatorname{Det}(g) = -1$, then $\operatorname{Det}(gg_-) = \operatorname{Det}(g) \cdot \operatorname{Det}(g_-) = +1$. Hence gg_- is a rotation and $g = (gg_-)g_-$ is a product of a rotation and a reflection in the origin.

§ 2. Parametrization of the Rotation Group

In the sequel we shall need a number of parametric representations of rotations. A rotation may be completely specified by giving the angle of rotation and the axis about which it takes place. We will frequently represent a rotation by a vector $\xi = (\xi_1, \xi_2, \xi_3)$, directed along the axis of rotation and equal in magnitude to the angle of rotation. The sense of the vector will be chosen in such a way that the angle of rotation does not exceed π. Thus the components of a vector ξ describing a rotation will satisfy the condition $\xi_1^2 + \xi_2^2 + \xi_3^2 \leqslant \pi^2$; and the points (ξ_1, ξ_2, ξ_3) will be contained in a sphere of radius π. It is clear that different rotations correspond to different interior points of this sphere, and that points at opposite ends of a diameter denote one and the same rotation through an angle π (since rotations through an angle π in two opposite directions give rise to one and the same result).

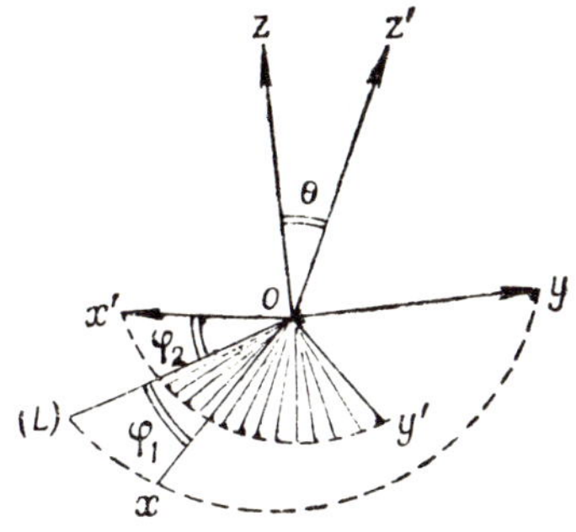

Fig. 1

This method of describing the rotation group exhibits the topological structure of the group, in that the group is topologically equivalent to a sphere in which diametrically opposite points (of the shell) are identified.

One important parametric form for the group of rotations uses the so-called Euler angles. Let the coordinate axes Ox, Oy, Oz be transformed as a result of the rotation g into the lines Ox', Oy', Oz'.

Denote the line of intersection of the planes xOy, $x'Oy'$ by (L). Suppose that (L) makes angles φ_1, φ_2 with Ox, Ox' respectively, and that θ is the angle between Oz and Oz'. It is then plain that the rotation g is equivalent to the product of three rotations, namely a rotation g_{φ_1} through angle φ_1 about Oz, carrying Ox to (L), then a rotation g_θ through an angle θ about (L) which takes Oz to Oz' and finally a rotation g_{φ_2} through angle φ_2 about Oz':

$$g(\varphi_1, \theta, \varphi_2) = g_{\varphi_2} g_\theta g_{\varphi_1}. \tag{8}$$

We are now in a position to determine explicitly the elements g_{ik} of the matrix of the rotation $g(\varphi_1, \theta, \varphi_2)$ in terms of the Euler angles. To do this we note that the matrices of g_{φ_1}, g_θ, g_{φ_2} with respect to the successive positions of the coordinate axes are respectively as follows

$$\begin{Vmatrix} \cos\varphi_1 & -\sin\varphi_1 & 0 \\ \sin\varphi_1 & \cos\varphi_1 & 0 \\ 0 & 0 & 1 \end{Vmatrix}, \qquad \begin{Vmatrix} 1 & 0 & 0 \\ 0 & \cos\theta & -\sin\theta \\ 0 & \sin\theta & \cos\theta \end{Vmatrix}$$

$$\left\| \begin{matrix} \cos\varphi_2 & -\sin\varphi_2 & 0 \\ \sin\varphi_2 & \cos\varphi_2 & 0 \\ 0 & 0 & 1 \end{matrix} \right\|.$$

The matrix of g is then obtained by multiplying these matrices in the reverse order. Hence:

$$g(\varphi_1, \theta, \varphi_2) = \left\| \begin{matrix} \cos\varphi_1\cos\varphi_2 - \cos\theta\sin\varphi_1\sin\varphi_2, & -\cos\varphi_1\sin\varphi_2 - \cos\theta\sin\varphi_1\cos\varphi_2 & \sin\varphi_1\sin\theta \\ \sin\varphi_1\cos\varphi_2 + \cos\theta\cos\varphi_1\sin\varphi_2, & -\sin\varphi_1\sin\varphi_2 + \cos\theta\cos\varphi_1\cos\varphi_2 & -\cos\varphi_1\sin\theta \\ \sin\varphi_2\sin\theta, & \cos\varphi_2\sin\theta & \cos\theta \end{matrix} \right\|. \tag{9}$$

The angles φ_1 and φ_2 lie between 0 and 2π, the angle θ between 0 and π. Different sets of three numbers lying within these ranges correspond to different rotations apart from the case for which $\theta = 0$ or π. If $\theta = 0$ the rotation is one about Oz through an angle $\varphi_1 + \varphi_2$ and if $\theta = \pi$ it is a rotation about Oz through an angle $\varphi_1 - \varphi_2$ so that in these cases different values for φ_1 and φ_2 may correspond to the same rotation.

Let us consider the rotation $g(\varphi_1, \theta, \varphi_2)$ whose matrix is given by (9). We see easily that the inverse transformation is given by the angles $\pi - \varphi_2$, θ, $\pi - \varphi_1$. For if we substitute $\pi - \varphi_2$ for φ_1 and $\pi - \varphi_1$ for φ_2 in the matrix (9), we obtain its transpose and we know from (5) that the inverse of an orthogonal matrix is the same as its transpose. Hence if the rotation g is given by the angles φ_1, θ, φ_2, then the rotation g^{-1} is given by the angles $\pi - \varphi_2$, θ, $\pi - \varphi_1$.

§ 3. Invariant Integration

For the study of functions of the elements of the group G (i.e. functions of φ_1, θ, φ_2) we need to consider the integrals of these functions over the group (i.e. over all permissible values of φ_1, θ, φ_2) $0 \leqslant \varphi_1 < 2\pi$, $0 \leqslant \theta \leqslant \pi$, $0 \leqslant \varphi_2 < 2\pi$). The most useful form for the theory is that known as invariant integration. The invariant integral of the function $f(g) \equiv f(\varphi_1, \theta, \varphi_2)$ is the integral*

$$\int f(g)\,dg = \int f(\varphi_1, \theta, \varphi_2)\, I(\varphi_1, \theta, \varphi_2)\, d\varphi_1\, d\theta\, d\varphi_2$$

over φ_1, θ, φ_2 where the "weight function" $I(\varphi_1, \theta, \varphi_2)$ is chosen so that

* We will sometimes write:

$$dg \equiv I\, d\varphi_1\, d\theta\, d\varphi_2.$$

$$\int f(gg_0)\,dg = \int f(g)\,dg. \tag{10}$$

Thus the invariant integral of the function $f(g)$ remains unaltered if we substitute gg_0 for the argument g, i.e. if we "displace" or "translate" the function $f(g)$. It may be proved that the weight function defined by the condition of invariance (10) is determined to within multiplication by a constant factor. This factor is chosen such that the integral of the function $f(g) \equiv 1$ is unity, i.e.

$$\int dg = 1.$$

Instead of the Euler angles we could have used any other parametric representation for the rotation group, for defining the invariant integral. The Euler-angles however are the most convenient.

Let us consider the rotation g, given by the angles φ_1, θ, φ_2. We denote by P the point on the unit sphere which is carried by the rotation g into the North Pole (i.e. the point where the sphere meets the axis Oz). By Q we denote the point on the same sphere which after the rotation g lies on the axis Ox. It is clear that the rotation is completely specified by the points P and Q. If the point P is given then Q will lie on the great circle whose plane is perpendicular to the radius OP.

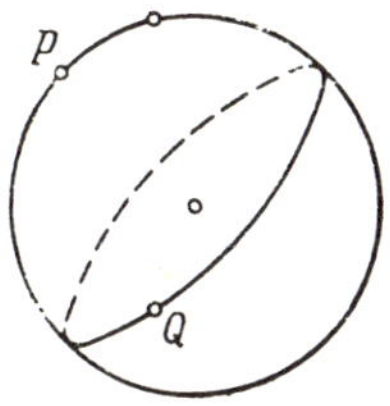

Fig. 2

The elements of the third row of the matrix (9) are the Cartesian coordinates of P. Hence its spherical coordinates are $\frac{\pi}{2} - \varphi_2$ and θ.

If g_0 is some other rotation and $\tilde{g} = gg_0$, then the points $\tilde{P}$ and $\tilde{Q}$ corresponding to the rotation $\tilde{g}$ are obtained from P and Q by the rotation g_0^{-1}.

We will show that the invariant integral is given by the formula

$$\int f(g)\,dg = \int_0^{2\pi}\int_0^{\pi}\int_0^{2\pi} f(\varphi_1, \theta, \varphi_2) \sin\theta \, d\varphi_1 \, d\theta \, d\varphi_2,$$

More precisely we show that, with this definition of the integral, the following formula hold

$$\int f(g)\,dg = \int f(gg_0)\,dg.$$

Denote by $\tilde{\varphi}_1$, $\tilde{\theta}$, $\tilde{\varphi}_2$ the Euler angles of the rotation $\tilde{g}$. We will show that

$$\sin\theta\, d\varphi_1\, d\theta\, d\varphi_2 = \sin\tilde{\theta}\, d\tilde{\varphi}_1\, d\tilde{\theta}\, d\tilde{\varphi}_2,$$

that is

$$dg = d\tilde{g}.$$

The expression $\sin\theta\, d\varphi_1\, d\theta\, d\varphi_2$ has a simple geometrical significance. In fact, $\sin\theta\, d\varphi_2\, d\theta$ is an infinitesimal square on the unit sphere at the point P, and $d\varphi_1$ defines an infinitesimal arc on the great circle which contains Q. If, then, we change the angle φ_1 by an amount $d\varphi_1$ without changing φ_2 and θ (i.e. fixing the position of the point P) this will give rise to an additional rotation about P through an angle $d\varphi_1$, i.e. a translation of Q a distance $d\varphi_1$. As was shown above the replacement of g by $\tilde{g} = gg_0$ results in an additional rotation of the pair P, Q through g_0. But this rotation will alter neither the infinitesimal square nor the infinitesimal arc; hence we have

$$\sin\theta\, d\varphi_1\, d\theta\, d\varphi_2 = \sin\tilde{\theta}\, d\tilde{\varphi}_1\, d\tilde{\theta}\, d\tilde{\varphi}_2.$$

It remains to prove the equation

$$\int f(\tilde{g})\, dg = \int f(g)\, dg,$$

or, what is the same thing,

$$\int f(\tilde{g})\, dg = \int f(\tilde{g})\, d\tilde{g}.$$

But this follows from the result just proved: $d\tilde{g} = dg$. If we use the remaining normalizing condition that

$$\int 1\, dg = 1,$$

then, since

$$\int_0^{2\pi}\int_0^{\pi}\int_0^{2\pi} \sin\theta\, d\varphi_1\, d\theta\, d\varphi_2 = 8\pi^2,$$

we find

$$\int f(g)\, dg = \frac{1}{8\pi^2}\int_0^{2\pi}\int_0^{\pi}\int_0^{2\pi} f(\varphi_1, \theta, \varphi_2)\sin\theta\, d\varphi_1\, d\theta\, d\varphi_2. \qquad (11)$$

§ 4. The Connexion between the Rotation Group and the Group of 2 by 2 Unitary Matrices

At this point we show that the group of rotations in three-dimensional space can be represented by complex matrices of the second order. In order to do this we consider the stereographic projection of a sphere on to a plane, setting up a correspondence between the points of the plane and the points of the sphere such that each point P of the sphere corresponds to the point M of the plane which lies on the line $O'P$ (O' is the North Pole). Every rotation about the centre of the sphere of three-dimensional space carries points on the sphere to points on the sphere, and therefore corresponds to a transformation in the plane. Our problem is to examine this transformation in more detail. We will consider a sphere with diameter 1. From the properties of similar triangles we easily obtain the relation between the coordinates x, y, z of the point P of the sphere and the coordinates ξ, η of the point M of the plane:

$$\xi = \frac{x}{\frac{1}{2} - z}, \qquad \eta = \frac{y}{\frac{1}{2} - z}.$$

It is convenient to introduce here the complex notation $\zeta = \xi + i\eta$. Then

$$\zeta = \xi + i\eta = \frac{x + iy}{\frac{1}{2} - z}. \tag{12}$$

Since $x^2 + y^2 = \frac{1}{4} - z^2$ for points on the sphere, we may also write this in the form:

$$\zeta = \frac{x^2 + y^2}{\left(\frac{1}{2} - z\right)(x - iy)} = \frac{\frac{1}{2} + z}{x - iy}. \tag{12'}$$

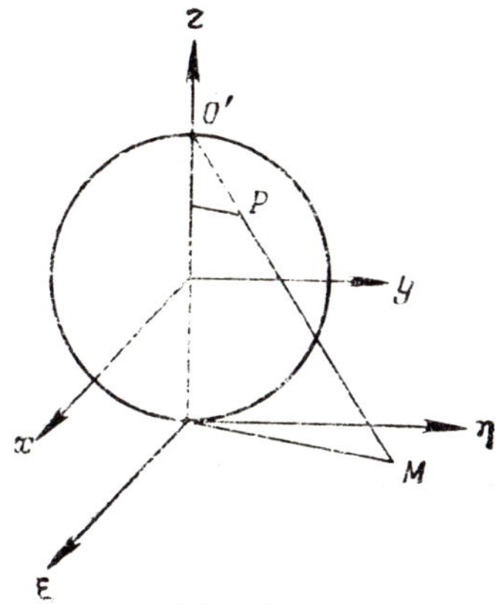

Fig. 3

We will determine the transformation of the plane corresponding to a rotation about Oz through an angle φ. We have

$$x' = x\cos\varphi - y\sin\varphi,$$

$$y' = x\sin\varphi + y\cos\varphi,$$

$$z' = z.$$

Hence

$$\zeta' = \frac{x' + iy'}{\frac{1}{2} - z} = \frac{e^{i\varphi}(x+iy)}{\frac{1}{2} - z} = e^{i\varphi}\zeta,$$

i.e. such a rotation corresponds to a plane transformation of the form

$$\zeta' = e^{i\varphi}\zeta. \tag{13}$$

We now consider a rotation about Ox through an angle θ. By an analogous argument we find that the effect of this rotation is to multiply the expression

$$w = \frac{y + iz}{\frac{1}{2} - x}$$

by $e^{i\theta}$, i.e.

$$w' = e^{i\theta} w. \tag{14}$$

It remains for us to express w in terms of ζ (and similarly w' in terms of ζ'). Consider the expression

$$\frac{w+i}{w-i} = \frac{\dfrac{y+iz}{\frac{1}{2}-x} + i}{\dfrac{y+iz}{\frac{1}{2}-x} - i} = \frac{-(x+iy) + \left(\frac{1}{2} + z\right)}{(x-iy) - \left(\frac{1}{2} - z\right)}.$$

Using formulae (12) and (12') we obtain,

$$\frac{w+i}{w-i} = \frac{-\zeta\left(\frac{1}{2} - z\right) + \left(\frac{1}{2} + z\right)}{\frac{1}{\zeta}\left(\frac{1}{2} + z\right) - \left(\frac{1}{2} - z\right)} = \zeta$$

and similarly

$$\frac{w'+i}{w'-i} = \zeta'.$$

Hence, expressing w and w' in terms of ζ and ζ' and substituting in formula (14) we find that

$$\frac{\zeta'+1}{\zeta'-1} = e^{i\theta}\frac{\zeta+1}{\zeta-1}.$$

Solving this equation for ζ' we obtain the transformation corresponding to a rotation through an angle θ about Ox:

$$\zeta' = \frac{\zeta(e^{i\theta}+1) + (e^{i\theta}-1)}{\zeta(e^{i\theta}-1) + (e^{i\theta}+1)} = \frac{\zeta\cos\frac{\theta}{2} + i\sin\frac{\theta}{2}}{i\zeta\sin\frac{\theta}{2} + \cos\frac{\theta}{2}}. \tag{15}$$

We see from this that rotations about the axes Ox and Oz correspond to bilinear transformations in the plane of ζ. It is clear also that to the product of such transformations in the plane there corresponds the product of the rotations defining them. Since every rotation may be obtained as a product of rotations about the axes Oz and Ox (see § 2) then to each rotation there corresponds a product of transformations of the forms (13) and (15) i.e. a bilinear transformation

$$\zeta' = \frac{\alpha\zeta + \beta}{\gamma\zeta + \delta}. \tag{16}$$

The bilinear transformation (16) is uniquely defined by the following complex 2 by 2 matrix:

$$\begin{Vmatrix} \alpha & \beta \\ \gamma & \delta \end{Vmatrix}. \tag{17}$$

Since ζ' in equation (16) does not change on multiplication of the numerator and denominator of the right-hand side by the same number, then by multiplying α, β, γ, δ by $\pm \frac{1}{\sqrt{\alpha\delta - \beta\gamma}}$ we may ensure that the determinant of (17) is equal to 1.* Thus to every rotation there corresponds a matrix of the form (17), defined except for a possible minus sign, for which

$$\alpha\delta - \beta\gamma = 1.$$

In particular, we can write down the matrices corresponding to the rotations g_φ and g_θ about the axes Oz and Ox. To the rotation g_θ corresponds the matrix

$$g_\theta \sim \begin{Vmatrix} \cos\frac{\theta}{2} & i\sin\frac{\theta}{2} \\ i\sin\frac{\theta}{2} & \cos\frac{\theta}{2} \end{Vmatrix}. \tag{18}$$

To the rotation g_φ there corresponds the transformation $\zeta' = e^{i\varphi}\zeta$. Writing this in the form $\zeta' = \frac{e^{i\frac{\varphi}{2}}\zeta}{e^{-i\frac{\varphi}{2}}}$, we obtain the matrix of the transformation:

$$g_\varphi \sim \begin{Vmatrix} e^{i\frac{\varphi}{2}} & 0 \\ 0 & e^{-i\frac{\varphi}{2}} \end{Vmatrix} \tag{19}$$

the determinant of which is equal to unity.

The rotation g with Euler angles φ_1, θ, φ_2 may be written as the product of rotations: $g = g_{\varphi_2} g_\theta g_{\varphi_1}$. Since the successive application of bilinear transformations the matrices must be multiplied in the reverse order, the matrix

---

* This condition determines the coefficients of the bilinear transformation to within a sign.

of the rotation g is given by

$$g \sim \begin{Vmatrix} e^{i\frac{\varphi_1}{2}} & 0 \\ 0 & e^{-i\frac{\varphi_1}{2}} \end{Vmatrix} \begin{Vmatrix} \cos\frac{\theta}{2} & i\sin\frac{\theta}{2} \\ i\sin\frac{\theta}{2} & \cos\frac{\theta}{2} \end{Vmatrix} \begin{Vmatrix} e^{i\frac{\varphi_2}{2}} & 0 \\ 0 & e^{-i\frac{\varphi_2}{2}} \end{Vmatrix} = \tag{20}$$

$$= \begin{Vmatrix} \cos\frac{\theta}{2} e^{i\frac{\varphi_1+\varphi_2}{2}} & i\sin\frac{\theta}{2} e^{-i\frac{\varphi_2-\varphi_1}{2}} \\ i\sin\frac{\theta}{2} e^{i\frac{\varphi_2-\varphi_1}{2}} & \cos\frac{\theta}{2} e^{-i\frac{\varphi_1+\varphi_2}{2}} \end{Vmatrix}.$$

The matrices (18) and (19) are unitary matrices and their determinants are equal to unity. Therefore the matrix (20), being a product of such matrices, is also unitary and has its determinant equal to unity.

We will now show conversely that every unitary matrix (17), whose determinant is equal to 1, corresponds to a rotation. Since the matrix is unitary

$$\alpha\bar{\gamma} + \beta\bar{\delta} = 0, \quad \alpha\bar{\alpha} + \beta\bar{\beta} = 1, \quad \gamma\bar{\gamma} + \delta\bar{\delta} = 1$$

and we have $\alpha\delta - \gamma\beta = 1$. From this it follows simply that $\delta = \bar{\alpha}$, $\gamma = -\bar{\beta}$. Therefore any matrix of the type under consideration may be written in the form

$$\begin{Vmatrix} \alpha & \beta \\ -\bar{\beta} & \bar{\alpha} \end{Vmatrix}, \tag{21}$$

where

$$|\alpha|^2 + |\beta|^2 = 1. \tag{21'}$$

It is easily seen that the matrix (21), with the condition 21') may be written in the form (20) by making the substitution

$$\cos\frac{\theta}{2} = |\alpha|, \quad \sin\frac{\theta}{2} = |\beta|^*$$

and defining the angles φ_1 and φ_2 by the equations

$$\frac{\varphi_1 + \varphi_2}{2} = \arg\alpha,$$

$$-\frac{\varphi_2 - \varphi_1}{2} + \frac{\pi}{2} = \arg\beta.$$

Thus every rotation can be determined by two complex numbers, satisfying the condition (21') or equivalently by four real numbers, the sum of the squares of which is equal to one.

We have shown, then, that to every unitary matrix of the second order with determinant unity there corresponds a rotation in three-dimensional space. Conversely, as was shown earlier, to each rotation there corresponds two such

* Recall that $0 \leqslant \theta \leqslant \pi$

matrices, differing only in sign. The correspondence established above between rotations and bilinear transformations is unique. On the other hand we see that every bilinear transformation is defined by two matrices with determinants equal to unity. So there correspond two matrices of the form (21) to each rotation. It does not follow however that we can rid ourselves of the dual nature of this correspondence by an appropriate choice of sign. Consider for example a rotation through an angle φ about Oz. To this there corresponds the matrix

$$\begin{Vmatrix} e^{i\frac{\varphi}{2}} & 0 \\ 0 & e^{-i\frac{\varphi}{2}} \end{Vmatrix}.$$

In particular, to the unit rotation ($\varphi = 2k\pi$) we have associated the two matrices

$$\pm \begin{pmatrix} 1 & 0 \\ 0 & 1 \end{pmatrix}. \tag{22}$$

If we substitute for this just the matrix

$$\begin{pmatrix} 1 & 0 \\ 0 & 1 \end{pmatrix},$$

then by changing the angle φ continuously from 0 to 2π we shall arrive at the position obtained from the matrix

$$\begin{pmatrix} -1 & 0 \\ 0 & -1 \end{pmatrix}.$$

Thus, if we do not wish to destroy the continuity, we must regard the unit rotation as being associated with the two matrices (22).

We write down the matrix of the rotation g in terms of the complex parameters α, β

$$\|g_{ik}\| = \begin{Vmatrix} \frac{1}{2}(\alpha^2 - \beta^2 + \bar{\alpha}^2 - \bar{\beta}^2), & \frac{i}{2}(-\alpha^2 - \beta^2 + \bar{\alpha}^2 + \bar{\beta}^2), & -\alpha\beta - \bar{\alpha}\bar{\beta} \\ \frac{i}{2}(\alpha^2 - \beta^2 - \bar{\alpha}^2 + \bar{\beta}^2), & \frac{1}{2}(\alpha^2 + \beta^2 + \bar{\alpha}^2 + \bar{\beta}^2), & -i(\alpha\beta - \bar{\alpha}\bar{\beta}) \\ \alpha\bar{\beta} + \bar{\alpha}\beta, & i(-\alpha\bar{\beta} + \bar{\alpha}\beta), & \alpha\bar{\alpha} - \beta\bar{\beta} \end{Vmatrix}. \tag{23}$$

To prove this we substitute for α and β the elements of the matrix (20). After a few simple transformations we obtain the expression (9) of the matrix of the rotation in terms of the Euler angles.

If we put $\alpha = \alpha_1 + i\alpha_2$, $\beta = \beta_1 + i\beta_2$, then since $|\alpha|^2 + |\beta|^2 = 1$ we may also consider the group of rotations in three-dimensional space as corresponding to the sphere in four-dimensional space

$$\alpha_1^2 + \beta_1^2 + \alpha_2^2 + \beta_2^2 = 1 \tag{24}$$

In this, points at opposite ends of a diameter represent one and the same rotation.

§ 5. The Notion of a Representation of the Group of Rotations

At this point we give some new definitions to be used later. We shall say that we have a (finite dimensional) representation $g \to T_g$ of the group if to each element g of the group there corresponds a linear transformation T_g of finite-dimensional linear space R*, such that the product of elements of the group corresponds to the product of their transformations, and to the unit element e of the group there corresponds the identity transformation, i.e.

if
$$T_{g_1} T_{g_2} = T_{g_1 g_2} \tag{25}$$

and
$$T_e = E. \tag{26}$$

Since in a finite dimensional space each linear transformation is given by a matrix, a finite dimensional representation may be defined as a correspondence between the elements g of the group and matrices T_g such that conditions (25) and (26) are satisfied.

The representation $g \to T_g$ is called continuous if the elements of the matrix T_g are continuous functions of g. In future we will be concerned only with continuous representations. A trivial example of a group representation is the correspondence in which every element is associated with the unit matrix. Such a representation is termed primary. Another example of a group representation, the fundamental representation, is obtained by assigning to each rotation its own matrix, determined in terms of some basis.

The subspace R_1 of the space R (in which the linear transformations T_g take place) is called invariant with respect to the representation $g \to T_g$, if R_1 is invariant with respect to all the transformations T_g.

If, in the space R there does not exist a subspace** which is invariant with respect to the representation $g \to T_g$, then this representation is called irreducible. We will see later that the study of any representation of the group of rotations leads us to a study of irreducible representations. These last will be classified in Sect.2, and the matrices T_g of irreducible representations will b given in Sect.7. The various special forms of representations of the group of rotations (i.e. the various special forms of the space R and the transformations T_g will be studied in Sects. 3, 5, 6, 8.

The representation $g \to T_g$ is called unitary if each of the linear transformations T_g is unitary with respect to some scalar product defined in the complex space R. We shall prove that each representation of the group of rotations is unitary in the sense that it is possible to define a scalar product

* The elements of the space R will in future be referred to as the quantities transformed by the representation $g \to T_g$.

** Excluding of course the whole space R and the null subspace, which may be considered formally as subspaces for any representation.

such that T_g is a unitary linear transformation for each g. To do this, we consider in R any scale product (ξ, η) where ξ, η are elements of R. In general, T_g will not be unitary with respect to this scalar product i.e. $(T_g\xi, T_g\eta)$ will not be the same as (ξ, η). We average the function $(T_g\xi, T_g\eta)$ over the group i.e. we consider the expression

$$\int (T_g\xi, T_g\eta)\, dg,$$

where the integration is carried out in the way defined in paragraph 3. Let us now define a new scalar product by the formula

$$(\xi, \eta)_1 = \int (T_g\xi, T_g\eta)\, dg.$$

We show that $(\xi, \eta)_1$ has all the properties of a scalar product. In fact

$$(\xi_1 + \xi_2, \eta)_1 = \int (T_g(\xi_1 + \xi_2), T_g\eta)\, dg =$$

$$= \int (T_g\xi_1, T_g\eta)\, dg + \int (T_g\xi_2, T_g\eta)\, dg = (\xi_1, \eta)_1 + (\xi_2, \eta)_1.$$

Similarly it may be shown that $(\xi, \eta)_1 = (\eta, \xi)_1$ and that $(\lambda\xi, \eta)_1 = \lambda(\xi, \eta)_1$. Moreover, $(\xi, \xi)_1 = \int (T_g\xi, T_g\xi)\, dg > 0$ for $\xi \neq 0$ since $(T_g\xi, T_g\xi) > 0$. We show that the transformation T_g is unitary with respect to the scalar product $(\xi, \eta)_1$ i.e. that $(T_{g_0}\xi, T_{g_0}\eta)_1 = (\xi, \eta)_1$. Let $T_{g_0}\xi = \xi'$ and $T_{g_0}\eta = \eta'$,

$$(T_{g_0}\xi, T_{g_0}\eta)_1 = (\xi', \eta')_1 = \int (T_g\xi', T_g\eta')\, dg =$$

$$= \int (T_g T_{g_0}\xi, T_g T_{g_0}\eta)\, dg = \int (T_{gg_0}\xi, T_{gg_0}\eta)\, dg.$$

But since the integration is invariant (i.e. $\int f(gg_0)\, dg = \int f(g)\, dg$) we also have:

$$\int (T_{gg_0}\xi, T_{gg_0}\eta)\, dg = \int (T_g\xi, T_g\eta)\, dg = (\xi, \eta)_1,$$

and, finally,

$$(T_{g_0}\xi, T_{g_0}\eta)_1 = (\xi, \eta)_1.$$

Thus we have shown that the representation $g \to T_g$ is unitary with respect to the scalar product $(\xi, \eta)_1$. The study of unitary representations of the group may be founded on the study of its irreducible unitary representations. Let us consider the unitary representation $g \to T_g$. If there exists in R no invariant subspace, then the representation is irreducible. If, however, there does exist in R an invariant subspace R_1, then the set of vectors orthogonal to all the vectors of R_1 also forms an invariant subspace. The whole space R has thus been resolved into a linear sum of two mutually orthogonal subspaces R_1 and R_2. If the representation is reducible either in R_1 or R_2 we resolve it further, until we obtained an irreducible representation.

So far we have been concerned with finite dimensional representations of the group of rotations. However, in the first part of this book (for example, in Sect. 3, § 5; Sect. 7, § 2) we will also meet with infinite dimensional unitary representations. We say that we have an infinite unitary representation if to each element of the group g there corresponds a unitary transformation T_g in Hilbert space (infinite dimensional Euclidean space) such that conditions (25) and (26) are satisfied. The representation is called continuous if for all vectors ξ, η $(T_g\xi, \eta)$ is a continuous function of g.

The theorem on the resolution of a representation into irreducible representations holds also for infinite dimensional representations of the group of rotations. We state this theorem without proof. Let there be given a unitary representation $g \rightarrow T_g$ of the group of rotations in a (separable) Hilbert space R. Then there exist finite dimensional subspaces $R_1, R_2, \ldots, R_n, \ldots,$ invariant with respect to T_g in each of which the representation T_g is irreducible. These subspaces R_i are mutually orthogonal and the sum of the R_i is the whole space R. This means that every vector in R may be expressed as a convergent series*

$$\xi = \xi_1 + \xi_2 + \ldots + \xi_n + \ldots,$$

where the vectors ξ_i belong to the invariant subspaces R_i.

Section 2. Infinitesimal Rotations and the Determination of the Irreducible Representations of the Group of Rotations

In this paragraph we consider the representations of the group of rotations. We already know that any such representation may be regarded as being unitary, and so we may limit ourselves to the study of the unitary representations of the group.

First we will determine all the irreducible representations and then show how an arbitrary representation may be resolved into irreducible representations.

§ 1. Definition of the Matrices A_k, Corresponding to Infinitesimal Rotations

Suppose we are given a unitary representation of the rotation group G. This means that there is associated with every element g a unitary matrix $T_g = \|a_{ik}(g)\|$ such that to the product of the rotations g_1 and g_2 there corresponds the product of the matrices T_{g_1} and T_{g_2}, i.e.

$$T_{g_1 g_2} = T_{g_1} T_{g_2}. \tag{1}$$

---

* The series $\xi_1 + \ldots + \xi_n \ldots$ is said to converge to ξ if the sequence $S_n = \xi_1 + \cdots + \xi_n$ has the limit ξ, i.e. if

$$(\xi - S_n, \xi - S_n) \rightarrow 0 \text{ as } n \rightarrow \infty.$$

In particular, to the identity rotation e there corresponds the unit matrix E. For the parameters defining the rotation g we take the coordinates ξ_1, ξ_2, ξ_3 of the vector, parallel to the axis of the rotation, whose length is equal to the angle of rotation (these parameters were introduced in a previous paragraph). Then the matrix T_g is a function of these parameters i.e. $T_g = T(\xi_1, \xi_2, \xi_3)$. It may be proved that T_g has derivatives of all orders with respect to ξ_1, ξ_2, ξ_3 *. Since the vector $\xi_1 = \xi_2 = \xi_3 = 0$ corresponds to a rotation through angle zero i.e. to the identity rotation,

$$T(0, 0, 0) = E. \tag{2}$$

Expand $T(\xi_1, \xi_2, \xi_3)$ about the point $\xi_1 = \xi_2 = \xi_3 = 0$ by Taylor's theorem. Then

$$T(\xi_1, \xi_2, \xi_3) = E + A_1\xi_1 + A_2\xi_2 + A_3\xi_3 + \dots, \tag{3}$$

when E is the unit matrix, and A_1, A_2, A_3 are constant matrices, the partial derivatives of the matrix $T(\xi_1, \xi_2, \xi_3)$ with respect to ξ_1, ξ_2, ξ_3 evaluated at the point $\xi_1 = \xi_2 = \xi_3 = 0$. The row of dots indicates that we have omitted the remaining terms in the Taylor expansion of order higher than the first, which are negligible by comparison with $\sqrt{\xi_1^2 + \xi_2^2 + \xi_3^2}$.

We will show that the representation, i.e. the function $T(\xi_1, \xi_2, \xi_3)$ is completely determined by A_1, A_2 and A_3, and then find the possible forms of A_1, A_2 and A_3.

The matrices A_1, A_2, A_3 have a simple meaning. To see this, consider a rotation about the axis Ox through an angle ξ_1. To this there corresponds the matrix

$$T(\xi_1, 0, 0) = E + A_1\xi_1 + \dots$$

It is evident from this formula that the matrix $T(\xi_1, 0, 0)$, corresponding to an infinitesimal rotation through an angle ξ_1 about Ox, is completely determined by the matrix A_1 if we neglect second order terms. The matrices A_1, A_2 and A_3 are called the matrices of the infinitesimal rotations about the coordinates axes.

We will now show that the matrices A_1, A_2, A_3 completely determine the representation; i.e. that given these three matrices we can determine $T(\xi_1, \xi_2, \xi_3)$ for all ξ_1, ξ_2, ξ_3. To do this we take an arbitrary vector (ξ_1, ξ_2, ξ_3) and consider two rotations about this vector: $g(t\xi_1, t\xi_2, t\xi_3)$ and $g(s\xi_1, s\xi_2, s\xi_3)$** The product of these two rotations is obviously a rotation

* The proof is given in the addendum to this paragraph. We note that a matrix is said to be a differentiable function of certain variables if each of its elements is a differentiable function of these variables.

** These are rotations through angles $t\sqrt{\xi_1^2 + \xi_2^2 + \xi_3^2}$ and $s\sqrt{\xi_1^2 + \xi_2^2 + \xi_3^2}$ respectively.

about the same axis, and is given by the parameters $(t+s)\xi_1$, $(t+s)\xi_2$, $(t+s)\xi_3$:

$$g((t+s)\xi_1,\ (t+s)\xi_2,\ (t+s)\xi_3)=g(t\xi_1,\ t\xi_2,\ t\xi_3)\,g(s\xi_1,\ s\xi_2,\ s\xi_3). \quad (4)$$

Since a representation preserves products, we have also

$$T((t+s)\xi_1,\ (t+s)\xi_2,\ (t+s)\xi_3)=T(t\xi_1,\ t\xi_2,\ t\xi_3)\,T(s\xi_1,\ s\xi_2,\ s\xi_3). \quad (4')$$

Differentiating both sides of equation (4') with respect to s and putting $s=0$, we get

$$\frac{d}{dt}T(t\xi_1,\ t\xi_2,\ t\xi_3)=\frac{d}{ds}T(s\xi_1,\ s\xi_2,\ s\xi_3)\Big|_{s=0}\cdot T(t\xi_1,\ t\xi_2,\ t\xi_3).$$

But from (3)

$$\frac{d}{ds}T(s\xi_1,\ s\xi_2,\ s\xi_3)\Big|_{s=0}=A_1\xi_1+A_2\xi_2+A_3\xi_3.$$

From this we obtain the following differential equation for the matrix $X(t)=T(t\xi_1,\ t\xi_2,\ t\xi_3)$

$$\frac{d}{dt}X(t)=(A_1\xi_1+A_2\xi_2+A_3\xi_3)\,X(t)^{*} \quad (5)$$

Apart from this the fundamental condition

$$X(0)=T(0,\ 0,\ 0)=E. \quad (5')$$

must be satisfied. From the theorem concerning the uniqueness of the solution of a system of differential equations, equation (5) and the fundamental condition (5') define $X(t)$ uniquely. In particular, $X(1)=T(\xi_1,\ \xi_2,\ \xi_3)$ is defined uniquely. Thus we have proved that a representation is uniquely defined by the matrices A_1, A_2, A_3 corresponding to infinitesimal rotations.

Equation (5) can be solved directly. Its solution, satisfying the fundamental condition (5') is evidently

$$X(t)=e^{t(A_1\xi_1+A_2\xi_2+A_3\xi_3)}\ {**}.$$

$$T(\xi_1,\ \xi_2,\ \xi_3)=e^{A_1\xi_1+A_2\xi_2+A_3\xi_3}. \quad (6)$$

Thus if the matrices A_1, A_2 and A_3 correspond to infinitesimal rotations about the axes of coordinates, then the matrix $T_g=T(\xi_1,\ \xi_2,\ \xi_3)$ of the representation is given in terms of A_1, A_2, A_3 by equation (6).

* Equation (5) is equivalent to a system of linear differential equations, each equation arising by equating corresponding matrix elements on the two sides.

** See the appendix to Sect. 2.

§. 2. Relations between the Matrices A_k

We will now obtain the conditions which must be satisfed by three matrices A_1, A_2, A_3 in order that the matrix T_g defined by equation (6) shall in fact give a unitary representation of the group of rotations. With this object in view we will derive equations, which must be satisfied by the matrices corresponding to infinitesimal rotations in the representation and then, in § 3 we will solve these equations.

Let g_0 be a fixed rotation, and g be an arbitrary one. We consider the rotation $\tilde{g}_0 = g g_0 g^{-1}$. The matrices of the rotations $\tilde{g}_0$ and g_0 are obtained from each other by a similarity transformation. This means that these two matrices represent rotations through the same angle φ. If the rotation g_0 is defined by the vector $\eta = (\eta_1, \eta_2, \eta_3)$ then $\tilde{g}_0$ is defined by the vector $\tilde{\eta}$ which is obtained from η by the operation g: $\tilde{\eta} = g\eta$. Obviously the vector η is unchanged by the rotation g_0 i.e. $g_0\eta = \eta$. Hence $\tilde{g}_0\tilde{\eta} = g g_0 g^{-1}\tilde{\eta} = \tilde{\eta}$. This shows that $\tilde{\eta}$ is unchanged by the rotation $\tilde{g}_0$, i.e. it lies along the axis of rotation. Since the magnitude of the vector η which is equal to the angle of rotation, is unchanged by the rotation g, we have $|\tilde{\eta}| = |\eta|$ and $\tilde{\eta}$ is the vector defining the rotation $\tilde{g}_0$.

Since to a product of rotations corresponds the product of their matrices T_g, in the representation then, from the equation $\tilde{g}_0 = g g_0 g^{-1}$ we obtain

$$T_g T_{g_0} T'_{g^{-1}} = T_{\tilde{g}_0}. \tag{7}$$

From this we now obtain the relations between the matrices A_1, A_2 and A_3. To do this we assume that the rotation g_0 and consequently $\tilde{g}_0$, is small, i.e. that η and $\tilde{\eta}$ are small vectors; equation (3) gives the following expressions for T_{g_0} and $T_{\tilde{g}_0}$:

$$T_{g_0} = E + A_1\eta_1 + A_2\eta_2 + A_3\eta_3 + \ldots,$$
$$T_{\tilde{g}_0} = E + A_1\tilde{\eta}_1 + A_2\tilde{\eta}_2 + A_3\tilde{\eta}_3 + \ldots$$

Substituting from these relations in equation (7) we obtain

$$T_g (E + A_1\eta_1 + A_2\eta_2 + A_3\eta_3 + \ldots)\, T_{g-1} = $$
$$= E + A_1\tilde{\eta}_1 + A_2\tilde{\eta}_2 + A_3\tilde{\eta}_3 + \ldots \tag{8}$$

We have put each small vector η in correspondence with a matrix $A_\eta = A_1\eta_1 + A_2\eta_2 + A_3\eta_3$, determined up to terms of the first order of magnitude in the components of these vectors. By equating terms of the first order in equation (8) we obtain the formula

$$T_g A_\eta T_{g-1} = A_{\tilde{\eta}}, \tag{9}$$

where $\tilde{\eta} = g\eta$.

Since on multiplying the vector η by a number A_η and $A_{\tilde{\eta}}$ are multiplied by the same number, we may evidently disregard the restriction applied to magnitude of the vector η after the derivation of (9).

In order to obtain from (9) the relations between A_1, A_2, A_3 we now assume that the rotation g has small angle and is about the axis Ox, and the vector η to be a unit vector directed along the axis Oy: $\eta = (0, 1, 0)$. Then $\tilde{\eta}$, the vector obtained from η by the rotation g is a unit vector lying in the plane yOz making an angle a say with the axis Oy. Consequently, the components of this vector will, to within terms of the second order of magnitude in a, be given by

$$\tilde{\eta}_1 = 0,$$
$$\tilde{\eta}_2 = 1,$$
$$\tilde{\eta}_3 = \alpha.$$

The expression (3) for T_g in our case (when $\xi_1 = \alpha$, $\xi_2 = \xi_3 = 0$,) gives

$$T_g = E_1 + A_1\alpha + \ldots, \quad T_{g-1} = E_1 - A_1\alpha + \ldots,$$

However,

$$A_\eta = A_2,$$
$$A_{\tilde{\eta}} = A_2 + \alpha A_3.$$

Hence, substituting these expressions in (9) and equating terms of the first order of magnitude in a, we obtain the equation

$$A_1A_2 - A_2A_1 = A_3.$$

The expression $AB - BA$ is called the <u>commutator</u> of the matrices A and B and is denoted by $[A, B]$: $AB - BA = [A, B]$. Our relation may then be written in the form

$$[A_1, A_2] = A_3.$$

Similarly we may obtain the other relations

$$[A_2, A_3] = A_1, \quad [A_3, A_1] = A_2.$$

<u>Thus it has been shown that if $g \to T_g$ is an arbitrary representation of the group of rotations, then the matrices A_1, A_2 and A_3, corresponding to infinitesimal rotations about the coordinate axes satisfy the relations</u>

$$\left.\begin{aligned} [A_1, A_2] &= A_3, \\ [A_2, A_3] &= A_1, \\ [A_0, A_1] &= A_2. \end{aligned}\right\} \qquad (10)$$

These relations will in future be referred to as the commutation relations.

The equations (10) also show that if the matrices A_1, A_2 and A_3 are interpreted as a basis of a three-dimensional space, then the commutators of the matrices correspond exactly to the vector products of the basic vectors. More precisely, consider the linear space of all matrices $A_\xi = A_1\xi_1 + A_2\xi_2 + A_3\xi_3$, where $\xi = (\xi_1, \xi_2, \xi_3)$ is an arbitrary vector. Then the relation expressing the commutator of two matrices of this space as a linear combination of the matrices A_k is easily derived from the formulae. In fact it follows from (10) that

$$[A_1\xi_1 + A_2\xi_2 + A_3\xi_3,\ A_1\eta_1 + A_2\eta_2 + A_3\eta_3] = \\ = A_3(\xi_1\eta_2 - \xi_2\eta_1) + A_1(\xi_2\eta_3 - \xi_3\eta_2) + A_2(\xi_3\eta_1 - \xi_1\eta_3).$$

Let ζ be the vector product of ξ and η: $\zeta = [\xi, \eta]$ We can now make the following statement: if the matrices A_1, A_2 and A_3 correspond in a representation to infinitesimal rotations about the coordinate axes then, for any two matrices $A_\xi = A_1\xi_1 + A_2\xi_2 + A_3\xi_3$ and $A_\eta = A_1\eta_1 + A_2\eta_2 + A_3\eta_3$ we have

$$[A_\xi, A_\eta] = A_1\zeta_1 + A_2\zeta_2 + A_3\zeta_3 = A_\zeta,$$

where $\zeta = (\zeta_1, \zeta_2, \zeta_3)$ is the vector product of ξ and

In deriving equations (10) we have not made use of the fact that T_g is unitary. We shall now proceed to determine the consequences of this condition on the A_k.

Put $\xi_2 = \xi_3 = 0$ in equation (3). Then

$$T(\xi_1, 0, 0) = E + \xi_1 A_1 + \ldots \qquad (3')$$

Since the matrix T is unitary,

$$T^*(\xi_1, 0, 0)\, T(\xi_1, 0, 0) = E.$$

Substituting in this for T as given by (3') we have

$$(E + \xi_1 A_1^* + \ldots)\ (E + \xi_1 A_1 + \ldots) = E.$$

Equating terms of the first order in ξ we see that $A_1 + A_1^* = 0$, i.e. $A_1^* = - A_1$.

Putting $H_1 = iA_1$ we have $H_1^* = H_1$, i.e. H_1 is a Hermitian matrix. Similarly, putting $H_k = iA_k$ we obtain the Hermitian matrices H_2 and H_3**

* The conclusion that the matrices A_k are skew Hermitian for a suitable choice of basis can be established directly from the commutation relations. The proof is rather unwieldy and is left as an exercise.

** We note that the converse also holds; if the H_k are Hermitian, then T_g is unitary by formula (6)

$$T_g = e^{i(H_1\xi_1 + H_2\xi_2 + H_3\xi_3)}$$

From the commutation relations (10) there follow analogous relations for H_1, H_2 and H_3:

$$\left.\begin{aligned} [H_1, H_2] &= iH_3, \\ [H_2, H_3] &= iH_1, \\ [H_3, H_1] &= iH_2. \end{aligned}\right\} \tag{10'}$$

The problem of determining all possible representations of the group of rotations therefore reduces to the following two problems:

1) to find all possible sets of three matrices H_k for which the relations (10') hold;

2) to select from these triads those that do in fact give rise to a representation of the group of rotations.

The second problem is solved by two different methods in Sects. 6, 7. From the results of these paragraphs it follows that any three skew-Hermitian matrices satisfying the relations (10) correspond to infinitesimal rotations about the axes of coordinates, in some representation of the rotation group.

Thus we arrive at the conclusion that any three Hermitian matrices satisfying conditions (10') yield a representation $T_g = e^{i(H_1\xi_1 + H_2\xi_2 + H_3\xi_3)}$ of the group of rotations.

§ 3. The Form of an Irreducible Representation

Instead of the matrices H_1, H_2 and H_3 in the form just defined, it will be more convenient for us to consider the following linear combinations of them:

$$\begin{aligned} H_+ &= H_1 + iH_2, \\ H_- &= H_1 - iH_2, \\ H_3 &= H_3. \end{aligned}$$

The commutators of these three matrices are easily determined. Thus

$$[H_+, H_3] = [H_1 + iH_2, H_3] = [H_1, H_3] + i[H_2, H_3] = \\ = -iH_2 - H_1 = -H_+.$$

Similarly we can determine $[H_-, H_3]$ and $[H_+, H_-]$. The results are as follows

$$\left.\begin{aligned} [H_+, H_3] &= -H_+, \\ [H_-, H_3] &= H_-, \\ [H_+, H_-] &= 2H_3. \end{aligned}\right\} \tag{11}$$

We have further

$$H_+^* = (H_1 + iH_2)^* = H_1 - iH_2 = H_-. \tag{12}$$

The problem is now reduced to that of finding the matrices H_+, H_-, H_3, which satisfy conditions (11) and (12). We determine these matrices in terms of a basis consisting of eigenvectors of the matrix H_3. In this basis the matrices H_+, H_-, H_3 take their simplest form. In preparation we will now prove the following lemma.

Lemma. Let f be an eigenvector of the transformation H_3 corresponding to an eigenvalue λ:

$$H_3 f = \lambda f.$$

Then the vector $f_1 = H_+ f$ is either the zero-vector or it is also an eigenvector of H_3 corresponding to an eigenvalue $\lambda + 1$. Similarly the vector $f_2 = H_- f$ is either the zero vector or an eigenvector of H_3 corresponding to the eigenvalue $\lambda - 1$.

In fact,

$$H_3 f_1 = H_3 H_+ f = [H_3, H_+] f + H_+ H_3 f =$$
$$= H_+ f + H_+ \lambda f = (\lambda + 1) H_+ f = (\lambda + 1) f_1.$$

Similarly we obtain $H_3 f_2 = (\lambda - 1) f_2$.

We now turn to the determination of the matrices H_+, H_-, H_3. Since H_3 is a Hermitian matrix its eigenvalues must be real. We denote by l the greatest of the eigenvalues of the matrix H_3 and by f_l a corresponding normalized eigenvector:

$$H_3 f_l = l f_l, \quad (f_l, f_l) = 1.$$

If $H_- f_l \neq 0$ we can put

$$H_- f_l = \alpha_l f_{l-1},$$

where the positive number α_l has been chosen such that $(f_{l-1}, f_{l-1}) = 1$. From the lemma f_{l-1} is a normalized eigenvector of H_3 corresponding to the eigenvalue $l - 1$. If $H_- f_{l-1} \neq 0$ then we introduce f_{l-2} similarly, putting $H_- f_{l-1} = \alpha_{l-1} f_{l-2}$ where $\alpha_{l-1} > 0$ and $(f_{l-2}, f_{l-2}) = 1$. Continuing this process we have

$$H_- f_{l-2} = \alpha_{l-2} f_{l-3}$$

and so on.

Now from the lemma the vectors $f_l, f_{l-1}, \ldots$ which have been constructed in this way are eigenvectors of the matrix H_3 corresponding to eigenvalues $l, l - 1, \ldots$ Since the matrix H_3 has but a finite number of distinct eigenvalues, then the sequence of vectors $f_l, f_{l-1}, f_{l-2}, \ldots$ terminates, i.e. for some values of k we obtain $H_- f_k = 0$.

We thus obtain a system of (mutually orthogonal) normalized eigenvectors of the transformation H_3:

$$H_3 f_m = m f_m. \tag{13}$$

We also have

$$H_- f_m = \alpha_m f_{m-1}. \tag{14}$$

In order that equation (14) holds for the last of the sequence of eigenvectors (for which $m = k$) we must put $\alpha_k = 0$.

It is now clear how the vectors f_m are affected by the transformation H_+. In the first place, in virtue of the lemma $H_+ f_m$ is either zero or an eigenvector of H_3 corresponding to an eigenvalue $m+1$. Since l is the greatest eigenvalue of H_3 then $H_+ f_l = 0$.

Next we determine $H_+ f_{l-1}$. We have

$$H_+ f_{l-1} = \frac{1}{\alpha_l} H_+ H_- f_l = \frac{1}{\alpha_l} [H_+, H_-] f_l + \frac{1}{\alpha_l} H_- H_+ f_l = \frac{2}{\alpha_l} H_3 f_l = \frac{2l}{\alpha_l} f_l$$

Thus the vector $H_+ f_{l-1}$ is proportional to f_l i.e. $H_+ f_{l-1} = \beta_l f_l$, $\beta_l > 0$. We will show that $H_+ f_m$ is proportional to f_{m+1}, i.e. $H_+ f_m = \beta_{m+1} f_{m+1}$. Let this equation hold for the vectors $f_l, f_{l-1}, \ldots, f_{m+1}$. We prove the assertion for the vector f_m:

$$H_+ f_m = \frac{1}{\alpha_{m+1}} H_+ H_- f_{m+1} = \frac{1}{\alpha_{m+1}} [H_+, H_-] f_{m+1} +$$

$$+ \frac{1}{\alpha_{m+1}} H_- H_+ f_{m+1} = \frac{2}{\alpha_{m+1}} H_3 f_{m+1} + \frac{\beta_{m+2}}{\alpha_{m+1}} H_- f_{m+2}.$$

Using the equations (13) and (14) we obtain

$$H_+ f_m = \frac{2(m+1) + \alpha_{m+2}\beta_{m+2}}{\alpha_{m+1}} f_{m+1}.$$

Putting

$$\frac{2(m+1) + \alpha_{m+2}\beta_{m+2}}{\alpha_{m+1}} = \beta_{m+1}, \tag{15}$$

we have:

$$H_+ f_m = \beta_{m+1} f_{m+1}. \tag{16}$$

Since $H_+ f_l = 0$, it follows from this that for $m = l$ we must put $\beta_{l+1} = 0$. For the determination of the transformations H_+ and H_- we will need in the first place to obtain the coefficients α_m and β_m. From the fact that $H_+^* = H_-$ it follows that

$$(H_+ f_{m-1}, f_m) = (f_{m-1}, H_- f_m).$$

Using equations (14) and (16) we obtain $\beta_m(f_m, f_m) = \alpha_m(f_{m-1}, f_{m-1})$, and as both vectors are normalized, $\alpha_m = \beta_m$.

Replacing β by α and $m+1$ by m in (15) we get $\alpha_m^2 - \alpha_{m+1}^2 = 2m$. To find α_m^2, we sum these equations for $m = l$ to an arbitrary value m. We obtain

$$\alpha_m^2 - \alpha_{l+1}^2 = 2l + 2(l-1) + 2(l-2) + \ldots + 2m.$$

Using the fact that $\beta_{l+1}^2 = 0$ and consequently that $\alpha_{l+1}^2 = 0$, we find

$$\alpha_m^2 = (l+m)(l-m+1). \tag{17}$$

This formula enables us to find the number of vectors f_m in the series $f_l, \ldots, f_k$. We put $\alpha_k = 0$ if f_k is the last of these vectors ($H_- f_k = 0$). From formula (17) we obtain $k = -l$. Since the number m is reduced in the process each time by unity, then the difference $l - (-l) = 2l$ is the required number. Consequently this l is either an integer or half an odd integer. The number of vectors obtained is obviously in either case $2l+1$.

Until now we have placed no limitations on the representation T_g, i.e. it could be either reducible or irreducible. We will now restrict ourselves to irreducible representations. This means that in the space R, in which the representation operates, there does not exist a subspace invariant with respect to all transformations T_g. It follows that there does not exist in R a subspace invariant with respect to the matrices A_1, A_2, A_3, that is to the matrices H_+, H_-, H_3. For it follows in fact from formula (6) that such a subspace would be invariant with respect to T_g. We show that in such a case the vectors $f_l, f_{l-1}, \ldots, f_{-l}$ for a basis for the space R. Since H_+, H_- and H_3 carry a vector f_m into another vector of the same system, then the subspace spanned by the vectors f_m ($m = -l, \ldots, l$) is invariant with respect to H_+, H_-, H_3 and consequently, since the representation is irreducible must coincide with the whole space R.

Thus we have found that for any irreducible representation the transformations H_+, H_- and H_3 define an orthogonal basis consisting of the normalized eigenvectors of H_3 by the equations

$$\left.\begin{aligned} H_+ f_m &= \alpha_{m+1} f_{m+1}, \\ H_- f_m &= \alpha_m f_{m-1}, \\ H_3 f_m &= m f_m, \end{aligned}\right\} \tag{18}$$

Here $m = -l, -l+1, \ldots, l$; l being an integer or half an odd integer; and $\alpha_m = \sqrt{(l+m)(l-m+1)}$.

This basis $f_l, f_{l-1}, \ldots, f_{-l}$ consisting of the normalized eigenvectors of the transformation H_3 will be called, in the case of an irreducible representation, the canonical basis of the representation.

Returning to the transformations A_1, A_2 and A_3 we will prove the following statement: each irreducible representation of the group of rotations is defined by a number l which is either an integer or half an odd integer. The transformations A_1, A_2 and A_3 corresponding in this representation to infinitesimal rotations about the axes of coordinates are given in terms of the canonical basis f_m $(m=-l, -l+1, \ldots, l)$ by the expressions

$$\left.\begin{aligned} A_1 f_m &= -iH_1 f_m = \\ &= -\frac{i}{2}\sqrt{(l+m+1)(l-m)}\, f_{m+1} - \frac{i}{2}\sqrt{(l+m)(l-m+1)}\, f_{m-1}, \\ A_2 f_m &= -iH_2 f_m = \\ &= -\frac{1}{2}\sqrt{(l+m+1)(l-m)}\, f_{m+1} + \frac{1}{2}\sqrt{(l+m)(l-m+1)}\, f_{m-1}, \\ A_3 f_m &= -iH_3 f_m = -imf_m. \end{aligned}\right\} \quad (19)$$

The number l is called the weight of the corresponding irreducible representation. Every irreducible representation is defined uniquely by its weight. We remark that the matrices, corresponding to an arbitrary rotation g in the representation, are given in terms of the matrices A_k by equations (6) of this paragraph.

We have found the form of the representation T_g on the assumption that it exists. The fact, that the matrices T_g given by formula (6) in terms of the matrices A_k determined by the equations (19), do actually form a representation is difficult to prove directly. We will prove later that it is possible to construct for every l a corresponding irreducible representation. For example in Sects. 2, 3, such representations will be constructed for any integer l and in Sects. 6 7 for all integers and half odd integers. At present we will simply show that if there exists a representation defined by the formula (6), where the A_k have the form (19), then it is irreducible. It is sufficient to show that there does not exist a subspace R_1, invariant with respect to H_+, H_-, H_3 and not coinciding with the whole space of $2l+1$ dimensions. We consider the transformation H_3 in this space. This transformation has an eigenvector $h = \sum_{m=-l}^{l} c_m f_m$, corresponding to the greatest eigenvalue of H_3 in the subspace R_1. From the lemma on p. 23 it follows that the transformation H_+ carries the vector corresponding to the greatest eigenvalue of H_3 into the null-vector. From formula (18) we have, consequently,

$$H_+ h = \sum_{m=-l}^{l} c_m H_+ f_m = \sum_{m=-l}^{l} c_m \alpha_{m+1} f_{m+1} = 0.$$

Since the vectors f_m are linearly independent the coefficient of each f_m must be zero. Hence, if $m < l$ $\alpha_{m+1} \neq 0$ and hence $c_m = 0$. This means that $h = c_l f_l$, i.e. the subspace R_1 contains the vector f_l. Hence it contains $H_- f_l$, $H_-^2 f_l$ ... and so on, i.e. $f_{l-1}, f_{l-2}, \ldots, f_{-l}$. Therefore R_1 coincides with R. We have proved that there does not exist a subspace, distinct from R and the null space, which is invariant with respect to H_+, H_- and H_0 i.e., with respect to A_1, A_2, A_3. But it follows from this that there does not exist a subspace distinct from R and the null-space, which is invariant with respect to all T_g, since any such subspace would have to be invariant with respect to all $A_k = \frac{\partial T(\xi_1, \xi_2, \xi_3)}{\partial \xi_k}\Big|_{\xi_1=\xi_2=\xi_3=0}$. Hence it follows that the representation $g \to T_g$ is irreducible. We note that our proof implies the following useful result: in the space R in which we have the irreducible representation there exists a vector f such that the vectors f_0 satisfying $H_+ f = 0$ are precisely the scalar multiples of f

§ 4. The Resolution of a Representation into Irreducible Constituents

We now consider a reducible representation of the group of rotations. Most of the reasoning up to now has not depended on the irreducibility of the representation. The irreducibility was used only at the very end, where it was proved that the vectors f_m $(m = -l, -l+1, \ldots, l)$ form a basis of the space R. If it had not been assumed that the representation was irreducible then from our argument it would have followed that the system of vectors obtained would at least have formed a basis of some invariant subspace R_0. Consider the orthogonal complement R' of this subspace i.e. the set of all vectors orthogonal to $f_l, f_{l-1}, \ldots, f_{-l}$. Since the transformations H_k are hermitian, then R' being the orthogonal complement of the invariant subspace R_0 is itself invariant with respect to H_1, H_2, H_3. We may now repeat the same argument in the invariant subspace R'; i.e. we take the greatest eigenvalue l_1 of the transformation H_3 in this space, then again construct a set of vectors f'_m $(-l_1 \leqslant m \leqslant l_1)$ from the corresponding eigenvector, take the orthogonal complement and proceed until the space R has been reduced as far as possible. Thus we arrive finally at the following conclusion.

Let there be given some unitary representation of the group of rotations. Then there exists an orthonormal basis in which the matrices $A_k = -iH_k$ have the following form

$$A_k = \left\| \begin{matrix} A_k^{(0)} & & & & & \\ & A_k^{(1)} & & & & \\ & & \cdot & & & \\ & & & \cdot & & \\ & & & & \cdot & \\ & & & & & A_k^{(s)} \end{matrix} \right\|, \tag{20}$$

where $A_1^{(j)}$, $A_2^{(j)}$, $A_3^{(j)}$ are matrices determined by equations (19) of the previous paragraph for $l = l_j$, that is,

$$\left.\begin{aligned}
A_1^{(j)} &= -\frac{i}{2}\begin{Vmatrix}
0 & \alpha_{-l_j+1} & 0 & \ldots & 0 & 0 \\
\alpha_{-l_j+1} & 0 & \alpha_{-l_j+2} & \ldots & 0 & 0 \\
0 & \alpha_{-l_j+2} & 0 & \ldots & 0 & 0 \\
\cdot & \cdot & \cdot & \cdot & \cdot & \cdot \\
\cdot & \cdot & \cdot & \cdot & \cdot & \cdot \\
0 & 0 & 0 & \ldots & 0 & \alpha_{l_j} \\
0 & 0 & 0 & \ldots & \alpha_{l_j} & 0
\end{Vmatrix}; \\
A_2^{(j)} &= \frac{1}{2}\begin{Vmatrix}
0 & \alpha_{-l_j+1} & 0 & \ldots & 0 & 0 \\
-\alpha_{-l_j+1} & 0 & \alpha_{-l_j+2} & \ldots & 0 & 0 \\
0 & -\alpha_{-l_j+2} & 0 & \ldots & 0 & 0 \\
\cdot & \cdot & \cdot & \cdot & \cdot & \cdot \\
\cdot & \cdot & \cdot & \cdot & \cdot & \cdot \\
0 & 0 & 0 & \ldots & 0 & \alpha_{l_j} \\
0 & 0 & 0 & \ldots & -\alpha_{l_j} & 0
\end{Vmatrix}; \\
A_3^{(j)} &= \begin{Vmatrix}
il_j & 0 & 0 & \ldots & 0 & 0 \\
0 & i(l_j-1) & 0 & \ldots & 0 & 0 \\
0 & 0 & i(l_j-2) & \ldots & 0 & 0 \\
\cdot & \cdot & \cdot & \cdot & \cdot & \cdot \\
\cdot & \cdot & \cdot & \cdot & \cdot & \cdot \\
0 & 0 & 0 & \ldots & -i(l_j-1) & 0 \\
0 & 0 & 0 & \ldots & 0 & -il_j
\end{Vmatrix}
\end{aligned}\right\} \tag{21}$$

$$\left(\alpha_m = \sqrt{(l_j+m)(l_j-m+1)}\right).$$

We will now comment on some of the frequently used forms into which a given representation can be resolved. It is easily seen that every vector f for which $H_+ f = 0$ has the form

$$\alpha f_l + \alpha' f'_{l_1} + \ldots \tag{22}$$

In particular such vectors will, of course, be f_l, f'_{l_1}, ... These are characterized by the fact that they are the eigenvectors of H_3: $H_3 f_l = l f_l$, $H_3 f'_{l_1} = l_1 f'_{l_1}$, ... From what has been said above we are able to give the following rule for the constructed basis, in which the matrices A_1, A_2, A_3 have the form (20), or as we will say, for the resolution of a representation into irreducible representations: we seek all solutions of the equation $H_+ f = 0$. The set of all such solutions is invariant with respect to H_3. We consider the transformation H_3 in the subspace of vectors f for which $H_+ f = 0$ and find its complete orthonormal system of eigenvectors.

For each one of these vectors, f'_{l_1} for example, we may construct the part of the basis corresponding to one of the invariant subspaces. This consists of the vectors $H\ f'_{l_1}$, $H^2 f'_{l_1}$, ..., $H_-^{2l_1} f'_{l_1}$.

From the preceding we are able to make the following deduction: if an arbitrary representation is given, then the irreducible representations into which it is resolved are given by the values of l, for which there exist solutions of the following simultaneous equations

$$H_+f = 0, \qquad H_3f = lf. \tag{23}$$

The representation corresponding to a given l appears in the resolution as many times as there are linearly independent solutions of the equations (23). We note further that if in the resolution of the representation into irreducible representations there are found to be more than one with a given weight l then this resolution is not unique. In fact in our construction of the corresponding chain of vectors of the basis we began with an orthonomal system of vectors for which $H_+f = 0$ and $H_3f = lf$. But, if one and the same value of l occurs several times then H_3 has a repeated eigenvalue in the subspace of vectors f for which $H_+f = 0$. This means that this system can be chosen in more ways than one.

We now consider a new method of resolution into invariant subspaces. For this the representation in each invariant subspace is either irreducible, or isotypic, i.e. can be broken down into irreducible representations with the same weight. Unlike the resolution into irreducible representations this resolution is unique. To perform this resolution consider the transformation

$$H^2 = H_1^2 + H_2^2 + H_3^2. \tag{24}$$

The transformation H^2 commutes with H_1, H_2, H_3, that is

$$[H^2, H_1] = 0, \quad [H^2, H_2] = 0, \quad [H^2, H_3] = 0.$$

As an illustration we verify that $[H^2, H_3] = 0$. We have

$$[H_1^2, H_3] = H_1^2H_3 - H_3H_1^2 = H_1^2H_3 - H_1H_3H_1 + H_1H_3H_1 - H_3H_1^2 =$$
$$= H_1[H_1, H_3] + [H_1, H_3]H_1 = -iH_1H_2 - iH_2H_1.$$

Similarly

$$[H_2^2, H_3] = H_2[H_2, H_3] + [H_2, H_3]H_2 = iH_2H_1 + iH_1H_2$$

and since obviously

$$[H_3^2, H_3] = 0.$$

we have

$$[H_1^2 + H_2^2 + H_3^2, H_3] = [H^2, H_3] = 0.$$

In the same way it may be proved that

$$[H^2, H_1] = 0, \quad [H^2, H_2] = 0.$$

In the case of an irreducible representation, when H_1, H_2, H_3 are given by the equations (19) a direct calculation gives us

$$H^2 = l(l+1)E,$$

where l is the number defining the representation.

In order to carry out this calculation, we note that

$$H_+H_- = (H_1 + iH_2)(H_1 - iH_2) =$$
$$= H_1^2 + H_2^2 + i(H_2H_1 - H_1H_2) = H_1^2 + H_2^2 + H_3,$$

Hence

$$H^2 = H_+H_- - H_3 + H_3^2.$$

From formula (18) we find

$$H_+H_-f_m = H_+\alpha_m f_{m-1} = \alpha_m^2 f_m,$$
$$H_3 f_m = mf_m,$$
$$H_3^2 f_m = m^2 f_m.$$

Since $\alpha_m^2 - m + m^2 = l(l+1)$, we have $H^2 f_m = l(l+1)f_m$ *. Thus we see that all vectors f in the space R of an irreducible representation with a given weight l satisfy the equation

$$H^2 f = l(l+1)f. \tag{25}$$

It follows that the number of linearly independent solutions of this equation is a multiple of $2l+1$.

With the help of the transformation H^2 we are able to resolve an arbitary representation into isotypic representations. This means that we are able to choose in the space R a basis consisting of separate groups of vectors which are such that, firstly, each group gives rise to an invariant subspace, and secondly, the representation in this subspace is either irreducible or isotypic. Obviously to obtain the resolution into isotypic representations it is necessary to find for each l a complete set of linearly independent solutions of equation (25). Such a set of solutions will give a basis, in which the representation is resolved into isotypic representations. As we have seen, this resolution is distinguished from that into irreducible representations by the fact that it is unique.

§ 5. Examples of Representations

In conclusion let us consider various examples of irreducible representations

---

* This fact, that for an irreducible representation $H^2 = aE$, where a is a scalar, can be obtained from the fact that H^2 commutes with all the H_k. The value of the constant a is easily found by applying both sides of the equation $H^2 = aE$ to the vector f_{-l}

of the group of rotations.

Take $l=0$. In this case the representation is one dimensional and the matrices T_g are simply numbers. Obviously we obtain this representation if we put $T_g \equiv 1$ (the identity representation). The matrices A_k will in this case be null, as is seen from the equation (19).

Next let $l=1$. Then $2l+1=3$, i.e. we obtain a representation of the group of rotations by transformations in three-dimensional space. We obtain such a representation by putting each rotation in correspondence with its matrix (the fundamental representation). Since every orthogonal matrix, if regarded as the matrix of a transformation in a complex space, is at the same time unitary, our representation is unitary. The canonical basis of the fundamental representation has the form

$$f_{-1}=\frac{e_x-ie_y}{\sqrt{2}},\quad f_0=e_z,\quad f_1=\frac{-e_x-ie_y}{\sqrt{2}}*,$$

where e_x, e_y and e_z are unit vectors along the axes of coordinates. It is easily verified that the transformations A_1, A_2 and A_3 have, in the canonical basis, the form given by equations (19) for $l=1$.

We will find the matrix in this basis corresponding to a finite rotation about the axis Oz through an angle φ. In the basis e_x, e_y, e_z this transformation is given by

$$\begin{aligned} e'_x &= e_x\cos\varphi + e_y\sin\varphi,\\ e'_y &= -e_x\sin\varphi + e_y\cos\varphi,\\ e'_z &= e_z. \end{aligned}$$

Hence

$$f'_{-1}=\frac{e'_x-ie'_y}{\sqrt{2}}=\frac{(e_x\cos\varphi+e_y\sin\varphi)-i(-e_x\sin\varphi+e_y\cos\varphi)}{\sqrt{2}}=$$

$$=\frac{(e_x-ie_y)e^{i\varphi}}{\sqrt{2}}=e^{i\varphi}f_{-1}$$

$$f'_0=e'_z=e_z=f_0,$$

$$f'_1=\frac{-e'_x-ie'_y}{\sqrt{2}}=\frac{-e_x-ie_y}{\sqrt{2}}e^{-i\varphi}=e^{-i\varphi}f_1.$$

---

* The elements of the canonical basis are the normalized eigenvectors of the transformation H_3. From equation (7) Sect.1 it is easily found that the matrix H_3 in the basis e_x, e_y, e_z has the form $\begin{Vmatrix} 0 & -i & 0\\ i & 0 & 0\\ 0 & 0 & 0\end{Vmatrix}$.

The matrix of the rotation g_φ consequently has the form

$$\begin{Vmatrix} e^{i\varphi} & 0 & 0 \\ 0 & 1 & 0 \\ 0 & 0 & e^{-i\varphi} \end{Vmatrix}.$$

We find the components a_-, a_0 and a_+ of the vector a in terms of the canonical basis. Substituting the above expression for f_{-1}, f_0, f_1 in the equation

$$a_x e_x + a_y e_y + a_z e_z = a_- f_{-1} + a_0 f_0 + a_+ f_1,$$

we find that

$$a_x = \frac{a_- - a_+}{\sqrt{2}}, \quad a_y = -\frac{i}{\sqrt{2}}(a_- + a_+), \quad a_z = a_0$$

or

$$a_+ = \frac{-a_x + ia_y}{\sqrt{2}},$$

$$a_0 = a_z,$$

$$a_- = \frac{a_x + ia_y}{\sqrt{2}}.$$

Before we consider other examples, we will find the matrix T_g which corresponds to a finite rotation about the axis Oz in a representation with any value of the weight l. From (19) we have that the matrix A_3 corresponding to an infinitesimal rotation about Oz has the form

$$A_3 = \begin{Vmatrix} il & 0 & 0 & \dots & 0 \\ 0 & i(l-1) & 0 & \dots & 0 \\ 0 & 0 & i(l-2) & \dots & 0 \\ \cdot & \cdot & \cdot & \cdot & \cdot \\ \cdot & \cdot & \cdot & \cdot & \cdot \\ 0 & 0 & 0 & \dots & -il \end{Vmatrix}.$$

As was shown in paragraph 1 of this section, a rotation g through an angle φ about the axis Oz corresponds to the matrix $T_g = e^{A_3\varphi}$, i.e.

$$T_g = \begin{Vmatrix} e^{il\varphi} & 0 & 0 & \dots & 0 \\ 0 & e^{i(l-1)\varphi} & 0 & \dots & 0 \\ 0 & 0 & e^{i(l-2)\varphi} & \dots & 0 \\ \cdot & \cdot & \cdot & \cdot & \cdot \\ \cdot & \cdot & \cdot & \cdot & \cdot \\ 0 & 0 & 0 & & e^{-il\varphi} \end{Vmatrix}.$$

If the number l is half an odd integer we must associate with the identity rotation the two matrices E and $-E$. For if the angle of the rotation changes continuously from 0 to 2π then the matrix T_g changes continuously from E to $-E$. Thus for such l we do not obtain a unique continuous representation. We obtain a situation analogous to that which arose in Paragraph 4 of Sect.1. There we put every rotation in correspondence with two unitary matrices of the second order, distinguished only by their sign. The product of two rotations corresponds to the product of their matrices, also taken with sign . In this way we obtain what is called a two-valued representation of the group of rotations. It will be more convenient for us to consider this further in Sect. 6. There it will be shown that an analogous correspondence can be set up for any value of l which is half an odd integer.

An example, taken from Sect.1, is the case $l=\frac{1}{2}$ we will find the matrices A_k corresponding to infinitesimal rotations about the coordinate axes. These matrices are determined uniquely, if we put the matrix $\begin{Vmatrix} 1 & 0 \\ 0 & 1 \end{Vmatrix}$ in correspondence with the identity rotation, and for other rotations define the sign so as to maintain continuity (see paragraph 4, Sect.1). In Sect.1 we determined the matrices of order two corresponding to infinitesimal rotations about Ox and Oz (see equations (18) and (19), Sect.1). Differentiating these matrices with respect to θ and φ respectively, and substituting zero for the values of these parameters, we find the matrices A_1 and A_3:

$$A_1=\frac{i}{2}\begin{Vmatrix} 0 & 1 \\ 1 & 0 \end{Vmatrix}, \quad A_3=\frac{i}{2}\begin{Vmatrix} 1 & 0 \\ 0 & -1 \end{Vmatrix}.$$

Further,

$$A_2=[A_3,\ A_1]=\frac{1}{2}\begin{Vmatrix} 0 & -1 \\ 1 & 0 \end{Vmatrix}.$$

The matrix A_3 thus determined is diagonal and coincides with the matrix A_3 of formula (19) of this paragraph, for the value $l=\frac{1}{2}$. At the same time, A_1 and A_2 do not coincide with the matrices (19). This means that although a basis in two-dimensional space consists of the eigenvectors of H_3, it is not a canonical basis in that it is not normalized. We multiply the first of the vectors of the basis by i and the second by $-i$, i.e. we put

$$\left.\begin{aligned} \tilde{A}_1&=\begin{Vmatrix} i & 0 \\ 0 & -i \end{Vmatrix}\frac{i}{2}\begin{Vmatrix} 0 & 1 \\ 1 & 0 \end{Vmatrix}\begin{Vmatrix} -i & 0 \\ 0 & i \end{Vmatrix}=-\frac{i}{2}\begin{Vmatrix} 0 & 1 \\ 1 & 0 \end{Vmatrix}, \\ \tilde{A}_2&=\begin{Vmatrix} i & 0 \\ 0 & -i \end{Vmatrix}\frac{1}{2}\begin{Vmatrix} 0 & -1 \\ 1 & 0 \end{Vmatrix}\begin{Vmatrix} -i & 0 \\ 0 & i \end{Vmatrix}=\frac{1}{2}\begin{Vmatrix} 0 & 1 \\ -1 & 0 \end{Vmatrix}, \\ \tilde{A}_3&=\begin{Vmatrix} i & 0 \\ 0 & -i \end{Vmatrix}\frac{i}{2}\begin{Vmatrix} 1 & 0 \\ 0 & -1 \end{Vmatrix}\begin{Vmatrix} -i & 0 \\ 0 & i \end{Vmatrix}=\frac{i}{2}\begin{Vmatrix} 1 & 0 \\ 0 & -1 \end{Vmatrix}. \end{aligned}\right\} \qquad (26)$$

We see that the matrices $\tilde{A}_k$ are the same as those defined in equation (19) for $l=\frac{1}{2}$, i.e. we have a new basis which is canonical. The matrix of second order corresponding, in the canonical basis, to an arbitrary rotation with Euler angles φ_1, θ, φ_2, has the following form:

$$\left\| \begin{matrix} \cos\frac{\theta}{2} e^{i\frac{\varphi_1+\varphi_2}{2}} & -i\sin\frac{\theta}{2} e^{-i\frac{\varphi_2-\varphi_1}{2}} \\ -i\sin\frac{\theta}{2} e^{i\frac{\varphi_2-\varphi_1}{2}} & \cos\frac{\theta}{2} e^{-i\frac{\varphi_1+\varphi_2}{2}} \end{matrix} \right\|. \qquad (27)$$

So far we have been concerned with the irreducible representations of the group of rotations. We now discuss briefly the representations of the entire orthogonal group, i.e. the group of rotations and reflections. Since each element of the entire orthogonal group is either a rotation g or the product gg_- of a rotation and a reflection in the origin of coordinates (see paragraph 1, Sect.1) it is sufficient for the study of the irreducible representations of the entire orthogonal group to specify the transformation representing a reflection in the origin of coordinates. Since $g_-^2 = e$, $T_{g_-}^2 = E$, it is possible to deduce from this that, if l is an integer then $T_{g_-} = \pm E$. Thus for each integer l there exist two different irreducible representations of weight l for the entire orthogonal group. For the first of these $T_{g_-} = E$ and for the second $T_{g_-} = -E$.

As an example, we take the case $l = 0$. We have our two possibilities $T_{g_-} = \pm E$. In the first case $T_{g_-} = E$ one has only one kind of quantity, namely "scalars", which remain unaltered by the transformations of the representation. In the second case $T_{g_-} = -E$ one has "pseudo-scalars" which are quantities unchanged by rotations and whose sign changes under reflection. An example of a pseudo-scalar is the determinant whose rows are three given vectors.

For the case $l = 1$ the quantities transformed under the representation are vectors of three-dimensional space. Under reflection, ordinary vectors change their sign, that is $T_{g_-} = -E$. If $T_{g_-} = +E$ then a quantity undergoing this transformation is known as a pseudo-vector. In textbooks on vectors, vectors of the usual type are called polar vectors, and pseudo-vectors (for example, the vector product of two polar vectors) are called axial vectors.

In the general case the quantities operated upon by an irreducible representation of the whole orthogonal group are called l-vectors if $T_{g_-} = (-1)^l E$, and the corresponding pseudo-quantities if $T_{g_-} = (-1)^{l+1} E$.

Appendix to Section 2. Proof that the Matrix T_g is Differentiable

For the derivation of the matrices A_1, A_2, A_3 we assumed that the matrix T_g was differentiable with respect to the parameters ξ_1, ξ_2, ξ_3. We now prove that this is so. To do this we show that for any vector η the function $T_g\eta$ is a differentiable function of g. Take an arbitrary element $\xi \in R$ and put:

$$\eta = \int f(g)\, T_g \xi\, dg,$$

where $f(g)$ is some function on G, differentiable with respect to g (i.e. with respect to the parameters ξ_1, ξ_2, ξ_3). Here, by integration we mean the invariant integration introduced in Sect. 1, §. 3. We observe that the integral of the vector $f(g)\,T_g\xi$ of R is the vector whose components (with respect an arbitrary basis) are the integrals of the components of $f(g)\,T_g\xi$.

We show that the function $T_{g_0}\eta$ is differentiable with respect to g_0. For this we must determine $T_{g_0}\eta$.

We have:

$$T_{g_0}\eta = T_{g_0}\int f(g)\,T_g\xi\,dg = \int f(g)\,T_{g_0}T_g\xi\,dg = \int f(g)\,T_{g_0 g}\xi\,dg.$$

Since the integration is invariant we have:

$$T_{g_0}\eta = \int f(g_0^{-1}g)\,T_g\xi\,dg.$$

Thus if the function $f(g_0^{-1}g)$ is differentiable with respect to the parameters then so also is $T_{g_0}\eta$. But we commenced by assuming that this function was differentiable. Hence $T_{g_0}\eta$ is also differentiable.

We now show that $T_{g_0}\eta$ is differentiable for any vector η. It is clear that if the function $T_{g_0}\eta$ is differentiable with respect to g_0 for $\eta = \eta_1$ and $\eta = \eta_2$ then it is differentiable for any vector η which is a linear combination of these. We have proved that $T_g\eta$ is differentiable for vectors of the form

$$\eta = \int f(g)\,T_g\xi\,dg. \quad *$$

Hence it is sufficient to show that the whole space R is spanned by linear combinations of vectors of the form $(*)$; this we do by proving that a vector η_0 which is orthogonal to all vectors of the form $(*)$ is necessarily the zero-vector. Let $(\eta_0, \eta) = 0$, whenever $\eta = \int f(g)\,T_g\xi\,dg$ and $f(g)$ is an arbitrary differentiable function, i.e.

$$\left(\eta_0, \int f(g)\,T_g\xi\,dg\right) = 0,$$

i.e.

$$\int f(g)\,(\eta_0, T_g\xi)\,dg = 0.$$

Since $f(g)$ is an arbitrary differentiable function we have:

$$(\eta_0, T_g\xi) = 0 \text{ for any } g \text{ and any } \xi.$$

In particular, putting $g = e$, we obtain:

$$(\eta_0, \xi) = 0 \text{ for any } \xi,$$

i.e. $\eta_0 = 0$, as required. Hence the function $T_g\eta$ is differentiable for any vector η. Taking the case when the vectors η are the vectors of a basis we obtain the result that the elements of the matrix T_g are differentiable.

Section 3. Spherical Functions and Representations of the Group of Rotations

In Sect. 2 we obtained the possible irreducible representations of the group of rotations of three-dimensional space. We now turn our attention to a realization of these representations which is frequently met with in analysis, namely that in which the operators T_g are regarded as transforming sets of functions. In this interpretation there arise systems of functions which are invariant with respect to rotations, the so-called spherical functions.

One of the results of this section will be the proof of the existence of irreducible representations of weight l, for any integral value of l. The method used to determine the spherical functions will be a fairly general one. In Sect. 7 we shall use a similar method to obtain another class of special functions.

§ 1. Definition of Spherical Functions

Consider the function $f(x) = f(x_1, x_2, x_3)$. Any rotation g may be written in the form

$$x' = gx \qquad x'_i = \sum g_{ik} x_k. \tag{1}$$

If we substitute in $f(x_1, x_2, x_3)$ for the x_k their values as given by (1) in terms of the x'_i we obtain a new function $f_1(x'_1, x'_2, x'_3)$. We say that the rotation g transforms the function f into f_1. We denote the transformation which takes f into f_1 by T_g. Thus we have associated with each rotation g a transformation T_g of the functions f, which transforms a function f into a function f_1, obtained from f by the substitution of x' for $x = g^{-1}x'$ i.e.

$$T_g f(x) = f_1(x), \text{ where } f_1(x) = f(g^{-1}x). \tag{2}$$

It is clear that the transformation T_g is linear: a sum of functions is transformed into a sum, a scalar multiple into a scalar multiple.

We show that a product of transformations T_g corresponds to a product of rotations. Take two successive rotations g_1 and g_2

$$x' = g_1 x,$$

$$x'' = g_2 x'.$$

As a result of the first rotation $f(x)$ is transformed into

$$T_{g_1} f(x) = f(g_1^{-1} x),$$

and as a result of the second into

$$T_{g_2} T_{g_1} f(x) = T_{g_1} f(g_1^{-1} x) = f((g_1^{-1} g_2^{-1} x)) = f((g_2 g_1)^{-1} x) = T_{g_2 g_1} f(x).$$

This means that

$$T_{g_2 g_1} = T_{g_2} T_{g_1}. \tag{3}$$

Since a sphere with centre the origin transforms into itself under a rotation, we limit ourselves to functions defined on the surface of such a sphere. We may thus assume that $x_1^2+x_2^2+x_3^2=1$, i.e. that x lies on the surface of the unit sphere. It will also be often convenient for us to consider the vector x as given by its spherical coordinates θ and φ, where $x_1=\sin\theta\cos\varphi$, $x_2=\sin\theta\sin\varphi$ and $x_3=\cos\theta$.

We may limit ourselves to the study of functions $f(\theta, \varphi)$, the square of the modulus of which is integrable over the surface of the sphere. We define the scalar product of two such functions by the equation

$$f, g)=\int_0^{2\pi}\int_0^{\pi} f(\theta, \varphi)\overline{g(\theta, \varphi)}\sin\theta\, d\theta\, d\varphi.$$

In the metric defined by this scalar product the transformations T_g are unitary. This is so because a rotation does not alter the element of surface of the unit sphere and so the integral of a product of transformed functions is equal to the integral of the product of the original functions. If we express the vector x in terms of θ, φ and x' in terms of θ', φ', we have

$$(T_g f, T_g g)=\int_0^{2\pi}\int_0^{\pi} f(\theta', \varphi')\overline{g(\theta', \varphi')}\sin\theta\, d\theta\, d\varphi=$$

$$=\int_0^{2\pi}\int_0^{\pi} f(\theta', \varphi')\overline{g(\theta', \varphi')}\sin\theta'\, d\theta'\, d\varphi'=(f, g),$$

i.e. the transformation T_g is unitary.

With the aid of the transformations just introduced we can obtain an irreducible representation corresponding to each integer l. To this end we construct a finite-dimensional space, consisting of functions, in which the transformations T_g provide an irreducible representation of the group of rotations with the given weight l. For given l, this finite dimensional space will consist of linear combinations of $2l+1$ functions $f_m(x)$ $(-l\leqslant m\leqslant l)$*. The functions $f_m(x)$ are chosen so as to form a canonical basis for this representation. The functions on the sphere, belonging to the space of the irreducible representation of weight l, are called the spherical functions of the l-th order. The functions $f_m(x)$ forming the canonical basis in the space are called the basic spherical functions of the l-th order.

Since a canonical basis is determined with the aid of the transformations

---

* Strictly speaking when the transformations T_g are specified so as to satisfy (3) and are proved unitary, a unitary representation is obtained in the space of all square-integrable functions, defined on the surface of the sphere. It is natural to consider the problem of determining the decomposition of this representation into irreducible components. By constructing a system of functions $f_m(x)$ for each l we solve the main part of this problem i.e. we separate the invariant subspaces corresponding to the irreducible representations of the rotation group. The problem in its entirety will be solved in paragraph 5 of this section.

corresponding to infinitesimal rotations, to find the $f_m(x)$ we firstly derive the transformations A_1, A_2, A_3, corresponding to infinitesimal rotations.

§ 2. The Differential Operators Corresponding to Infinitesimal Rotations

In paragraph 1 we introduced the linear transformations T_g in the space of functions on the surface of the sphere. We now obtain the transformations A_1, A_2 and A_3 which correspond in this space to infinitesimal rotations about the coordinate axes. The functions f, on which the transformations T_g operate, will be assumed to be differentiable.*

We first find the operator A_3 corresponding to an infinitesimal rotation about the axis Oz. Consider a rotation g through an angle α. Since (see Sect. 2) $T_g = E + \alpha A_3 + \ldots$, in order to determine $A_3 f$ we must expand $T_g f$ in terms of α and take the coefficient of the first term in α.

We have $T_g f(x) = f(g^{-1}x)$. Therefore, for a rotation g about Oz we have $T_g f(\theta, \varphi) = f(\theta, \varphi - \alpha)$.

Expanding $f(\theta, \varphi - \alpha)$ as a power series in α, we obtain

$$f(\theta, \varphi - \alpha) = f(\theta, \varphi) - \alpha \frac{\partial f(\theta, \varphi)}{\partial \varphi} + \ldots \tag{4}$$

Hence

$$A_3 f = -\frac{\partial f(\theta, \varphi)}{\partial \varphi},$$

and so the operator A_3 is the differential operator given by

$$A_3 = -\frac{\partial}{\partial \varphi}.$$

It is easily seen that generally, each A_k $(k = 1, 2, 3)$ for the transformations T_g of the functions f defined in paragraph 1, will be differential operators of the first order. For a rotation g through a small angle α about some fixed axis $T_g f(\theta, \varphi) = f(\theta', \varphi')$, where θ', φ' depend on the angle of rotation α and coincide with θ and φ, respectively when $\alpha = 0$.

Expanding $T_g f = f(\theta', \varphi')$ as a power series in α we have

$$f(\theta', \varphi') = f(\theta, \varphi) + \left(\frac{\partial f}{\partial \theta'} \frac{d\theta'}{d\alpha} + \frac{\partial f}{\partial \varphi'} \frac{d\varphi'}{d\alpha} \right) \Bigg|_{\alpha=0} \alpha + \ldots,$$

Consequently the operator A corresponding to an infinitesimal rotation has the form

---

* The fact that the functions forming an irreducible invariant subspace are differentiable follows from the theorem on complete reducibility of a unitary representation together with the supplement of Sect. 2.

$$A = a(\theta, \varphi)\frac{\partial}{\partial\theta} + b(\theta, \varphi)\frac{\partial}{\partial\varphi},$$

where

$$a(\theta, \varphi) = \frac{d\theta}{d\alpha}\Big|_{\alpha=0}, \quad b(\theta, \varphi) = \frac{d\varphi}{d\alpha}\Big|_{\alpha=0}. \tag{5}$$

We now find the differential operator A_1, corresponding to an infinitesimal rotation about Ox. To obtain the functions $a(\theta, \varphi)$ and $b(\theta, \varphi)$, i.e. $\frac{d\theta}{d\alpha}\Big|_{\alpha=0}$ and $\frac{d\varphi}{d\alpha}\Big|_{\alpha=0}$ for this rotation, it will be easier for us to obtain the derivatives with respect to α from the Cartesian coordinates x_1, x_2, x_3 of the vector x. If g is a rotation through an angle α about Ox then g^{-1} is also a rotation about this axis through an angle $-\alpha$ Therefore the vector $x' = g^{-1}x$ has the components $x'_1 = x_1$, $x'_2 = x_2\cos\alpha + x_3\sin\alpha$, $x'_3 = -x_2\sin\alpha + x_3\cos\alpha$. The functions $\frac{dx_k}{d\alpha}\Big|_{\alpha=0}$ are given by

$$\frac{dx_1}{d\alpha}\Big|_{\alpha=0} = 0, \quad \frac{dx_2}{d\alpha}\Big|_{\alpha=0} = x_3, \quad \frac{dx_3}{d\alpha}\Big|_{\alpha=0} = -x_2. \tag{6}$$

Differentiating with respect to α the equations $x_1 = \sin\theta\cos\varphi$, $x_2 = \sin\theta\sin\varphi$, $x_3 = \cos\theta$, connecting the Cartesian coordinates with the spherical coordinates, and using (6) we have for $\alpha = 0$,

$$\cos\theta\cos\varphi\frac{d\theta}{d\alpha} - \sin\theta\sin\varphi\frac{d\varphi}{d\alpha} = 0,$$

$$\cos\theta\sin\varphi\frac{d\theta}{d\alpha} + \sin\theta\cos\varphi\frac{d\varphi}{d\alpha} = \cos\theta$$

$$-\sin\theta\frac{d\theta}{d\alpha} = -\sin\theta\sin\varphi.$$

from which we obtain

$$\frac{d\theta}{d\alpha} = \sin\varphi \quad \text{and} \quad \frac{d\varphi}{d\alpha} = \cot\theta\cos\varphi.$$

Substituting these values of $\frac{d\theta}{d\alpha}\Big|_{\alpha=0}$ and $\frac{d\varphi}{d\alpha}\Big|_{\alpha=0}$ in formula (5) we find that A_1 is the differential operator, given by the equation

$$A_1 = \sin\varphi\frac{\partial}{\partial\theta} + \cot\theta\cos\varphi\frac{\partial}{\partial\varphi}. \tag{7}$$

The operator A_2 corresponding to an infinitesimal rotation about Oy may be obtained in a similar manner. However, since the interchange of φ and $\varphi - \frac{\pi}{2}$ interchanges the axes Oy and Ox, we may obtain A_2 by putting $\varphi - \frac{\pi}{2}$ for φ in equation (7). Thus,

$$A_2 = -\cos\varphi\frac{\partial}{\partial\theta} + \cot\theta\sin\varphi\frac{\partial}{\partial\varphi}. \tag{8}$$

We are now in a position to find the operators H_+, H_- and H_3 which were introduced in Sect.2. Using the expressions above for A_1, A_2 and A_3 we obtain

$$\left.\begin{aligned} H_+ &= H_1 + iH_2 = iA_1 - A_2 = e^{i\varphi}\left(\frac{\partial}{\partial\theta} + i\cot\theta\frac{\partial}{\partial\varphi}\right), \\ H_- &= H_1 - iH_2 = iA_1 + A_2 = e^{-i\varphi}\left(-\frac{\partial}{\partial\theta} + i\cot\theta\frac{\partial}{\partial\varphi}\right), \\ H_3 &= iA_3 = -i\frac{\partial}{\partial\varphi}. \end{aligned}\right\} \quad (9)$$

§ 3. The Differential Equation of Spherical Functions

As previously stated, the functions on the sphere which belong to the invariant subspace of the irreducible representation of weight l are called the spherical functions* of order l. The functions forming the canonical basis in this subspace (i.e. the eigenvectors of the transformation H_3) are called the basic spherical functions. The basic spherical functions will be denoted by $Y_l^m(\theta, \varphi)$ where m is the number of the corresponding vector of the basis i.e. the corresponding eigenvalue of H_3 $(-l \leqslant m \leqslant l)$. Thus each spherical function of order l is a linear combination of the $2l+1$ basic spherical functions $Y_l^m(\theta, \varphi)$. Simple expressions for these functions will be found in the next paragraph. Meanwhile we consider how they depend on φ, and obtain their differential equation.

The function $Y_l^m(\theta, \varphi)$ is the eigenfunction of H_3 corresponding to the eigenvalue m. Using the expression (9) for H_3, we obtain:

$$H_3 Y_l^m(\theta, \varphi) = -i\frac{\partial Y_l^m(\theta, \varphi)}{\partial\varphi} = mY_l^m(\theta, \varphi).$$

Hence we have:

$$Y_l^m(\theta, \varphi) = e^{im\varphi}F_l^m(\theta). \quad (10)$$

The way Y_l^m depends on φ is now clear; its dependence on θ will be considered in more detail later.

From equation (10) it is clear that the functions thus obtained are single-valued on the surface of the sphere, corresponding to any integer l.**

Thus we obtain at least the irreducible representations corresponding to integral values of l. Since the $Y_l^m(\theta, \varphi)$ are normalized eigenfunctions of H_3, we have

---

* It will be shown in paragraph 4 that there is exactly one such invariant subspace for every weight l.

** If l is half of an odd integer, we would arrive at the same point with the sign of the value of the function reversed by starting at the point (θ, φ) and varying φ continuously to $\varphi + 2\pi$.

$$\int_0^{2\pi}\int_0^{\pi} |Y_l^m(\theta, \varphi)|^2 \sin\theta \, d\theta \, d\varphi = 1. \tag{11}$$

Since $\int_0^{2\pi} |e^{im\varphi}| \, d\varphi = 2\pi$, we shall replace (10) by

$$Y_l^m(\theta, \varphi) = \frac{1}{\sqrt{2\pi}} F_l^m(\theta) e^{im\varphi}. \tag{10'}$$

We may now write condition (11) in the form

$$\int_0^{\pi} |F_l^m(\theta)|^2 \sin\theta \, d\theta = 1. \tag{11'}$$

We now derive differential equations for the spherical functions and for the functions $F_l^m(\theta)$. As was shown in Sect. 2, the vectors transformed by the irreducible representation with the given weight l satisfy the equation $H^2 f = l(l+1) f$, where $H^2 = H_1^2 + H_2^2 + H_3^2$.

We now find an explicit form of this equation for the case we are considering. We note that $H_1^2 + H_2^2 = \frac{1}{2}(H_+ H_- + H_- H_+)$. Substituting for H_+ and H_- from equations (9) we find that

$$H_1^2 + H_2^2 = -\frac{\partial^2}{\partial\theta^2} - \cot\theta \frac{\partial}{\partial\theta} - \cot^2\theta \frac{\partial^2}{\partial\varphi^2}.$$

Adding to this the equation $H_3^2 = -\frac{\partial^2}{\partial\varphi^2}$ and simplifying we obtain:

$$-H^2 = \frac{1}{\sin\theta}\frac{\partial}{\partial\theta}\left(\sin\theta \frac{\partial}{\partial\theta}\right) + \frac{1}{\sin^2\theta}\frac{\partial^2}{\partial\varphi^2}.$$

Thus the equation $[-H^2 + l(l+1)E] f = 0$ has in this case the form

$$\frac{1}{\sin\theta}\frac{\partial}{\partial\theta}\left(\sin\theta \frac{\partial f}{\partial\theta}\right) + \frac{1}{\sin^2\theta}\frac{\partial^2 f}{\partial\varphi^2} + l(l+1) f = 0. \tag{12}$$

This is known as the differential equation of the spherical functions of the l-th order. The number of linearly independent solutions of this equation (we are, naturally, only interested in those solutions which are continuous and differentiable over the whole sphere) must be a multiple of $2l+1$. In paragraph 4 we show that in fact the number is precisely $2l+1$.

Let us now consider the basic spherical functions. Substituting $Y_l^m(\theta, \varphi)$ from equation (10) in equation (12) we obtain the usual differential equation for the function $F_l^m(\theta)$. It has the form

$$\frac{1}{\sin\theta}\frac{d}{d\theta}\left(\sin\theta \frac{dF_l^m}{d\theta}\right) + \left[l(l+1) - \frac{m^2}{\sin^2\theta}\right] F_l^m(\theta) = 0 \tag{13}$$

Or, introducing a new independent variable $\mu = \cos\theta$ and putting $P_l^m(\mu)$ for $F_l^m(\theta)$, the form

$$\left[(1-\mu^2)P_l^{m\prime}(\mu)\right]' + \left[l(l+1) - \frac{m^2}{1-\mu^2}\right]P^m(\mu) = 0. \tag{13'}$$

Thus finally we obtain the result that the basic spherical functions have the form

$$Y_l^m(\theta, \varphi) = \frac{1}{\sqrt{2\pi}} e^{im\varphi} P_l^m(\cos\theta), \tag{14}$$

where $P_l^m(\mu)$ satisfies condition (13'). In the next paragraph we obtain a simple expression for $P_l^m(\mu)$.

§ 4. An Explicit Expression for Spherical Functions

We now obtain a simple expression for the basic spherical functions. We shall at the same time prove that for every integer l there exists one, and only one, invariant subspace of functions in which there is a realization of a representation of weight l. To find the canonical basis for the irreducible representation of weight l we begin (as in Sect.2, p.29) with the solution of the simultaneous equations

$$H_3 f = lf,$$
$$H_+ f = 0,$$

i.e. with the determination of $Y_l^l(\theta, \varphi)$ (the eigenvector of H_3 corresponding to the greatest eigenvalue).

The first of these equations, as in paragraph 3, implies that $Y_l^l(\theta, \varphi)$ has the form

$$Y_l^l(\theta, \varphi) = \frac{1}{\sqrt{2\pi}} e^{il\varphi} F_l^l(\theta).$$

Substituting this in the second equation and cancelling $e^{i(l+1)\varphi}$, we obtain the following differential equation for $F_l^l(\theta)$:

$$\frac{dF_l^l(\theta)}{d\theta} - l\cot\theta F_l^l(\theta) = 0.$$

The general solution of this equation has the form

$$F_l^l(\theta) = C\sin^l\theta. \tag{15}$$

Hence it is clear that of all the eigenfunctions of the operator H_3 corresponding to the eigenvalue l only one satisfies the equation $H_+ f = 0$. Consequently for every l there exists only one irreducible representation of weight l since, if this were not so, the equation $H_+ Y_l^l = 0$ would have for some l at least two linearly independent solutions of the form $e^{il\varphi}\Phi(\theta)$.

Before we find the $Y_l^m(\theta, \varphi)$ for $m < l$, we first normalize the function

$F_l^l(\theta) = C \sin^l \theta$, i.e. find a constant C such that, according to (11'),

$$\int_0^\pi |F_l^l(\theta)|^2 \sin\theta \, d\theta = 1.$$

Evaluating the integral we find that

$$C^2 \int_0^\pi \sin^{2l+1}\theta \, d\theta = C^2 \cdot 2^{2l+1} \frac{[l!]^2}{(2l+1)!} = 1,$$

whence

$$C = \pm \frac{1}{2^l l!} \sqrt{\frac{2l+1}{2}} \sqrt{(2l)}.$$

We shall take

$$C = (-1)^l \frac{1}{2^l l!} \sqrt{\frac{2l+1}{2}} \sqrt{(2l)}.$$

Thus

$$Y_l^l(\theta, \varphi) = \frac{(-1)^l}{\sqrt{2\pi} \cdot 2^l \cdot l!} \sqrt{\frac{2l+1}{2}} \sqrt{(2l)}\, e^{il\varphi} \sin^l \theta = \frac{C}{\sqrt{2\pi}} e^{il\varphi} \sin^l \theta.$$

We now find the other functions of the canonical basis $f_m = Y_l^m(\theta, \varphi)$. We make use of the equation $H_- f_m = \alpha_m f_{m-1}$ where $\alpha_m = \sqrt{(l+m)(l-m+1)}$. Since $H_- = e^{-i\varphi}\left(-\frac{\partial}{\partial\theta} + i \cot\theta \frac{\partial}{\partial\varphi}\right)$, we have

$$e^{-i\varphi}\left(-\frac{\partial Y_l^m}{\partial\theta} + i\cot\theta \frac{\partial Y_l^m}{\partial\varphi}\right) = \alpha_m Y_l^{m-1}.$$

Substituting $Y_l^m(\theta, \varphi) = \frac{1}{\sqrt{2\pi}} e^{im\varphi} F_l^m(\theta)$ and cancelling $\frac{1}{\sqrt{2\pi}} e^{i(m-1)\varphi}$, we obtain the following recurrence relation for $F_l^m(\theta)$

$$-\frac{dF_l^m(\theta)}{d\theta} - m\cot\theta F_l^m(\theta) = \alpha_m F_l^{m-1}(\theta).$$

Putting, as before, $\mu = \cos\theta$ and denoting $F_l^m(\theta)$ by $P_l^m(\mu)$ we obtain the relation

$$\sqrt{1-\mu^2}\left(\frac{dP_l^m(\mu)}{d\mu} - m\frac{\mu}{1-\mu^2} P_l^m(\mu)\right) = \alpha_m P_l^{m-1}(\mu). \tag{16}$$

Since $P_l^l(\mu)$ is known, this equation makes it possible to determine $P_l^m(\mu)$. With this object in view we make the substitution

$$P_l^m(\mu) = (1-\mu^2)^{-\frac{m}{2}} u_m(\mu). \tag{17}$$

We then obtain from (16) a relation, which after the cancellation of

$(1-\mu^2)^{-\frac{m-1}{2}}$ takes the simple form

$$u_{m-1}(\mu) = \frac{1}{\alpha_m}\frac{du_m}{d\mu}. \tag{18}$$

From (15) we see that $P_l^l(\mu) = C(1-\mu^2)^{\frac{l}{2}}$. This means that $u_l(\mu) = C(1-\mu^2)^l$. Hence by equation (18)

$$u_{l-1}(\mu) = \frac{C}{\alpha_l}\frac{d(1-\mu^2)^l}{d\mu}, \quad u_{l-2}(\mu) = \frac{C}{\alpha_l\alpha_{l-1}}\frac{d^2(1-\mu^2)^l}{d\mu^2}, \ldots$$

$$\ldots, \; u_m(\mu) = \frac{C}{\alpha_l\alpha_{l-1}\ldots\alpha_{m+1}}\frac{d^{l-m}(1-\mu^2)^l}{d\mu^{l-m}}$$

If we substitute the value determined for $u_m(\mu)$ in equation (17) we obtain the expression

$$P_l^m(\mu) = \frac{C}{\alpha_l\alpha_{l-1}\ldots\alpha_{m+1}}(1-\mu^2)^{-\frac{m}{2}}\frac{d^{l-m}(1-\mu^2)^l}{d\mu^{l-m}}.$$

Here $m = l, l-1, l-2, \ldots$ We note that, as we might expect, $P_l^m(\mu) \equiv 0$ for $m \leqslant -l-1$. Substituting for C and α_m and taking $(-1)^l$ into the derivative we obtain finally

$$P_l^m(\mu) = \sqrt{\frac{(l+m)!}{(l-m)!}}\sqrt{\frac{2l+1}{2}}\frac{1}{2^l\cdot l!}(1-\mu^2)^{-\frac{m}{2}}\frac{d^{l-m}(\mu^2-1)^l}{d\mu^{l-m}}. \tag{19}$$

In particular the function $P_l^0(\mu)$ which will frequently be denoted just by $P_l(\mu)$ has the form

$$P_l(\mu) = \sqrt{\frac{2l+1}{2}}\frac{1}{2^l\cdot l!}\frac{d^l(\mu^2-1)^l}{d\mu^l}. \tag{20}$$

The polynomial $P_l(\mu)$ is called the normalized Legendre polynomial of the l-th order, and the functions $P_l^m(\mu)$ are called the normalized associated Legendre functions.

Thus the following theorem has been proved: the basic spherical functions of the l-th order have the form

$$Y_l^m(\theta,\varphi) = \frac{1}{\sqrt{2\pi}}e^{im\varphi}P_l^m(\cos\theta),$$

where $P_l^m(\mu)$ is given by (19). The linear combinations of the functions $Y_l^m(\theta,\varphi)$ form, for a given l, a function space of dimension $(2l+1)$ which is invariant with respect to the rotation group and provides a realization of the irreducible representation of weight l.

In conclusion we obtain a recurrence formula for the Legendre polynomials and functions of a given value of l. Two recurrence relations, in terms of the functions $P_l^m(\mu)$ and their first derivatives, are contained in the transformation formulae of the basic spherical functions $H\ Y_l^m(\theta,\varphi) =$

$= \alpha_m Y_l^{m-1}(\theta, \varphi)$ and $H_+ Y_l^m(\theta, \varphi) = \alpha_{m+1} Y_l^{m+1}(\theta, \varphi)$ The first of these equations was used above for the derivation of a recurrence formula. For this we substituted $Y_l^m(\theta, \varphi) = \frac{e^{im\varphi}}{\sqrt{2\pi}} P_l^m(\cos\theta)$ and obtained the equation

$$\sqrt{1-\mu^2}\frac{dP_l^m}{d\mu} = m\frac{\mu}{\sqrt{1-\mu^2}} P_l^m(\mu) + \sqrt{(l+m)(l-m+1)}\, P_l^{m-1}(\mu). \tag{21}$$

Similarly the relation $H_+ Y_l^m = \alpha_{m+1} Y_l^{m+1}$ gives another recurrence formula

$$-\sqrt{1-\mu^2}\frac{dP_l^m}{d\mu} = m\frac{\mu}{\sqrt{1-\mu^2}} P_l^m(\mu) + \sqrt{(l+m+1)(l-m)}\, P_l^{m+1}(\mu). \tag{22}$$

Combining (21) and (22) we obtain a relation between three consecutive normalized Legendre functions.

$$\sqrt{(l+m+1)(l-m)}\, P_l^{m+1}(\mu) + 2m\frac{\mu}{\sqrt{1-\mu^2}} P_l^m(\mu) + \\ + \sqrt{(l+m)(l-m+1)}\, P_l^{m-1}(\mu) = 0. \tag{23}$$

In Sect. 7 recurrence relations will be derived for the associated Legendre functions and Legendre polynomials for different values of l.

From equation (22) it is possible to find another more general expression for the associated Legendre functions. Putting $m = 0$ in this equation we find that

$$P_l^1(\mu) = -\frac{1}{\alpha_1}(1-\mu^2)^{\frac{1}{2}}\frac{dP_l}{d\mu}.$$

In general we put:

$$P_l^m(\mu) = (1-\mu^2)^{\frac{m}{2}} V_m(\mu). \tag{24}$$

Then equation (22) gives a simple relation between the functions $V_m(\mu)$, namely

$$V_{m+1}(\mu) = -\frac{1}{\alpha_{m+1}}\frac{dV_m(\mu)}{d\mu}.$$

Since $V_0(\mu) = \sqrt{\frac{2l+1}{2}}\frac{1}{2^l \cdot l!}\frac{d^l(\mu^2-1)^l}{d\mu^l}$ this relation implies that

$$V_m(\mu) = (-1)^m \frac{1}{\alpha_1\alpha_2\ldots\alpha_m}\sqrt{\frac{2l+1}{2}}\frac{1}{2^l \cdot l!}\frac{d^{l+m}(\mu^2-1)^l}{d\mu^{l+m}}.$$

Putting this value of $V_m(\mu)$ in (24) and substituting for the α_m, we obtain the following expression for the associated Legendre functions:

$$P_l^m(\mu) = (-1)^m \sqrt{\frac{(l-m)!}{(l+m)!}} \sqrt{\frac{2l+1}{2}} \frac{1}{2^l \cdot l!} (1-\mu^2)^{\frac{m}{2}} \frac{d^{m+l}(\mu^2-1)^l}{d\mu^{m+l}}. \tag{25}$$

A comparison of (19) and (25) shows that the one arises from the other by replacing m by $-m$. Thus we may conclude that

$$P_l^m(\mu) = (-1)^m P_l^{-m}(\mu),$$

i.e. the normalized associated Legendre functions corresponding to values of m equal in magnitude but opposite in sign, are proportional.

Finally we observe that if we substitute $\sqrt{\frac{2l+1}{2}} \frac{1}{2^l \cdot l!} \frac{d^l(\mu^2-1)^l}{d\mu^l}$ for $P_l(\mu)$, in equation (25) we obtain

$$P_l^m(\mu) = (-1)^m \sqrt{\frac{(l-m)!}{(l+m)!}} (1-\mu^2)^{\frac{m}{2}} \frac{d^m P_l(\mu)}{d\mu}. \tag{26}$$

This equation gives the normalized associated Legendre functions in terms of of the normalized Legendre polynomials of the same order.

§ 5. The Expression of Functions Defined on the Sphere in Terms of the Spherical Functions

From Sect.1, paragraph 5, we know that any unitary representation of the group of rotations, even if it is infinite dimensional, may be resolved into irreducible representations. At the beginning of this section we considered the infinite dimensional function space consisting of functions, the squared moduli of which were integrable over the surface of the sphere. The transformation $g \to T_g$ gives a unitary representation of the group of rotations in this space.

The general theorem concerning the resolution of a representation into irreducible representations means in this case that every function, the squared modulus of which is integrable over the surface of the sphere, may be expanded as a series of the form

$$f = \varphi_0 + \varphi_1 + \varphi_2 + \ldots + \varphi_l + \ldots,$$

where each term belongs to the invariant subspace of an irreducible representation.

We have already shown that these invariant subspaces consist of the spherical functions of the appropriate order, i.e.

$$\varphi_l = \sum_{m=-l}^{l} C_l^m Y_l^m(\theta, \varphi).$$

From the general theorems of Sect.3 it follows that $Y_l^m(\theta, \varphi)$ form an orthogonal system on the surface of the sphere. The spherical functions corresponding to different values of l are orthogonal, since they belong to subspaces of non-equivalent irreducible representations. For the same l and different m they are orthogonal since they are elements of a canonical basis.

Thus it has been proved that an arbitrary function, the square of whose modulus is integrable over the surface of the sphere, may be expanded in a convergent series whose terms are multiples of the elements of an orthogonal system of spherical functions.

The expansion of functions in series of spherical functions is particularly useful in physics because of the invariance of such an expansion under rotation. For this reason the spherical functions play, in problems associated with the sphere, a role similar to that of trigonometrical functions in problems connected with the circle.

Section 4. Multiplication of Representations

In this section we describe a method of constructing from two representations of the rotation group a third representation known as their product. From this it will appear that many important representations may be expressed as a product of very simple representations. Thus for example tensor representations, with which we shall be concerned in the following section, may be expressed as products of irreducible representations for which $l=1$, and spinor representations (cf.Sect.6) as products of irreducible representations for which $l=\frac{1}{2}$.

We show further, how to resolve the product of two irreducible representations into the direct sum of irreducible representations.

§ 1. Definition of the Product of Representations

Before we define the product of representations we must define the product of spaces. Let R_1 be a p-dimensional Euclidean space with an orthonormal basis $e_1, e_2, \ldots, e_p$, let R_2 be a q-dimensional Euclidean space with an orthonormal basis $f_1, f_2, \ldots, f_q$. We consider all formal linear combinations of all possible pairs $e_i f_k$ with arbitrary coefficients:

$$h=\sum_{i,k=1}^{\substack{k=q\\ i=p}} a_{ik}e_i f_k$$

to be vectors in a new space R. This new space R we call the product space of R_1 and R_2 and we denote it by $R_1\times R_2$. Thus the vector h of the space R is determined by the pq numbers

$$a_{ik} \quad (i=1, 2, \ldots, p;\ k=1, 2, \ldots, q),$$

i.e. this space has pq dimensions. If an arbitrary element $e=\sum_i \mu_i e_i$ is taken from R_1 and an arbitrary element $f=\sum_k \lambda_k f_k$ is taken from R_2, then by ef we understand the element of the space $R_1\times R_2$, which is $\sum_{i,k}\mu_i\lambda_k e_i f_k$.

The scalar product of the two vectors $h'=\sum_{i,k} a'_{ik} f_i e_k$ and $h''=$ $=\sum_{i,k} a''_{ik} e_i f_k$ in the space $R_1\times R_2$ is defined by the equation

$$(h', h'') = \sum_{i,k} a'_{ik}\bar{a}''_{ik}, \tag{1}$$

This ensures that the basis $e_i f_k$ of the space $R = R_1 \times R_2$ is orthonormal. Similarly we may define the product of three, four or more, spaces.

In what follows we shall encounter the product of a three-dimensional space with itself. Each element is given by a set of nine numbers a_{ik} $(i, k = 1, 2, 3)$.

The product of three three-dimensional spaces is a 27-dimensional space, each element being given by a set a_{ikl} $(i, k, l = 1, 2, 3)$.

Similarly an element of the product of r three-dimensional spaces is given by a set of 3^r numbers $a_{i_1 i_2 \ldots i_r}$ $(i_1, i_2, i_3, \ldots, i_r = 1, 2, 3)$.

We now turn to the definition of the product of representations. Let there be given two representations of the rotation group in three-dimensional space: a representation with matrices U_g in a p-dimensional space R_1, and a representation with matrices V_g in a q-dimensional space R_2. Consider the product R of the spaces R_1 and R_2. From the representations in the spaces R_1 and R_2 we are able to construct a representation in R. Since the vector e_i of the space R_1 is transformed into $U_g e_i$, and the vector f_k of the space R_2 into the vector $V_g f_k$, we may define the transformation T_g in the space $R_1 \times R_2$ which corresponds to the rotation g by the equation

$$T_g e_i f_k = U_g e_i V_g f_k. \tag{2}$$

Since the vectors $e_i f_k$ form a basis in the space $R_1 \times R_2$ then, a knowledge of the $T_g e_i f_k$ determines a linear transformation T_g in R. If the matrix $U_g = \|u_{si}\|$ corresponds to the rotation g for the representation in the space R_1, and the matrix $V_g = \|v_{rk}\|$ corresponds to g for the representation in R_2, that is, if

$$U_g e_i = \sum_{s=1}^{p} u_{si} e_s, \quad \text{and} \quad V_g f_k = \sum_{r=1}^{q} v_{rk} f_r,$$

then by definition the transformation T_g is given by

$$T_g e_i f_k = \sum_{s=1}^{p} \sum_{r=1}^{q} u_{si} v_{rk} e_s f_r.$$

Hence it follows that an arbitrary element of the space $R_1 \times R_2$ having the coordinates a_{ik} in the basis $e_i f_k$ is transformed into the element with coordinates

$$a'_{sr} = \sum_{i,k} u_{si} v_{rk} a_{ik}.$$

From (2) it is easily proved that the transformations T_g so defined do in fact form a representation of the rotation group in the space R, i.e. that to a product of rotations g_1 and g_2 there corresponds the product of the transformations T_{g_1} and T_{g_2}

Hence, by the product of the representations $g \to U_g$ and $g \to V_g$, in the spaces R_1 and R_2 respectively, we mean the representation $g \to T_g$ in the space $R = R_1 \times R_2$. If the matrix of U_g in the basis e_i is $\|u_{si}\|$ and the matrix of V_g in the basis f_k is $\|v_{rk}\|$ then the vector h of $R_1 \times R_2$ having the components a_{ik} in the basis $e_i f_k$ is transformed by the rotation g into the vector $h' = T_g h$, whose components are defined by the equation

$$a'_{sr} = \sum_{i=1}^{p} \sum_{k=1}^{q} u_{si} v_{rk} a_{ik}. \tag{3}$$

Similarly we may define the product of any number of representations.

It is possible to express this result in another form. The element $h = \sum a_{ik} e_i f_k$ of the product space may be put in correspondence with the matrix $\|a_{ik}\|$ with p rows and q columns, consisting of the components of h. We denote this matrix also by h: $h = \|a_{ik}\|$. Then equation (3), which defines the transformation $h \to T_g h$, may be rewritten in the form

$$T_g h = U_g h V_g^T, \tag{3'}$$

where V_g^T is the transpose of V_g.

We now show that the product of unitary representations U_g and V_g is also unitary. First we note that, in virtue of the definition of the scalar product in a product space (equation (1)) we have:

$$(ef, e'f') = (e, e')(f, f').$$

To show that T_g is unitary, it will be sufficient to show that T_g transforms the orthonormal basis $e_i f_k$ into an orthonormal basis. This is obvious, for

$$(T_g e_i f_k, T_g e_{i'} f_{k'}) = (U_g e_i V_g f_k, U_g e_{i'} V_g f_{k'}) =$$
$$= (U_g e_i, U_g e_{i'})(V_g f_k, V_g f_{k'}) = \delta_{ii'} \cdot \delta_{kk'},$$

i.e. $T_g e_i f_k$ is an orthonormal basis; it follows that T_g is unitary.

Consider an important example. The simplest representation of the rotation group is the identical representation in three-dimensional space, in which to each rotation g there corresponds the matrix of this rotation $g = \|g_{si}\|$. We find the product of this representation with itself. The product of two three-dimensional spaces is a nine-dimensional space the elements of which have components a_{ik} $(i, k = 1, 2, 3)$. Since the rotation g transforms three-dimensional space by the matrix $\|g_{is}\|$, the numbers a_{ik} will, in in accordance with (2), be transformed by the rotation g according to the equations:

$$a'_{sr} = \sum_{i=1}^{3} \sum_{k=1}^{3} g_{si} g_{rk} a_{ik}.$$

Similarly the product of r representations is defined. The elements of the product of r three-dimensional spaces are sets of 3^r numbers $a_{i_1 i_2 \ldots i_r}$ $(i_1, i_2, \ldots, i_r = 1, 2, 3)$. The rotation g transforms these numbers according to the equations

$$a'_{i'_1 i'_2 \ldots i'_r} = \sum_{i_1, i_2, \ldots i_r = 1}^{3} g_{i'_1 i_1} g_{i'_2 i_2} \ldots g_{i'_r i_r} a_{i_1 i_2 \ldots i_r}. \qquad (3'')$$

Such a representation is called a tensor representation.

The representation just introduced is closely linked with the notion of a tensor. Suppose that some magnitude is defined in every system of coordinates by means of a set of 3^r numbers $a_{i_1 \ldots i_r}$. We wish to find how the sets of numbers corresponding to the various systems of coordinates are related. In the case of a vector in three-dimensional space the relation is clear, namely, if $g = \|g_{ik}\|$ is the matrix of the transformation from one orthogonal system of coordinates to another, then

$$a'_i = \sum g_{ik} a_k.$$

If a set of numbers $a_{i_1 \ldots i_r}$ $(i_1, \ldots, i_r = 1, 2, 3)$ is given in each orthogonal coordinate system, and the transformation from one system to another with matrix $g = \|g_{ik}\|$ takes the form

$$a'_{i'_1 i'_2 \ldots i'_r} = \sum_{i_1, i_2, \ldots i_r = 1}^{3} g_{i'_1 i_1} g_{i'_2 i_2} \ldots g_{i'_r i_r} a_{i_1 i_2 \ldots i_r},$$

then we say that we have a tensor of rank r in three-dimensional Euclidian space. In other words, a tensor of rank r is an element of the product of x three-dimensional Euclidean spaces in which the above defined product of representations of the rotation group acts. Tensors and tensor representations will be considered further in the sequel.

§ 2. The Transformations which Correspond to Infinitesimal Rotations in the Product Representation

We now find the transformations corresponding to infinitesimal rotations about the coordinate axes in the product representation. To do this we must, in accordance with Sect.2, take the transformation T_g, corresponding to a rotation through an angle a about the k-th coordinate axis, and determine the coefficient of a in its expansion in powers of a. This will give us A_k, the transformation corresponding to an infinitesimal rotation about the given axis.

We know that $T_g e_i f_k = U_g e_i V_g f_k$ but for a rotation through a about a fixed axis

$$U_g e_i = e_i + \alpha A e_i + \ldots,$$

$$V_g f_k = f_k + \alpha A f_k + \ldots,$$

where we have used the same letter A to denote the transformation corresponding to an infinitesimal rotation in both the representations. Substituting these expansions in the expression for $T_g e_i f_k$, we obtain:

$$T_g e_i f_k = (e_i + \alpha A e_i + \dots)(f_k + \alpha A f_k + \dots) = e_i f_k + \alpha (A e_i f_k + e_i A f_k) + \dots,$$

where the row of dots indicates, as usual, that the remaining terms are higher order than the first in α. Since this equation is the expansion for its own representation of $T_g e_i f_k$ in terms of α we must have:

$$A(e_i f_k) = A e_i f_k + e_i A f_k. \tag{4}$$

Equation (4) provides a rule for finding the transformation corresponding in the product representation to an infinitesimal rotation about any axis, which is analogous to the rule for the differentiation of a product.

A similar rule holds for the product of several representations.

§ 3. The Product of Two Irreducible Representations

We now assume that the representations $g \to U_g$ and $g \to V_g$ in the spaces R_1 and R_2 are irreducible. The product of two irreducible representations is in general reducible. We now show how it may be resolved into a product of irreducible representations.

Consider the irreducible representations $g \to U_g$ with weight l_1, in the space R_1 and the representation $g \to V_g$ with weight l_2, in the space R_2. Let us take, for the spaces R_1 and R_2, canonical bases $e_{-l_1}, e_{-l_1+1}, \dots, e_{l_1}$ and $f_{-l_2}, f_{-l_2+1}, \dots, f_{l_2}$ (i.e. bases consisting of normalized eigenvectors of the transformation $H_3 = iA_3$ in these spaces). Then we have

$$H_3 e_{m_1} = m_1 e_{m_1} \qquad (-l_1 \leqslant m_1 \leqslant l_1),$$
$$H_3 f_{m_2} = m_2 f_{m_2} \qquad (-l_2 \leqslant m_2 \leqslant l_2)$$

(see equation (19), Sect.2). Consider the transformation H_3 for the product of the representations U_g and V_g. From the result proved in paragraph 2 we have

$$\begin{aligned} H_3(e_{m_1} f_{m_2}) = H_3 e_{m_1} f_{m_2} + e_{m_1} H_3 f_{m_2} = m_1 e_{m_1} f_{m_2} + m_2 e_{m_1} f_{m_2} = \\ = (m_1 + m_2)\, e_{m_1} f_{m_2}. \end{aligned} \tag{5}$$

Thus the basis $e_{m_1} f_{m_2}$ in the space $R_1 \times R_2$ is itself a system of orthogonal normalized eigenvectors of the transformation H_3, with eigenvalues $m = m_1 + m_2$.

The irreducible representations into which the representation may be reduced are now obvious. We first determine the eigenvalues of H_3 in the space $R_1 \times R_2$ and their multiplicities.

Since $-l_1 \leqslant m_1 \leqslant l_1, -l_2 \leqslant m_2 \leqslant l_2$, it follows from equation (5) that each eigenvalue m of the transformation H_3 in $R_1 \times R_2$ is one of the numbers $l_1+l_2, l_1+l_2-1, \ldots, -l_1-l_2$.

To determine the multiplicity of the eigenvalue m it is necessary to find the number of vectors $e_{m_1}f_{m_2}$ for which $m_1+m_2=m$. Since the vectors $e_{m_1}f_{m_2}$ and $e_{-m_1}f_{-m_2}$ both belong to the basis, it is clear that the number of solutions to the problem is unaltered by interchanging m and $-m$. It is enough therefore to determine the multiplicity of the non-negative eigenvalues m.

Let us assume that $l_1 \leqslant l_2$. Obviously the greatest eigenvalue $m=l_1+l_2$ has associated with it at least one eigenvector $e_{l_1}f_{l_2}$. To the eigenvalue $m=l_1+l_2-1$ there corresponds two linearly independent eigenvectors $e_{l_1-1}f_{l_2}$ and $e_{l_1}f_{l_2-1}$, to the eigenvalue $m=l_1+l_2-2$ there corresponds three linearly independent eigenvectors and so on. Finally to the eigenvalue $m=l_2-l_1$ there corresponds the $2l_1+1$ eigenvectors

$$e_{-l_1}f_{l_2},\ e_{-l_1+1}f_{l_2-1},\ \ldots,\ e_{l_1}f_{l_2-2l_1}. \tag{6}$$

The number of eigenvectors cannot be increased by a further reduction of eigenvalues because (6) contains all the vectors e_{m_1}. Thus we obtain the result that $2l_1+1$ eigenvectors correspond to each of the eigenvalues $m=l_2-l_1$, $m=l_2-l_1-1, \ldots, m=-(l_2-l_1)$. For example the $2l_1+1$ eigenvectors corresponding to the eigenvalue $m=-(l_2-l_1)$ are

$$e_{-l_1}f_{-l_2+2l_1},\ e_{-l_1+1}f_{-l_2+2l_1-1},\ \ldots,\ e_{l_1}f_{-l_2}.$$

If we now reduce the eigenvalues from $-l_2+l_1$ to $-l_2-l_1$ we shall reduce the number of eigenvectors from $2l_1+1$ to 1, so that to the eigenvalue $-l_2-l_1$ there will correspond at least one eigenvector, namely $e_{-l_1}f_{-l_2}$. In the case $l_2<l_1$, the result is exactly the same except that instead of l_2-l_1 we now have l_1-l_2. Thus we obtain the following proposition:

The transformation H_3 in the space $R_1 \times R_2$ has eigenvalues l_1+l_2, $l_1+l_2-1, \ldots, -l_1-l_2$. The eigenvalue l_1+l_2 has multiplicity 1, the eigenvalue l_1+l_2-1 multiplicity 2, and so on until $|l_2-l_1|$. Eigenvalues between $|l_2-l_1|$ to $-|l_2-l_1|$ have the same multiplicity $l_2+l_1-|l_2-l_1|+1$. Further, from $-|l_2-l_1|$ to $-l_2-l_1$, a reduction of the eigenvalue by 1 reduces its multiplicity by 1, so that the smallest eigenvalue again has multiplicity 1.*

* This can also be seen from Fig.4. The rectangular array of points $-l_1 \leqslant x \leqslant l_1$, $-l_2 \leqslant y \leqslant l_2$ correspond to the pairs of values m_1, m_2. The straight line $x+y=m$ contains one of these points for $m=l_1+l_2$ When m is reduced by unity, the number of points occurring increases by unity every time, to the value $m=|l_2-l_1|$ (i.e. till the line contains the top left corner). From this value up to $m=-|l_2-l_1|$ the same number of points occurs, and then as m decreases the number drops to zero.

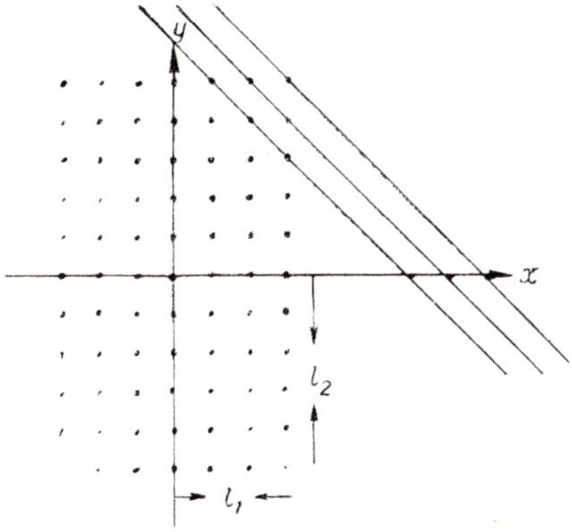

Fig. 4

Into what irreducible representations may the representation be resolved? Since the eigenvalues m of the transformation H_3 in the space $R_1 \times R_2$ satisfy the inequalities $-(l_1+l_2) \leqslant m \leqslant (l_1+l_2)$, the weight l of any such irreducible representation can not exceed l_1+l_2.

The transformation H_3 has one eigenvector $e_{l_1}f_{l_2}$ corresponding to the eigenvalue l_1+l_2. As we know from Sect. 2, the normalized vectors $e_{l_1}f_{l_2}$, $H_-(e_{l_1}f_{l_2})$, $H_-^2(e_{l_1}f_{l_2})$, ..., $H_-^{2l}(e_{l_1}f_{l_2})$ form a basis of the subspace $R^{(1)}$, in which we have the irreducible representation with weight $l=l_1+l_2$. Among these vectors are the eigenvectors of H_3 corresponding to all the eigenvalues from l_1+l_2 to $-(l_1+l_2)$. Therefore, the eigenvalues of H_3 in the orthogonal complement R' of $R^{(1)}$, have multiplicity one less than in $R=R_1\times R_2$. In particular, the eigenvalue l_1+l_2 of H_3 will not be contained in R' and the greatest eigenvalue of H_3 in R', namely l_1+l_2-1 will not be repeated.

We take in R' the eigenvector corresponding to the eigenvalue l_1+l_2-1* and beginning with this, construct a sequence of eigenvectors forming a basis for the subspace $R^{(2)}$ of an irreducible representation with weight l_1+l_2-1.

Continuing this process we construct from R subspaces which determine irreducible representations with weights l_1+l_2, l_1+l_2-1, l_1+l_2-2 and so on, until we reach a subspace in which the greatest eigenvalue of H_3

---

* It should not be assumed that this eigenvector is necessarily one of the basis vectors $e_{l_1-1}f_{l_2}$ or $e_{l_1}f_{l_2-1}$, corresponding to this eigenvalue. In fact, the vector $H_-(e_{l_1}f_{l_2})$ of $R^{(1)}$ is equal to $H_-e_{l_1}f_{l_2}+e_{l_1}H_-f_{l_2}=\sqrt{2l_1}e_{l_1-1}f_{l_2}+$ $+\sqrt{2l_2}e_{l_1}f_{l_2-1}$. For this reason the vector corresponding to the same eigenvalue l_1+l_2-1 and orthogonal to $R^{(1)}$ will also be a linear combination of $e_{l_1-1}f_{l_2}$ and $e_{l_1}f_{l_2-1}$.

is exactly $|l_2 - l_1|$ and has multiplicity 1. In this subspace H_3 has eigenvalues $|l_2 - l_1|$, $|l_2 - l_1| - 1$, $|l_2 - l_1| - 2$, ..., $-|l_2 - l_1|$ and, as is easily seen, all these eigenvalues have the same multiplicity 1. Hence the representation in this subspace is irreducible and has the weight $|l_2 - l_1|$.

Thus we have obtained the following result: the product of irreducible representations with weights l_1 and l_2 may be resolved into irreducible representations with weights $l_1 + l_2$, $l_1 + l_2 - 1$, ..., $|l_2 - l_1|$, and each of these irreducible representations occurs in the resolution only once.

§ 4. The Resolution of the Product of Irreducible Representations one of which has the Weight 1 or $\frac{1}{2}$

In the preceding paragraph we have obtained the resolution into irreducible representations of the product of two irreducible representations. To carry out this resolution in practice it is necessary to express the canonical bases of the subspaces of $R_1 \times R_2$, as linear combinations of the vectors $e_{m_1} f_{m_2}$. The coefficients of the corresponding linear combinations can be determined in the general case but the result is rather unwieldy. We shall here determine these coefficients in a frequently-encountered simple case, namely when one of the irreducible representations has the weight $l = 1$ or $l = \frac{1}{2}$. The general case will be examined in Sect. 10.

Let the irreducible representation $g \to U_g$ have the weight $l = 1$ and let $e_{-1} = e_-$, e_0 and $e_1 = e_+$ be a canonical basis in three-dimensional space, R_1.* Suppose the representation $g \to V_g$ has weight $l \geqslant 1$, and denote the vectors of the canonical basis of this representation by f_m $(-l \leqslant m \leqslant l)$. We now consider the product of these representations. The basis of the representation space $R_1 \times R_2$, consists of the $3(2l+1)$ vectors $e_+ f_m$, $e_0 f_m$, $e_- f_m$ $(-l \leqslant m \leqslant l)$. According to subsection 3 the space $R_1 \times R_2$ is the sum of three invariant subspaces, the spaces of the irreducible representations with weights $l+1, l, l-1$ respectively. We denote the canonical bases consisting of the eigenvectors of H_3 in these subspaces by g_m^{l+1}, g_m^l and g_m^{l-1}; the upper indices indicate the weight of the irreducible representation in the given subspace, and m, as usual, denotes the eigenvalue of H_3, corresponding to the given eigenvector.

For a fixed value of m each of the vectors g_m^{l+1}, g_m^l, and g_m^{l-1} is an eigenvector of H_3 in the space $R_1 \times R_2$, corresponding to the eigenvalue m. The vectors g_m^{l+1}, g_m^l, g_m^{l-1} are therefore linear combinations of those basic vectors $e_{m_1} f_{m_2}$, which are themselves eigenvectors of H_3 with the same eigenvalue m, i.e. $e_- f_{m+1}$, $e_0 f_m$, $e_+ f_{m-1}$. In other words g_m^{l+1}, g_m^l, g_m^{l-1} and $e_- f_{m+1}$, $e_0 f_m$, $e_+ f_{m-1}$ are different bases of the three-dimensional** subspace

---

* For the relation between this basis and the ordinary basis in three-dimensional space see paragraph 5 of Sect. 2.

** If $m = l + 1$ this space is one-dimensional, and if $m = +l$, it is two-dimensional.

space of $R_1 \times R_2$, consisting of all the eigenvectors of H_3 corresponding to the eigenvalue m. We need to determine the matrix of the transformation from either of these bases into the other.

Put:

$$\left.\begin{aligned} e_-f_{m+1} &= c_{11}^m g_m^{l+1} + c_{12}^m g_m^l + c_{13}^m g_m^{l-1}, \\ e_0 f_m &= c_{21}^m g_m^{l+1} + c_{22}^m g_m^l + c_{23}^m g_m^{l-1}, \\ e_+ f_{m-1} &= c_{31}^m g_m^{l+1} + c_{32}^m g_m^l + c_{33}^m g_m^{l-1}. \end{aligned}\right\} \tag{7}$$

Since the vectors e_-f_{m+1}, e_0f_m, e_+f_{m-1} are mutually orthonormal, and the vectors g_m^{l+1}, g_m^l, g_m^{l-1} are also orthonormal (they belong to distinct, mutually orthogonal subspaces of $R_1 \times R_2$), the matrix $C^{(m)} = \|c_{ik}^m\|$ is a unitary matrix of the third order.

For $m = -l - 1$ equations (7) reduce to a single equation:

$$e_-f_{-l} = c_{11}^{-l-1} g_{-l-1}^{l+1}. \tag{7'}$$

For $m = -l$ we have the two equations

$$\left.\begin{aligned} e_-f_{-l+1} &= c_{11}^{-l} g_{-l}^{l+1} + c_{12}^{-l} g_{-l}^l, \\ e_0 f_{-l} &= c_{21}^{-l} g_{-l}^{l+1} + c_{22}^{-l} g_{-l}^l. \end{aligned}\right\} \tag{7''}$$

Similarly we may note the form of (7) for $m = l+1$ and for $m = l$.

To determine the matrix $C^{(m)}$, apply the transformations H_- and H_+ to both sides of the equations (7). We then obtain recurrence relations which enable us to determine $C^{(m)}$.

We find the first row of the matrix $C^{(m)}$ by applying the transformation H_- to the first of the equations (7). Remembering that $H_-f_m = \alpha_m^l f_{m-1}$, where $\alpha_m^l = \sqrt{(l+m)(l-m+1)}$, we obtain from the left-hand side of (7):

$$\begin{aligned} H_-(e_-f_{m+1}) &= e_-H_-f_{m+1} = \alpha_{m+1}^l e_-f_m = \\ &= \alpha_{m+1}^l \left(c_{11}^{m-1} g_{m-1}^{l+1} + c_{12}^{m-1} g_{m-1}^l + c_{13}^{m-1} g_{m-1}^{l-1}\right). \end{aligned}$$

On the other hand applying H_- to the right-hand side, we have

$$c_{11}^m \alpha_m^{l+1} g_{m-1}^{l+1} + c_{12}^m \alpha_m^l g_{m-1}^l + c_{13}^m \alpha_m^{l-1} g_{m-1}^{l-1}.$$

Equating the coefficients of the same vector g and substituting for the α_m^l we find the recurrence relations:

$$\left.\begin{aligned} c_{11}^{m}\sqrt{l-m+2} &= c_{11}^{m-1}\sqrt{l-m}, \\ c_{12}^{m}\sqrt{(l+m)(l-m+1)} &= c_{12}^{m-1}\sqrt{(l+m+1)(l-m)}, \\ c_{13}^{m}\sqrt{l+m-1} &= c_{13}^{m-1}\sqrt{l+m+1}. \end{aligned}\right\} \quad (8)$$

Since the vectors $e_{-}f_{-l}$ and g_{-l-1}^{l+1} are normalized it follows from formula (7') that $|c_{11}^{-l-1}|=1$. Putting $c_{11}^{-l-1}=1$* we find from the first of equations (8) that

$$c_{11}^{m}=\sqrt{\frac{l-m}{l-m+2}}\,c_{11}^{m-1}=\sqrt{\frac{(l-m)(l-m+1)}{(l-m+2)(l-m+3)}}\,c_{11}^{m-2}=\dots$$

$$\dots=\sqrt{\frac{(l-m)(l-m+1)}{(l-n)(l-n+1)}}\,c_{11}^{n}=\dots=\sqrt{\frac{(l-m)(l-m+1)}{(2l+1)(2l+2)}}\,c_{11}^{-l-1},$$

Hence

$$c_{11}^{m}=\sqrt{\frac{(l-m)(l-m+1)}{(2l+1)(2l+2)}}.$$

Taking $m=-l$ and using the fact that the matrix of (7") is unitary, in particular that $|c_{11}^{-l}|^2+|c_{12}^{-l}|^2=1$, we find that $|c_{12}^{-l}|=\sqrt{\frac{1}{l+1}}$. Again, putting $c_{12}^{-l}=\sqrt{\frac{1}{l+1}}$ (see the footnote on this page) we find from the second of equations (8) that

$$c_{12}^{m}\sqrt{\frac{(l+m+1)(l-m)}{2l(l+1)}}.$$

In the same way using the third of equations (8) we find that

$$c_{13}^{m}=\sqrt{\frac{(l+m)(l+m+1)}{2l(l+1)}}$$

(for this we use the fact that $|c_{13}^{-l+1}|^2=1-|c_{12}^{-l+1}|^2-|c_{11}^{-l+1}|^2=\frac{1}{l(2l+1)}$, and, as above, we have put $c_{13}^{-l+1}=\sqrt{\frac{1}{l(2l+1)}}$). Thus we have obtained all the elements of the first row of the matrix $C^{(m)}$.

To find the second row, we apply the transformation H_{+} to the first of the equations (7); we obtain:

$$\sqrt{2}e_0f_{m+1}+\alpha_{m+2}^{l}e_{-}f_{m+2}=c_{11}^{m}\alpha_{m+1}^{l+1}g_{m+1}^{l+1}+c_{12}^{m}\alpha_{m+1}^{l}g_{m+1}^{l}+c_{13}^{m}\alpha_{m+1}^{l-1}g_{m+1}^{l-1},$$

---

* Since each of the vectors g_m^{l+1}, g_m^{l}, g_m^{l-1} is determined uniquely to within a numerical factor of unit modulus, we may choose this factor such that one coefficient in each column of the matrix $C^{(m)}$ is equal to any preassigned number.

By substituting for each of the vectors on the left-hand side as given by the equations (7) we obtain simple expressions for the elements of the second row of the matrix $C^{(m)}$ in terms of the elements of the first row already determined. Hence c_{21}^m, c_{22}^m, c_{23}^m. Similarly by applying the transformation H_+ to the second of equations (7) we can express the elements of the third row of the matrix $C^{(m)}$ in terms of the elements of the second row and thus find c_{31}^m, c_{32}^m, c_{33}^m.

Finally the following result is obtained:

$$C^{(m)} = \left\| \begin{matrix} \sqrt{\frac{(l-m)(l-m+1)}{(2l+1)(2l+2)}} & \sqrt{\frac{(l+m+1)(l-m)}{2l(l+1)}} & \sqrt{\frac{(l+m)(l+m+1)}{2l(2l+1)}} \\ \sqrt{\frac{(l+m+1)(l-m+1)}{(2l+1)(l+1)}} & \frac{m}{\sqrt{l(l+1)}} & -\sqrt{\frac{(l+m)(l-m)}{l(2l+1)}} \\ \sqrt{\frac{(l+m)(l+m+1)}{(2l+1)(2l+2)}} & -\sqrt{\frac{(l+m)(l-m+1)}{2l(l+1)}} & \sqrt{\frac{(l-m)(l-m+1)}{2l(2l+1)}} \end{matrix} \right\| . \tag{9}$$

Since this matrix is orthogonal its inverse is equal to its transpose and so the vectors g_m^{l+1}, g_m^l, g_m^{l-1} may be expressed in terms of e_-f_{m+1}, e_0f_m, e_+f_{m-1} with the aid of the columns of this matrix.

We now consider the product of an irreducible representation with $l=\frac{1}{2}$ and an arbitrary irreducible representation with weight l. This may be resolved into two representations with weights $l+\frac{1}{2}$ (the basis in the corresponding subspace being denoted by $g_m^{l+\frac{1}{2}}\left(-l-\frac{1}{2}\leqslant m\leqslant l+\frac{1}{2}\right)$) and a representation with weight $l-\frac{1}{2}$ $\left(\text{the basis being } g_m^{l-\frac{1}{2}}\left(-l+\frac{1}{2}\leqslant \right.\right.$ $\left.\left.\leqslant m\leqslant l-\frac{1}{2}\right)\right)$. By a method similar to the preceding we have:

$$e_{\frac{1}{2}}f_{m-\frac{1}{2}} = c_{11}^m g_m^{l+\frac{1}{2}} + c_{12}^m g_m^{l-\frac{1}{2}},$$

$$e_{-\frac{1}{2}}f_{m+\frac{1}{2}} = c_{21}^m g_m^{l+\frac{1}{2}} + c_{22}^m g_m^{l-\frac{1}{2}}.$$

Applying H_+ to the first equation and H_- to the second and putting $c_{11}^{l+\frac{1}{2}} = c_{21}^{-l-\frac{1}{2}} = 1$, we find that the matrix $C^{(m)}$ for the remaining eigenvalues m has the form

$$C^{(m)} = \left\| \begin{matrix} \sqrt{\frac{l+m+\frac{1}{2}}{2l+1}} & -\sqrt{\frac{l-m+\frac{1}{2}}{2l+1}} \\ \sqrt{\frac{l-m+\frac{1}{2}}{2l+1}} & \sqrt{\frac{l+m+\frac{1}{2}}{2l+1}} \end{matrix} \right\| . \tag{10}$$

Section 5. Tensors and Tensor Representations

In the preceding section, in connexion with the definition of the product of representations, we introduced the notion of a tensor. We stated that we had a tensor of rank r in three-dimensional Euclidean space if sets of 3^r numbers $a_{i_1 i_2 \ldots i_r}$ were given which transformed according to the equations

$$a'_{i'_1 i'_2 \ldots i'_r} = \sum_{i_1=1}^{3} \sum_{i_2=1}^{3} \cdots \sum_{i_r=1}^{3} g_{i'_1 i_1} g_{i'_2 i_2} \cdots g_{i'_r i_r} a_{i_1 i_2 \ldots i_r}, \quad (1)$$

where $\|g_{ik}\|$ is the matrix of transformation from one coordinate system to another. The matrix of transformation is given in terms of the old and new vector bases by the equations

$$e'_i = \sum_{k=1}^{3} g_{ik} e_k.$$

We consider the set of all tensors of rank r in a given coordinate system, i.e. all sets of 3^r numbers $a_{i_1 i_2 \ldots i_r}$. They form a 3^r dimensional space R, in which the operations of addition and multiplication (of numbers of the form (1)) by a constant are defined in the usual way. Equation (1) gives a linear transformation in a 3^r dimensional space, taking $a_{i_1 i_2 \ldots i_r}$ into $a'_{i'_1 i'_2 \ldots i'_r}$. If we put the rotation $g = \|g_{ik}\|$ in correspondence with the linear transformation (1) we obtain the representation of the rotation group which was considered in sect.4 (a tensor representation). This representation as we have seen, is the product of three-dimensional representations of the group *.

We now consider the effect of reflection in the origin of coordinates on a vector, i.e. a tensor of the first rank; it changes its sign. Hence a reflection in the origin leaves the components of a tensor of the second rank unchanged, changes the sign of the components of a tensor of rank r by $(-1)^r$.

The tensor representations of rank $r > 1$ are reducible. In paragraph 2 of this section we reduce these representations into isotypic components. Before this, however, we shall consider in paragraph 1 some operations on tensors, and the connexion these have with the problem of resolving tensor spaces into invariant subspaces, i.e. the resolution of tensor representations.

§ 1. The Fundamental Algebraic Operations on Tensors and Invariant Subspaces

We define the operation on tensors known as "contraction". Consider a

---

* We note that equation (1) can be interpreted in two ways: as the law describing the effect of a change of coordinates on the tensor and as the equation defining the effect of a linear transformation on the given tensor. Both points of view are equally tenable. This is clearly seen in the case of tensors of rank 1, i.e. vectors a_i. If the matrix $\|g_{ik}\|$ is understood as a rotation of the coordinate system the equation $a'_{i'} = \sum g_{i'i} a_i$ gives the components of the given vector in the new system. If, however, $\|g_{ik}\|$ is interpreted as a rotation of the whole space then this equation gives the components of the new position of the given vector in a fixed coordinate system.

tensor of the r-th rank $a_{i_1 i_2 \dots i_r}$. We form the sum of all those components whose first two indices are equal:

$$\sum_{i=1}^{3} a_{iii_3 i_4 \dots i_r} = b_{i_3 i_4 \dots i_r}. \tag{2}$$

The expression (2) is called the trace of the tensor for the first two indices and the operation of forming the trace is called contraction. Similarly we define contraction for any other pair of indices.

From formula (1) and the orthogonality of the matrix $\|g_{ik}\|$ it follows that the trace for any two indices of a tensor of rank r is a tensor of rank $r-2$. Thus *

$$b'_{i'_3 i'_4 \dots i'_r} = a_{i'i'i'_3 \dots i'_r} = g_{i'i_1} g_{i'i_2} g_{i'_3 i_3} \dots g_{i'_r i_r} a_{i_1 i_2 \dots i_r} =$$

$$= \delta_{i_1 i_2} g_{i'_3 i_3} \dots g_{i'_r i_r} a_{i_1 i_2 \dots i_r} = g_{i'_3 i_3} g_{i'_4 i_4} \dots g_{i'_r i_r} a_{iii_3 i_4 \dots i_r} =$$

$$= g_{i'_3 i_3} g_{i'_4 i_4} \dots g_{i'_r i_r} b_{i_3 i_4 \dots i_r}.$$

Hence

$$b'_{i'_3 i'_4 \dots i'_r} = g_{i'_3 i_3} g_{i'_4 i_4} \dots g_{i'_r i_r} b_{i_3 i_4 \dots i_r}, \tag{3}$$

which shows that the trace of a tensor is transformed according to the tensor rule and is therefore a tensor of rank $r-2$. In particular the trace of a tensor of the second rank is a tensor of rank zero, i.e. a scalar (a number which is independent of the coordinate system). A tensor a_{ijk} of the third rank has three traces: a_{iik}, a_{iki}, a_{kii}, each of which is a vector.

The operation of forming the trace enables us to find subspaces in the space R which are invariant for tensor representations. Thus tensors $a_{i_1 i_2 \dots i_r}$, whose trace, for a given pair of indices, is the null tensor, form a subspace invariant with respect to tensor representation. We consider elements

---

* In future we will, as is usual when dealing with tensors, omit the summation sign in the expressions involving a sum over any pair of indices. Using this convention equation (1) for example, may be written in the form

$$a'_{i'_1 i'_2 \dots i'_r} = g_{i'_1 i_1} g_{i'_2 i_2} \dots g_{i'_r i_r} a_{i_1 i_2 \dots i_r},$$

and the definition of the trace in the form

$$a_{iii_3 \dots i_r} = b_{i_3 \dots i_r}.$$

The equation

$$g_{i'i_1} g_{i'i_2} = \delta_{i_1 i_2}$$

is in this notation the condition that the matrix $\|g_{ik}\|$ be orthogonal.

$a_{i_1i_2\ldots i_r}$ in R for which the trace for, say, the first two indices is zero, i.e. $b_{i_3i_4\ldots i_r} = a_{iii_3i_4\ldots i_r} = 0$. Then, from equation (3), it follows that $a'_{i'i'i'_3\ldots i'_r} = 0$, i.e. $a'_{i'_1i'_2\ldots i'_r}$ belongs to the same subspace. Hence the subspace is invariant.

We now define multiplication of tensors. Consider two tensors $a_{i_1i_2\ldots i_r}$ and $b_{j_1j_2\ldots j_p}$ of rank r and p respectively. The product of these tensors is the tensor $c_{i_1i_2\ldots i_rj_1j_2\ldots j_p}$ of rank $r+p$, whose components are the products of the various components of $a_{i_1i_2\ldots i_r}$ and $b_{j_1j_2\ldots j_p}$, i.e.

$$c_{i_1i_2\ldots i_rj_1j_2\ldots j_p} = a_{i_1\ldots i_r}b_{j_1\ldots j_p}.$$

It is not difficult to show that under a transformation of coordinates these elements transform in the same manner as those of a tensor of rank $r+p$ i.e. that a product of tensors is in fact a tensor. From equation (1) it is clear also that under transformations of the tensor representations, the product of tensors $a_{i_1i_2\ldots i_r}$ and $b_{j_1j_2\ldots j_p}$ goes into the product of $a'_{i'_1i'_2\ldots i'_r}$ and $b'_{j'_1j'_2\ldots j'_p}$.

To obtain invariant subspaces of another type, we introduce what is known as the unit tensor of the second rank δ_{ik}, the components of which in any system of coordinates are given by the equations

$$\delta_{ik} = \begin{cases} 0, & i \neq k, \\ 1, & i = k. \end{cases}$$

Since,

$$\delta'_{i'k'} = g_{i'i}g_{k'k}\delta_{ik} = g_{i'k}g_{k'k} = \begin{cases} 0, & i' \neq k', \\ 1, & i' = k', \end{cases}$$

it follows that δ_{ik} is indeed a tensor whose components do not change if the coordinate system is altered.

We now consider the tensors of the r-th rank which are products of the unit tensor and tensors of rank $r-2$, i.e. tensors of the form

$$a_{i_1i_2i_3\ldots i_r} = \delta_{i_1i_2}b_{i_3\ldots i_r}. \tag{4}$$

Since the transformations of a representation take the unit tensor into the unit tensor, and a product into a product, tensors of the form (4) constitute an invariant subspace of the space R of tensors of rank r.

We show that every tensor of rank r may be expressed as the sum of two tensors, the first of which has the form $\delta_{i_1i_2}b_{i_3\ldots i_r}$, and the second satisfies $a_{iii_3\ldots i_r} = 0$. Take an arbitrary tensor $a_{i_1i_2\ldots i_r}$ and subtract from it the tensor $\frac{1}{3}\delta_{i_1i_2}b_{i_3\ldots i_r}$, where $b_{i_3\ldots i_r}$ is the trace of $a_{i_1i_2i_3\ldots i_r}$ for the first two indices. We obtain the tensor $c_{i_1i_2i_3\ldots i_r}$ whose trace for the first two indices is the null tensor (since $\delta_{ii} = 3$) and this proves our assertion. Thus the space R of tensors of the r-th rank may be resolved into a sum of invariant subspaces R' and R'', where R' is the space of tensors whose trace for the first two indices is null, and R'' is the space of tensors of the form (4).

A similar resolution may be obtained by contraction with respect to any other pair of indices.

There is yet another method for constructing invariant subspaces in R. Let s denote an arbitrary permutation of the numbers $i_1 i_2 \ldots i_r$, taking these numbers into $j_1 j_2 \ldots j_r$. We put the tensor $a_{i_1 i_2 \ldots i_r}$ in correspondence with the tensor $a_{j_1 j_2 \ldots j_r}$ obtained by rearranging the indices. We denote this correspondence by S, i.e. we put

$$Sa_{i_1 i_2 \ldots i_r} = a_{j_1 j_2 \ldots j_r}.$$

It is obvious that S is a linear transformation in the space of tensors of rank r. We call them permutation transformations. Since it does not matter whether we first carry out this operation and then change to new coordinates or vice-versa, it follows that the operation is unaltered by transformation by the tensor representation T_a.

Consider the sequence of "permutation" transformations $S_1, \quad S_2, \ldots, S_p$ and take the linear combination $\lambda_1 S_1 + \lambda_2 S_2 + \ldots + \lambda_p S_p$. Tensors satisfying the equation

$$(\lambda_1 S_1 + \lambda_2 S_2 + \ldots + \lambda_p S_p)\, a_{i_1 i_2 \ldots i_r} = 0, \tag{5}$$

form a linear subspace of the space R.

Since permutation transformations commute with the transformations of the representation it follows that this subspace is invariant with respect to the tensor representation. *

We show how the resolution of a tensor representation may be carried out with the help of eigenvalues of the permutation transformation.** Since any permutations if carried out a certain number of times will restore the initial order, the transformation S of tensor space will give the identical transformation E, if it is repeated a sufficient number of times. But this means that the eigenvalues of the transformation S raised to the appropriate power are equal to the eigenvalues of E, namely unity.

---

* Let $a_{i_1 i_2 \ldots i_r}$ belong to the subspace defined by equation (5), i.e., let $\sum_j \lambda_j S_j a_{i_1 i_2 \ldots i_r} = 0.$ Then

$$\sum_j \lambda_j S_j T_g a_{i_1 \ldots i_r} = T_g \left(\sum_j \lambda_j S_j a_{i_1 i_2 \ldots i_r} \right) = 0,$$

i.e. $T_j a_{i_1 i_2 \ldots i_r}$ belongs to the same subspace.

** Of course, in general this resolution yields neither isotypic nor irreducible representations.

Consider the permutation s and let $s^n = s_0$, where s_0 is the identical permutation. We denote by $\varepsilon_0 = 1$, ε_1, $\varepsilon_2 = \varepsilon_1^2, \ldots, \varepsilon_{n-1}$ the roots of the equation $\varepsilon^n - 1$ and consider the subspace of tensors satisfying the equation:

$$Sa_{i_1 i_2 \ldots i_r} = \varepsilon_k a_{i_1 i_2 \ldots i_r}.$$

From the general theory of linear transformations it is known that these subspaces form a complete resolution of the tensor space. As has been mentioned above they are invariant with respect to the tensor representation; we have thus obtained a resolution of tensor space.

For example, if s_1 is the permutation involving an interchange of the first two indices, taking $i_1 i_2 \ldots i_r$ into $i_2 i_1 \ldots i_r$, then $S_1^2 = S_0$ and the eigenvalues of the corresponding transformation S_1 are ± 1.

Every tensor may therefore be expressed as the sum of two tensors $b_{i_1 i_2 \ldots i_r}$ and $c_{i_1 i_2 \ldots i_r}$

$$a_{i_1 i_2 \ldots i_r} = b_{i_1 i_2 \ldots i_r} + c_{i_1 i_2 \ldots i_r},$$

where $b_{i_1 i_2 \ldots i_r}$ satisfies the condition $b_{i_2 i_1 \ldots i_r} = -b_{i_1 i_2 \ldots i_r}$ and $c_{i_1 i_2 \ldots i_r}$ satisfies the equation $c_{i_2 i_1 i_3 \ldots i_r} = c_{i_1 i_2 \ldots i_r}$. To see this, it is in fact sufficient to put:

$$b_{i_1 i_2 \ldots i_r} = \frac{1}{2}\left(a_{i_1 i_2 \ldots i_r} - a_{i_2 i_1 \ldots i_r}\right),$$

$$c_{i_1 i_2 \ldots i_r} = \frac{1}{2}\left(a_{i_1 i_2 \ldots i_r} + a_{i_2 i_1 \ldots i_r}\right).$$

We observe that the set of all conditions of the form (5) also defines an invariant tensor subspace.

Tensors which are unchanged by any rearrangement of the indices $i_1 i_2 \ldots \ldots i_r$, are called symmetric tensors of rank r. Tensors which are not changed by an even permutation of the indices and which change sign for an odd permutation are called skew-symmetric tensors. It is clear from what has been said that both symmetric and skew-symmetric tensors form invariant subspaces in the space R.

We will now consider in more detail skew-symmetric tensors in three-dimensional space. For an interchange of any two indices a skew-symmetric tensor changes sign. From this it will be seen that if a component of a skew-symmetric tensor is non-zero, then all of its indices must be different. Since the indices $i_1, i_2, \ldots, i_r$ may only take the values 1, 2 or 3, it follows that skew-symmetric tensors of rank higher than the third must be zero.

Let a_{ijk} be a skew-symmetric tensor of the third rank. For such a tensor the component a_{123} alone may be regarded as arbitrary since the remaining non-zero components are given in terms of it by the condition for skew-symmetry: $a_{231} = a_{312} = a_{123}$; $a_{213} = a_{321} = a_{132} = -a_{123}$. We now denote the tensor of the third rank for which $a_{123} = 1$, by ε_{ijk}. All other skew-symmetric tensors may be obtained from ε_{ijk} by multiplication by an arbitrary factor.

From equation (1) it is easily seen that for a transformation of coordinates each component of the tensor ε_{ijk} is multiplied by the determinant of the matrix of the transformation $\|g_{ik}\|$. Thus for rotations the components of this tensor are unchanged, and for a rotation combined with reflection they change sign. Hence a skew-symmetric tensor of the third rank represents a pseudo-scalar.

Consider a skew-symmetric tensor of the second rank a_{ij}. It is evident that this tensor has three independent components a_{12}, a_{23} and a_{13}, i.e. such tensors form a three-dimensional space. Multiply a_{ij} by $\varepsilon_{i'j'k'}$ and then contract with respect to the pairs of indices ii' and jj'. The result is a tensor of the first rank

$$x_k = \varepsilon_{ijk}a_{ij},$$

i.e. a vector. Clearly $\varepsilon_{ijk}a_{ij} = 2a_{ij}$ or $\varepsilon_{ijk}a_{ij} = -2a_{ij}$ depending on the choice of the coordinate system. Thus the components of the skew-symmetric tensor coincide with those of the vector x_k, to within multiplication by an arbitrary constant, the sign of which will depend on the coordinate system. In the transformation from one system to another the elements a_{23}, a_{31}, a_{12} will, under a rotation, transform as do the components of the vector x_1, x_2, x_3, and in the case of rotation combined with reflection they will also change sign. This means, in other words, that skew-symmetric tensors of the second rank behave, as far as representations are concerned as pseudo-vectors.

As an example we consider the resolution of a tensor representation of the second rank into irreducible representations. Each tensor of the second rank may be expressed as the sum of a symmetric tensor and a skew-symmetric tensor.

$$a_{ij} = b_{ij} + c_{ij}.$$

It is, in fact, enough to put

$$b_{ij} = \frac{1}{2}(a_{ij} - a_{ji}),$$

$$c_{ij} = \frac{1}{2}(a_{ij} + a_{ji}).$$

We have resolved a nine-dimensional tensor space of the second rank into the sum of a three-dimensional space of pseudo vectors (skew-symmetric tensors of the second rank) and a six-dimensional space of symmetric tensors. The representation in the space of pseudo-vectors is irreducible. The representation in the space of symmetric tensors may be resolved further. Every symmetric tensor may be represented as the sum of a symmetric tensor with zero trace and a multiple of the unit tensor. Thus $c_{ij} = \lambda\delta_{ij} + d_{ij}$, where $\lambda = \frac{1}{3}c_{ii}$, and the symmetric tensor $d_{ij} = c_{ij} - \lambda\delta_{ij}$ has trace zero.

We have, then, resolved the nine-dimensional space of tensors of the second rank into a sum of invarient subspaces: the one-dimensional subspace of tensors of the form $\lambda\delta_{ij}$, the three-dimensional subspace of skew-symmetric tensors, and the five-dimensional subspace of symmetric tensors with trace equal to zero.

The irreducibility of the representations in the first two cases is obvious. It is not difficult to show that for the last subspace the representation is irreducible, with weight $l = 2$. This will, however, be shown later. *

The reduction of a tensor representation of the third rank is more complicated. We obtain it in paragraph 3 as a particular case of a more general result.

§2. The Determination of the Weights of the Irreducible Representations into which the Tensor Representation may be Resolved

At this point we determine the possible irreducible constituents of an arbitrary tensor representation. The answer to this question is obtained from the results of the preceding section concerning the resolution into irreducible representations of the product of two irreducible representations.

We know that the tensor representation of rank 1 is an irreducible representation of weight $l = 1$ (the fundamental representation).

Consider the tensor representation of rank 2. It is the product of two irreducible representations of weight 1, and hence it may be resolved into three irreducible representations with weights 0, 1 and 2 respectively.

The next tensor representation of rank 3 is the product of a tensor representation of rank 2 and an irreducible three-dimensional representation. The space R of tensors of rank 2 is resolved, as shown above, into the sum of three invariant subspaces (denoted by R_0, R_1 and R_2) in which we have the irreducible representations of rank 0, 1 and 2 respectively. Hence the product $R \times R_1$ of this space and the three-dimensional space R_1, may be resolved into the sum of invariant subspaces $R_0 \times R_1$, $R_1 \times R_1$ and $R_2 \times R_1$. Since the representation in each of these subspaces is the product of two irreducible representations, we may apply to them the results of Sect.4. The product of the representations of weights 2 and 1 is resolved into three representations of weights 3, 2, 1.

$R_1 \times R_1$ (again in virtue of the results of 4) is resolved into representations of weight 2, 1, 0. $R_0 \times R_1$ is an irreducible representation of weight 1.

Combining these results we see that a tensor representation of rank 3 may be resolved into the sum of an irreducible representation of weight 0, three irreducible representations of weight 1, two irreducible representations of weight 2, and one irreducible representation of weight 3.

Since a tensor representation of rank 4 is the product of the tensor representations of ranks 3 and 1, by using the resolution just determined we can find the irreducible representations into which it may be resolved. The representation of rank 3 is first resolved into irreducible representations, each of these is then multiplied by the representation of rank 1; the products thus

* This result also follows from Sect.4. The tensor representation for $r = 2$ is the product of two irreducible representations with $r = 1$ (the fundamental representation). Hence it may be resolved into irreducible representations of weights $l = 0, 1, 2$.

obtained may then be resolved anew and the results added.

A simple calculation gives the following table:

TABLE 1

Rank of tensor	Weight of representation					
	$l=0$	$l=1$	$l=2$	$l=3$	$l=4$	$l=5$
$r=0$	1					
$r=1$	0	1				
$r=2$	1	1	1			
$r=3$	1	3	2	1		
$r=1$	3	6	6	3	1	
$r=5$	6	15	15	10	4	1

Each row corresponds to a tensor representation of the given rank. The columns correspond to the irreducible representations of various weights. The number at the intersection of the l-th column and r-th row indicates the number of representations of weight l in the resolution of the representation in the space of tensors of rank r. The following simple rule may be used to construct the table: in each position of the r-th row, beginning with the second column, is the sum of the numbers in the $(r-1)$-th row which are in the positions immediately to the left, above, and immediately to the right of the given position.

In fact the representations of weight l in the resolution of a representation of the rank r may be obtained by multiplying the representations of weights $l-1$, l $l+1$ in the resolution of the representation of rank $r-1$ by a representation of weight 1.

In the first column of the table we have the representation of weight $l=0$. The magnitude transformed by such a representation is either a scalar, or a pseudo-scalar which is transformed into minus itself by reflection. Since a tensor of rank r is multiplied by $(-1)^r$ on reflection in the origin of co-ordinates, the quantity corresponding to the first column is a scalar for r even and a pseudo-scalar for r odd. The magnitude transformed by the representation of weight l is a vector, if its components do not change sign under reflection in the origin and pseudo-vectors if this is not the case. The representations in the second column for r even apply to pseudo-vectors, and for r odd to vectors.

More generally the representations of the l-th column apply to l-vectors if $r+l$ is even and to the corresponding pseudo-magnitudes if $r+l$ is odd.

§ 3. The Resolution of a Tensor Representation into Isotypic Representations Tensors of the Third Rank *

In paragraph 2 we saw that in the reduction of a new tensor representation into

* For a more detailed account of the resolution of tensor representations into irreducible representations see "The Classical Groups" by H.Weyl (1947).

irreducible representations, the representation corresponding to a given weight l occurred, in general, several times. We now proceed to show how to resolve a tensor representation into isotypic representations (for the definition of an isotypic representation see Sect. 2).

The resolution is obtained by using the method indicated in Sec. 2. We know from this that vectors transformed by an irreducible representation of weight l satisfy the equation

$$H^2f - l(l+1)f = 0, \tag{6}$$

where $H^2 = H_1^2 + H_2^2 + H_3^2 = -(A_1^2 + A_2^2 + A_3^2)$ and the A_k correspond in the representation to infinitesimal rotations about the coordinate axes. Having found H^2 for the tensor representation we substitute in equation (6) all values in the r-th row of Table 1 and obtain a set of equations. The solutions of each of these equations form an invariant subspace which yields an isotypic representation with the corresponding weight l.

To find the form of the equations determining these invariant subspaces we must determine, for a tensor representation, the transformation $H^2 = -(A_1^2 + A_2^2 + A_3^2)$. It is necessary first of all to find the transformations A_1, A_2 and A_3. The tensor representation of rank r is the product of r three-dimensional representations. In Sect.4 we have shown that the transformation A_k, corresponding to an infinitesimal rotation in the product representation has the form

$$A_k(efg\ldots) = (A_k e)fg\ldots + e(A_k f)g\ldots + ef(A_k g)\ldots + \ldots,$$

where e, f, $g\ldots$ are vectors in the spaces of the representations whose product is being considered. Hence to find how the transformation A_k acts on the tensor $a_{i_1 i_2 \ldots i_r}$, we must apply this transformation to each index of the tensor in turn as on a three-dimensional vector, without changing any of the remaining indices, and add the results. Consequently it will be sufficient for us to know how the A_k act on vectors, i.e. to determine the matrices of these transformations in three-dimensional space. But this is already known. If a basis is chosen in three-dimensional space, consisting of the usual orthogonal system, it is easily seen that the matrices A_1, A_2 and A_3 take the form

$$A_1 = \begin{Vmatrix} 0 & 0 & 0 \\ 0 & 0 & -1 \\ 0 & 1 & 0 \end{Vmatrix}, \quad A_2 = \begin{Vmatrix} 0 & 0 & 1 \\ 0 & 0 & 0 \\ -1 & 0 & 0 \end{Vmatrix}, \quad A_3 = \begin{Vmatrix} 0 & -1 & 0 \\ 1 & 0 & 0 \\ 0 & 0 & 0 \end{Vmatrix}.$$

If the matrix of a rotation about the k-th axis is denoted by A_{st}, where the triad k, s, t is obtained from 1, 2, 3 by cyclic permutation, we may combine these three formulae into one:

$$A_{st} = \|\alpha_{ij}^{st}\|,$$

where

$$\alpha_{ij}^{st} = -\delta_{is}\delta_{jt} + \delta_{it}\delta_{js}.$$

Thus the transformation A_{st} acts on a vector in the following manner

$$A_{st}a_j = a_{ij}^{st}a_j = -\delta_{is}\delta_{jt}a_j + \delta_{it}\delta_{js}a_j = -\delta_{is}a_t + \delta_{it}a_s. \quad (7)$$

We take a tensor of the r-th rank $a_{i_1i_2\dots i_r}$. Applying the transformation A_{st} in turn to each index and adding the results we obtain:

$$A_{st}a_{i_1i_2\dots i_r} = \sum_{p=1}^{r} -\delta_{i_ps}a_{i_1i_2\dots i_{p-1}ti_{p+1}\dots i_r} + \delta_{i_dt}a_{i_1i_2\dots i_{p+1}si_{p+1}\dots i_r}. \quad (8)$$

Since we must also find the transformations $(A_{st})^2$ we again apply the transformation A_{st}. We obtain

$$(A_{st})^2a_{i_1i_2\dots i_r} = \sum_{q=1}^{r} -\delta_{i_qs}\Big[\sum_{\substack{p=1\\p\neq q}}^{r}(-\delta_{i_ps}a_{i_1\dots i_{q-1}ti_{q+1}\dots i_{p-1}ti_{p+1}\dots i_r} +$$
$$+\delta_{i_pt}a_{i_1i_2\dots i_{q-1}ti_{q+1}\dots i_{p-1}si_{p+1}\dots i_r}) - \delta_{ts}a_{i_1\dots i_{q-1}ti_{q+1}\dots i_r} +$$
$$+\delta_{tt}a_{i_1\dots i_{q-1}si_{q+1}\dots i_r}\Big] +$$
$$+\delta_{i_qt}\Big[\sum_{\substack{p=1\\p\neq q}}^{r}(-\delta_{i_ps}a_{i_1\dots i_{q-1}si_{q+1}\dots i_{p-1}ti_{p+1}\dots i_r} +$$
$$+\delta_{i_pt}a_{i_1\dots i_{q-1}si_{q+1}\dots i_{p-1}si_{p+1}\dots i_r}) - \delta_{ss}a_{i_1\dots i_{q-1}ti_{q+1}\dots i_r} +$$
$$+\delta_{st}a_{i_1\dots i_{q-1}si_{q+1}\dots i_r}\Big]. \quad (9)$$

Before substituting in equation (6) we have to find $H^2 = -A_{23}^2 - A_{31}^2 - -A_{12}^2$. But from equation (8) it is obvious that $A_{ss} = 0$ and $A_{ts} = -A_{st}$, hence $(A_{ts})^2 = (A_{st})^2$. By adding $(A_{st})^2$ for all possible values of s or t from 1 to 3 we obtain the transformation $-2H^2$. Carrying out the summation, using (9) and dividing by 2 we have:

$$-H^2 = \sum_{q=1}^{r}\sum_{\substack{p=1\\p\neq q}}^{r}(\delta_{i_pi_q}a_{i_1\dots i_{q-1}ti_{q+1}\dots i_{p-1}ti_{p+1}\dots i_r} -$$
$$-a_{i_1\dots i_{q-1}i_pi_{q+1}\dots i_{p-1}i_qi_{p+1}\dots i_r}) - 2ra_{i_1i_2\dots i_r}. \quad (10)$$

Thus we see that the operator H^2 represents the sum of all the possible traces of the tensor $a_{i_1i_2\dots i_r}$ multiplied by the unit tensor with the corresponding indices, minus the sum of the tensors obtained from $a_{i_1i_2\dots i_r}$ by all possible permutations of two indices and the tensor itself multiplied by 2r.

Consequently the equation $(-H^2 + l(l+1))a = 0$ has the form

$$\sum_{\substack{p,q=1\\p\neq q}}^{r} \delta_{i_p i_q} a_{i_1 i_2 \ldots i_{q-1} t i_{q+1} \ldots i_{p-1} t i_{p+1} \ldots i_r} -$$
$$- a_{i_1 \ldots i_{q-1} i_p i_{q+1} \ldots i_{p-1} i_q i_{p+1} \ldots i_r} + [l(l+1) - 2r]\, a_{i_1 i_2 \ldots i_r} = 0. \tag{11}$$

The solutions of this equation for any l form an invariant subspace, in which we have an isotypic representation of weight l.

For tensors of rank 2 this method yields, as was shown in paragraph 2, a resolution of these tensors into skew-symmetric tensors, multiples of the unit tensor and symmetric tensors of zero trace.

We consider the resolution of tensors of rank 3 in more detail. In accordance with the table in paragraph 2 we obtain a resolution into one representation of weight 0 (a pseudo-scalar) three of weight 1 (vectors) two of weight 2 and one of weight 3.

Equation (11) for the tensor a_{ijk} acquires the following form after division by 2

$$\delta_{ij} a_{ssk} + \delta_{jk} a_{iss} + \delta_{ki} a_{sjs} - a_{jik} - a_{ikj} - a_{kji} + \\ + \left[\frac{l(l+1)}{2} - 3\right] a_{ijk} = 0. \tag{12}$$

We must substitute the values $l = 0, 1, 2, 3$.

Take the trace of the left-hand side for the indices i and j. Then the second and third terms cancel with the sixth and fifth respectively and we obtain:

$$2a_{ssk} + \left[\frac{l(l+1)}{2} - 3\right] a_{ssk} = 0,$$

i.e.

$$\left[\frac{l(l+1)}{2} - 1\right] a_{ssk} = 0.$$

For $l(l+1) \neq 2$, i.e. $l \neq 1$, we have $a_{ssk} = 0$ and similarly $a_{sjs} = a_{iss} = 0$. Thus for tensors of rank 3 the subspaces of the isotypic representations with weight $l \neq 1$ consist of tensors with trace zero.

Put $l = 0$. Since all the traces of a_{ijk} are zero we obtain the equation:

$$a_{ijk} = -\frac{a_{jik} + a_{ikj} + a_{kji}}{3} \tag{13}$$

A cyclic permutation of the indices permutes the last three terms amongst themselves and so leaves their sum unaltered. Consequently the tensor is unaltered by a cyclic permutation of the indices, i.e.

$$a_{ijk} = a_{jki} = a_{kij}.$$

Interchanging the indices i and j in equation (13) we have also that

$$a_{jik} = a_{kij} = a_{ikj}.$$

Returning once more to (13) we obtain finally

$$a_{ijk} = -a_{jik}.$$

Comparing all these results we see that the solution of equation (13) is a skew-symmetric tensor of the third rank.

Put $l = 1$. The equation takes the form

$$\delta_{ij}a_{ssk} + \delta_{jk}a_{iss} + \delta_{ki}a_{sjs} - a_{jik} - a_{ikj} - a_{kji} - 2a_{ijk} = 0. \quad (14)$$

We substitute the tensor $\delta_{ij}x_k$ in this equation, x_k being an arbitrary vector. This gives:

$$\delta_{ij}\delta_{ss}x_k + \delta_{jk}\delta_{is}x_s + \delta_{ki}\delta_{sj}x_s - \delta_{ij}x_k - \delta_{ik}x_j - \delta_{kj}x_i - 2\delta_{ij}x_k = 0, \quad (15)$$

i.e. the tensor $\delta_{ij}x_k$ satisfies our equation. Similarly the tensors $\delta_{ik}y_j$ and $\delta_{jk}z_i$ also satisfy it, and hence also the tensor

$$a_{ijk} = \delta_{ij}x_k + \delta_{ik}y_j + \delta_{jk}z_i. \quad (16)$$

From paragraph 2 we know that in the resolution of a tensor representation of the third rank the irreducible representation of weight 1 appears three times. It follows that the solutions of equation (14) form a nine-dimensional space. But the three vectors x_k, y_j, z_i have nine components. We show that the tensor is zero only if these nine components are zero, i.e. that the space of tensors of the form (16) is in fact nine-dimensional. Equating the tensor a_{ijk} to 0 in equation (16) and taking traces we obtain nine equations in x_k, y_j and z_i, having only the zero solution.

Let $l = 2$. Since the traces of the solution of this equation have been shown to be zero, the equation must have the form

$$a_{jik} + a_{ikj} + a_{kji} = 0. \quad (17)$$

The dimensionality of the space of solutions of this equation is 10, in virtue of the results of paragraph 2. Later we will show how to resolve the representation in this space into two irreducible representations with weight $l = 2$.

Finally we put $l = 3$. Since the traces are zero we get:

$$a_{ijk} = \frac{a_{jik} + a_{ikj} + a_{kji}}{3}.$$

By reasoning similar to that for the case $l = 0$, we find that in this case the components of a_{ijk} are unchanged by any rearrangement of indices, i.e. a_{ijk} is a symmetric tensor of the third rank. It follows from the results of paragraph 2 that the dimensionality of the space of symmetric tensors of the third rank is 7.

We have therefore proved the following result: the 27-dimensional space of tensors of the third rank can be resolved into the sum of the following isotypic subspaces: a one-dimensional space of skew-symmetric tensors or pseudo-scalars, giving an irreducible representation of weight 0; a nine-dimensional subspace of tensors of the form $a_{ijk} = \delta_{ij}x_k + \delta_{ik}y_j + \delta_{jk}z_i$, giving an irreducible representation of weight 1, repeated three times; a ten-dimensional subspace of tensors, satisfying the equation $a_{jik} + a_{ikj} + a_{kji} = 0$, giving an irreducible representation of weight 2, repeated twice; and finally, a seven-dimensional subspace of symmetric tensors of trace zero. In this last subspace the irreducible representation is of weight 3. From the results of Sect.3 follows that this resolution into isotypic representations is unique.

To complete the resolution of tensor representation of the third rank into irreducible representations, it is necessary to resolve further the above isotypic representations when possible. This can be done in several ways. We indicate some of them.

The nine-dimensional space of tensors of the form $a_{ijk} = \delta_{ij}x_k + \delta_{ik}y_j + \delta_{jk}z_i$ may be broken down into three three-dimensional spaces $a^{(1)}_{ijk} = \delta_{ij}x_k$, $a^{(2)}_{ijk} = \delta_{ik}y_j$, $a^{(3)}_{ijk} = \delta_{jk}z_i$. in each of which there is an irreducible representation. This resolution, of course, is not unique. Any three linearly independent combinations of the vectors x_k, y_j and z_i gives rise to a resolution of this type.

Consider now the irreducible representation in the space of tensors satisfying equation (17): $a_{jik} + a_{ikj} + a_{kji} = 0$. As we have seen, the space of solutions of this equation is of dimension 10, and yields the irreducible representation of weight 2 twice. Hence any resolution of this representation must be a resolution into irreducible representations, and it will suffice to take any decomposition of this 10-dimensional space into non-zero invariant subspaces.

This can be done by making use of the method of resolution concerned with the eigenvalues of permutation transformations discussed in paragraph 1. Thus for example, it is possible to resolve the space of tensors satisfying equation (17) into subspaces of tensors, symmetrical and skew-symmetrical with respect to some pair of indices because among the non-zero solutions of equation (17) there exist tensors of both forms.

It is also possible to use this resolution, for example, for a cyclic rearrangements s of all three indices, taking ijk into jki. It is obvious that $s^3 = s_0$, and hence the eigenvalues of the corresponding transformation S are the cube roots of unity: 1, $\varepsilon = \frac{1}{2} + i\frac{\sqrt{3}}{2}$, $\varepsilon^2 = \frac{1}{2} - i\frac{\sqrt{3}}{2}$.

The equation $Sa_{ijk} = a_{ijk}$ combined with equation (17) gives $a_{ijk} = 0$, so that we are left with the two equations:

$$Sa_{ijk} = a_{jki} = \varepsilon a_{ijk}$$
$$Sa_{ijk} = a_{jki} = \varepsilon^2 a_{ijk},$$

giving the resolution into irreducible representations. A more detailed form of this can be applied to any other permutation transformation.

Thus we see that the resolution of the tensor representation of the third rank into irreducible representation is not unique. Every resolution from an isotypic representation to an irreducible one gives a resolution for tensor representations as a whole.

Section 6. Spinors and Spinor Representations

In the preceding sections we saw that it is possible to realize the irreducible representations of the rotation group of any integral weight l with the aid of tensor transformations. In this section we consider another method of representing the rotation group. This is the method of spinor representation which enables us to realize all the irreducible representations of the group without exception (this includes representations whose weights are halves of odd integers, i.e., two-valued representations).

§ 1. Definition of Spinors and Spinor Representations

We begin with the definition of spinors and spinor representations of rank 1. In Sect.2 we considered, among other examples, the irreducible representation of weight $l = \frac{1}{2}$. For this representation the rotation g with Euler angles φ_1, θ, φ_2 was set in correspondence with a complex matrix of the second order, which was defined to within a minus sign.

$$\left\| \begin{matrix} \alpha & \beta \\ -\bar{\beta} & \bar{\alpha} \end{matrix} \right\|, \tag{1}$$

where

$$\alpha = \pm \cos\frac{\theta}{2} e^{i\frac{\varphi_1+\varphi_2}{2}}, \quad \beta = \mp i \sin\frac{\theta}{2} e^{i\frac{\varphi_2-\varphi_1}{2}}, \quad |\alpha|^2 + |\beta|^2 = 1.$$

A product of rotations corresponds to the product of the corresponding complex matrices of the form (1).

We now postulate that in each system of coordinates we are given a pair (a^1, a^2) of complex numbers, defined to within sign, which, on transformation from one system to the other by the rotation $g = \|g_{ik}\|$ * are transformed by the matrix (1):

$$\left.\begin{aligned} a^{1'} &= \alpha a^1 + \beta a^2, \\ a^{2'} &= -\bar{\beta} a^1 + \bar{\alpha} a^2. \end{aligned}\right\} \tag{2}$$

* This is given by the equations $x'_i = \sum g_{ik} x_k$ of transformation of coordinates.

Such a system of numbers is called a spinor of the first rank in three-dimensional Euclidean space. The numbers a^1 and a^2 are called the components of the spinor.

If we regard equation (2) as giving, for a fixed system of coordinates, the transformation of one spinor with components $\{a^1, a^2\}$ to another spinor with components $\{a^{1'}, a^{2'}\}$, then the transformation (2) yields a two-dimensional representation of the rotation group. This representation we call a spinor representation of the first rank. From what has been said above it follows that it is an irreducible representation of weight $\frac{1}{2}$. In what follows it will be more convenient to denote the matrix $\begin{Vmatrix} \alpha & \beta \\ -\bar{\beta} & \bar{\alpha} \end{Vmatrix}$ by $\begin{Vmatrix} \alpha_{11} & \alpha_{12} \\ \alpha_{21} & \alpha_{22} \end{Vmatrix}$ i.e., to put $\alpha = \alpha_{11}$, $\beta = \alpha_{12}$, $-\bar{\beta} = \alpha_{21}$, $\bar{\alpha} = \alpha_{22}$. The equations transforming the components of a spinor now become

$$a^{i'} = \sum_{k=1}^{2} \alpha_{ik} a^k, \qquad \operatorname{Det} |\alpha_{ik}| = 1. \tag{2'}$$

Now let us be given in each orthogonal system of coordinates a set of 2^r complex numbers $a^{\lambda_1\lambda_2\ldots\lambda_r}$ $(\lambda_1, \lambda_2, \ldots, \lambda_r = 1, 2)$, defined to within sign, which, on transformation from one system to the other by the matrix $\|\alpha_{ik}\|$ are given by the equations

$$a'^{\lambda'_1\lambda'_2\ldots\lambda'_r} = \sum\sum\ldots\sum \alpha_{\lambda'_1\lambda_1}\alpha_{\lambda'_2\lambda_2}\ldots\alpha_{\lambda'_r\lambda_r}a^{\lambda_1\lambda_2\ldots\lambda_r}, \tag{3}$$

where $\alpha_{11} = \bar{\alpha}_{22} = \alpha$, $\alpha_{12} = -\bar{\alpha}_{21} = \beta$. Such a system is called a contravariant spinor of rank r in three-dimensional Euclidean space.

The set of all spinors of the r-th rank, in some system of coordinates, forms a 2^r-dimensional linear space.

Equation (3) of the transformation of spinors gives a representation of the rotation group by transformations in this space. It is called a spinor representation of rank r. Since it is clear from equation (3) that the components of a spinor of rank r are transformed as the product of the components of r spinors of the first rank, this representation is the product of r irreducible representations of weight $\frac{1}{2}$, just as a tensor representation is a product of irreducible representations of weight 1.

§ 2. Symmetric Spinors. The Existence of Irreducible Representations for any (Integral or Half an Odd Integer) Weight l.

At this point we prove that for any integer, or half of an odd integer, l there exist subspaces of spinors transforming under rotation in such a way as to give an irreducible representation of weight l. These are the subspaces of symmetric spinors.

The spinor $a^{i_1i_2\ldots i_r}$ of rank r is called symmetric if its components are unchanged by any permutation of the indices $i_1 i_2 \ldots i_r$. Since the indices i_μ

can only take the values 1 and 2, it is clearly possible to transform any component of a symmetric spinor into one of the following $r+1$ components.

$$a^{\overbrace{11\ldots 1}^{r}}, \ldots, a^{\overbrace{11\ldots 1}^{r-k},\overbrace{22\ldots 2}^{k}}, \ldots, a^{\overbrace{22\ldots 2}^{r}}.$$

Hence it is clear that symmetric spinors of the r-th rank form an $r+1$-dimensional subspace of the space of all spinors. It is easily proved that this subspace is invariant with respect to the spinor representation. In fact from the formula

$$a'^{i'_1 i'_2 \ldots i'_r} = \sum_{i_1=1}^{2} \sum_{i_2=1}^{2} \ldots \sum_{i_r=1}^{2} \alpha_{i'_1 i_1} \alpha_{i'_2 i_2} \ldots \alpha_{i'_r i_r} a^{i_1 i_2 \ldots i_r},$$

by which spinors of rank r are transformed, it will be seen that the same matrix α_{ik} acts on all the indices i_α and that therefore the symmetry of the spinor is not destroyed as a result of the transformation.

We show that the representation in the space of symmetric spinors of rank r is irreducible. It will be more convenient in what follows to denote r by $2l$, where l is an integer or half an odd integer. To prove the irreducibility of the representation in the space of symmetric tensors of rank $2l$ it is sufficient to show that the transformation H_3 has $2l+1$ distinct eigenvalues in this space (as many as the dimensionality of the space). To do this we determine the form of the transformation H_3 in the spinor representation. The spinor representation of rank one is irreducible of weight $\frac{1}{2}$ and the matrix $H_3 = iA_3$ for this representation has the form (see p.33)

$$\begin{Vmatrix} -\frac{1}{2} & 0 \\ 0 & \frac{1}{2} \end{Vmatrix}.$$

Since the spinor representation of rank r is the product of r representations of rank $\frac{1}{2}$, we find $H_3 a^{i_1 i_2 \ldots i_r}$, for this representation by operating on each of the indices in turn with the matrix, without changing the others, and then adding the results (see paragraph 2, Section 4). We obtain:

$$H_3 a^{i_1 i_2 \ldots i_r} = \left(-\frac{p_1}{2} + \frac{p_2}{2}\right) a^{i_1 i_2 \ldots i_r},$$

where p_1 is the number of times 1 appears among the indices i_α, and p_2 is the number of times 2 appears. From this equation it is clear that a non-zero spinor $a^{i_1 i_2 \ldots i_r}$, having p_1^0 indices equal to 1 and p_2^0 indices equal to 2 is an eigenvector of H_3 with the eigenvalue $\frac{1}{2}(p_2^0 - p_1^0)$.

Let f_m be a symmetric spinor of rank $2l$ for which $a^{\overbrace{11\ldots 1}^{l-m},\overbrace{22\ldots 2}^{l+m}} = 1$, and all other components are zero. Then

$$H_3 f_m = \left[-\frac{1}{2}(l-m) + \frac{1}{2}(l+m)\right] f_m = m f_m,$$

i.e. f_m corresponds to the eigenvalue m for the transformation H_3. Since m can take the values $-l, -l+1, \ldots, l$, we obtain $2l+1$ eigenvectors of the transformation H_3 in the space of symmetric spinors corresponding to the various eigenvalues. We have thus proved that the representation in this space is irreducible. The spinors f_m are simply scalar multiples of the elements of the canonical base for this representation.

Thus we have shown that it is possible to realize any irreducible representation of the rotation group in the space of symmetric spinors. In the space of symmetric spinors of rank $2l$ there exists an irreducible representation of dimensionality $2l+1$ and consequently of weight l. Hence symmetric spinors of even order provide irreducible representations of integral weight, which we have already met; and symmetric spinors of odd order provide representations with weights equal to an integer plus one half. In particular, a symmetric spinor of the second rank determines a representation with $l=1$. Thus a correspondence may be set up between symmetric spinors of the second rank and vectors of three-dimensional space, which is unique for rotations. To establish this connexion we observe that if a^{ik} is a symmetric spinor then its components a^{11}, $\sqrt{2}a^{12}$, a^{22} are its coordinates in a canonical basis. Since the components of a vector a_x, a_y, a_z in a canonical basis are $\frac{a_x+ia_y}{\sqrt{2}}$, a_z, $\frac{a_x+ia_y}{\sqrt{2}}$ (see paragraph 5, Sect.2) this connexion provides equations

$$a_x=-\frac{1}{\sqrt{2}}(a^{11}-a^{22}),\quad a_y=-\frac{i}{\sqrt{2}}(a^{11}+a^{22}),\quad a_z=\sqrt{2}a^{12}.$$

§ 3. Fundamental Operations on Spinors

Spinors and spinor representations are often used today in theoretical physics. In connexion with this we will study some operations in spinor algebra which are analogous to the operations on tensors discussed in Sect. 5.

We begin with the operation of contraction or forming the trace. First consider two spinors of the first rank $\{a^1, a^2\}$ and $\{b^1, b^2\}$. The effect of a transformation of coordinates on the determinant of the components of these spinors, namely

$$a^1b^2-a^2b^1, \tag{4}$$

is to multiply it by the determinant of the matrix $\|\alpha_{ik}\|$. Since $\alpha_{11}\alpha_{22}-\alpha_{12}\alpha_{21}=|\alpha|^2+|\beta|^2=1$, the expression $a^1b^2-a^2b^1$ is unaltered by a transformation of coordinates. Thus the bilinear form $a^1b^2-a^2b^1$ in the components of two spinors of the first rank is a scalar.

By introducing the matrix

$$\|\varepsilon_{\alpha\beta}\|=\begin{Vmatrix}0 & 1\\ -1 & 0\end{Vmatrix}, \tag{5}$$

this scalar may be written in the form

$$\sum_{\alpha=1}^{2}\sum_{\beta=1}^{2}\varepsilon_{\alpha\beta}a^{\alpha}b^{\beta}.$$

We define the number b_α by the equation

$$b_\alpha = \sum_{\beta=1}^{2} \varepsilon_{\alpha\beta} b^\beta, \tag{6}$$

i.e. we put $b_1 = b^2$, $b_2 = -b^1$. Obviously the numbers b_1 and b_2 define a spinor as well as b^1 and b^2 They are called the co-variant components of the spinor. From equations (4) and (2) it is easily seen that the co-variant components are transformed by the complex conjugate of the matrix (1) under a transformation of coordinates.

With the aid of the co-variant components, the bilinear form (4) may be written as the sum of products $a^1 b_1 + a^2 b_2 = \sum_{\lambda=1}^{2} a^\lambda b_\lambda$. In the following we shall omit the summation sign where the summation is carried out for an index in the upper position and one in the lower, i.e. we shall write the form (4) simply as $a^\lambda b_\lambda$.

Consider an arbitrary spinor $a^{\lambda_1 \cdots \lambda_r}$. We form from its components the system of numbers

$$b^{\lambda_3\lambda_4 \cdots \lambda_r} = \varepsilon_{\alpha\beta} a^{\alpha\beta\lambda_3 \cdots \lambda_r}. \tag{7}$$

The expression (7) is called the trace of the spinor $a^{\lambda_1 \cdots \lambda_r}$ for the first two indices. It follows from equation (4) that the trace of a spinor of the r-th rank is a spinor of rank $r-2$. For since a spinor of the r-th rank transforms as a product of r spinors of the first rank we have

$$b'^{\lambda'_3 \cdots \lambda'_r} = \alpha_{\lambda'_3\lambda_3} \alpha_{\lambda'_4\lambda_4} \cdots \alpha_{\lambda'_r\lambda_r} b^{\lambda_3 \cdots \lambda_r}.$$

If by analogy with what we did in the case of a spinor of the first rank we introduce for the spinor $a^{\lambda_1\lambda_2 \cdots \lambda_r}$ The spinor (co-variant for one index and contravariant for the rest)

$$a_{\lambda_1}^{\lambda_2 \cdots \lambda_r} = \varepsilon_{\lambda_1\alpha} a^{\alpha\lambda_2 \cdots \lambda_r},$$

then the trace of the spinor $a^{\lambda_1\lambda_2 \cdots \lambda_r}$ for the first two indices may be written simply as

$$\mathrm{Sp}\, a^{\lambda_1\lambda_2 \cdots \lambda_r} = a_\beta^{\beta\lambda_3 \cdots \lambda_r}.$$

The operation of forming the trace, as for the corresponding operation for tensors, is called contraction of the spinor with respect to the first two indices. The operation is defined similarly for any pair of indices.

In the process of defining the operation of contraction we began with the operation of lowering an index, replacing the contravariant components of of the spinor $a^{\lambda_1\lambda_2 \cdots \lambda_r}$ by the mixed (covariant for one index and contravariant for the rest) components $a_{\lambda_1}^{\lambda_2 \cdots \lambda_r}$. In the same way we may lower any number of indices and obtain mixed components, covariant for some indices and contravariant for the others. Under a transformation of coordinates the upper indices are transformed by the matrix $\|\alpha_{ik}\|$, and the lower by the complex conjugate of this matrix, so that the equation of the transformation of the components $a^{\lambda\mu\nu\cdots}_{\sigma\rho\tau\cdots}$ has the form

$$a'^{\lambda'\mu'\nu'\cdots}_{\sigma'\rho'\tau'\cdots} = \alpha_{\lambda'\lambda}\alpha_{\mu'\mu}\alpha_{\nu'\nu}\cdots\bar{\alpha}_{\sigma'\sigma}\bar{\alpha}_{\rho'\rho}\bar{\alpha}_{\tau'\tau}a^{\lambda\mu\nu\cdots}_{\sigma\rho\tau\cdots}.$$

The matrix $\|\varepsilon^{\alpha\beta}\|$, the inverse of the matrix $\|\varepsilon_{\alpha\beta}\|$, on the other hand, 'raises' the indices of a tensor and so, by using it, we are able to change any mixed spinor into contravariant components, in accordance with equation (3).

The product of two spinors $a^{i_1\cdots i_p}_{i_{p+1}\cdots i_s}$ and $b^{j_1\cdots j_{p'}}_{j_{p'+1}\cdots j_s}$, each of which is contravariant for some indices and covariant for the others, is the spinor $c^{i_1\cdots i_p j_1\cdots j_{p'}}_{i_{p+1}\cdots i_s j_{p'+1}\cdots j_s}$ of rank $r+s$, whose components are all the possible products of the components of the factors.

§ 4. The Irreducible Constituents of a Spinor Representation

Suppose we are given an arbitrary spinor of rank r. We know that the transformations of this spinor form a representation of the rotation group of dimensionality 2^r. At this point we determine the irreducible representations into which this representation may be resolved.

The solution of this problem may be achieved in a fashion similar to that used in the case of tensors. We already know that spinors of the first rank form an irreducible representation of weight $\frac{1}{2}$. An arbitrary spinor of the second rank is transformed exactly as is a product of two spinors of the first rank. Since each of the factors is transformed according to an irreducible representation, the product may be resolved into a sum of irreducible representations with weights $\frac{1}{2}-\frac{1}{2}=0$ and $\frac{1}{2}+\frac{1}{2}=1$. We turn now to spinors of rank 3. These are transformed as a product of spinors of the first and second rank. But a representation of spinors of the second rank has already been shown to be reducible; hence we may consider this representation as resolved into a sum of irreducible representations and multiply each factor by a representation of weight $\frac{1}{2}$ Then, from the representation of weight 0 we obtain a representation of weight $\frac{1}{2}$, and from the representation of weight 1 we obtain one representation of weight $\frac{1}{2}$ and one of weight $\frac{3}{2}$. (See Table 2). Hence in the resolution of a spinor representation of rank 3 there will be two representations of weight $\frac{1}{2}$ and one of weight $\frac{3}{2}$.

By continuing this process, i.e., expressing a spinor of rank r as the product of a spinor of rank $r-1$ and a spinor of rank 1, and using the results already established for the resolution of a spinor of rank $r-1$, we may obtain the irreducible constituents of any spinor representation.

TABLE 2

Order of Spinor	Weight of Representation						
	0	$\frac{1}{2}$	1	$\frac{3}{2}$	2	$\frac{5}{2}$	3
1		1					
2	1		1				
3		2		1			
4	2		3		1		
5		5		4		1	
6	5		9		5		1

Each entry in the r-th row is the sum of the entries immediately to the left and right of this entry in the $(r-1)$-th row.

CHAPTER II

FURTHER ANALYSIS OF THE REPRESENTATIONS OF THE ROTATION GROUP

Section 7. The Matrix Elements of an Irreducible Representation (the Generalized Spherical Functions)

In Sect.2 we found, using a particular basis, the matrices which correspond to infinitesimal rotations about the coordinate axes for an arbitrary irreducible representation. In this section we will find the matrix corresponding to an arbitrary rotation g in the same basis.

§ 1. The Operator U_g

Let there be given an irreducible representation of weight $l, g \rightarrow T_g$. The elements T_{mn} $(-l \leqslant m, \quad n \leqslant l)$ of the matrix T_g are functions of g, which we proceed to determine.

Multiply g by an arbitrary rotation g_1. Then $T_{mn}(g)$ is transformed into another function of g namely $T_{mn}(gg_1)$. This transformation of the functions T_{mn} depends on the rotation g_1, and, denoting it by U_{g_1}, we may write

$$U_{g_1} T_{mn}(g) = T_{mn}(gg_1). \tag{1}$$

It is easily proved that the transformation U_{g_1} satisfies the equation

$$U_{g_2} U_{g_1} = U_{g_2 g_1}. \tag{2}$$

For $U_{g_2} U_{g_1} T_{mn}(g) = U_{g_2} T_{mn}(gg_1) = T_{mn}(gg_2 g_1) = U_{g_2 g_1} T_{mn}(g)$.

Let us consider the function $T_{mn}(gg_1)$ in more detail. By the definition of the function T_{mn} this is an element of the matrix T_{gg_1}. Since the matrices T_g form a representation we have

$$T_{gg_1} = T_g T_{g_1}.$$

Equating the elements of the matrices on the left- and right-hand sides we obtain

$$T_{mn}(gg_1)=\sum_{s=-l}^{l} T_{ms}(g)\, T_{sn}(g_1). \tag{3}$$

From this equation it follows that

$$U_{g_1} T_{mn}(g)=\sum_{s=-l}^{l} T_{ms}(g)\, T_{sn}(g_1). \tag{4}$$

Thus we see that the transformation U_{g_1} takes an element of the m-th row of the matrix T_g into a linear combination of elements of the same row, the coefficients being functions of g_1.

Consider the $(2l+1)$-dimensional space R^m of functions of g generated by the elements of the m-th row of the matrix T_g, i.e. by the functions

$$T_{mn}(g) \quad (-l \leqslant n \leqslant l).$$

It follows from equations (2) and (4) that, for any m, the transformations U_{g_1} provide a $(2l+1)$-dimensional representation of the rotation group in the space R^m.

It follows also from equation (4) that the coefficients of the linear transformations in which U_{g_1} operates on the functions $T_{mn}(g)$, are the $T_{sn}(g_1)$ $(-l \leqslant s,\ n \leqslant l)$. This means that the matrices of the transformations U_{g_1} in the space R^m coincide with the matrices T_{g_1}.

From this it is clear that, firstly, the representation $g_1 \to U_{g_1}$ in the space R^m is irreducible; and secondly, the functions $T_{mn}(g)$ $(-l \leqslant n \leqslant l)$, on which the matrices U_{g_1} operate form a canonical basis in this space. Thus the transformations H_+ and H_- of the representation $g_1 \to U_{g_1}$, operate on these functions in accordance with the equations found earlier (see equation (19), Sect.2).

With the aid of these results we will find the functions $T_{mn}(g)$ and establish a set of recurrence formulae for them.

§ 2. The Differential Operators Corresponding to Infinitesimal Rotation

It is necessary first to find the operators A_k for the representation defined in paragraph 1 which correspond to infinitesimal rotations about the axes of coordinates. To do this we take, as in Sect.3, g_1 to be a rotation about a fixed axis through an angle a and expand $U_{g_1} T_{mn}(g) = T_{mn}(gg_1)$ as a power series in a.

The work is considerably simplified if we take the axis Oz as the axis of rotation. Consider an arbitrary rotation g with Euler angles φ_1, θ, φ_2, and let g_1 be a rotation about Oz through an angle a. Then the rotation gg_1 is given by the Euler angles φ_1, θ, $\varphi_2+\alpha$. Therefore,

$$T_{mn}(gg_1)=T_{mn}(\varphi_1,\ \theta,\ \varphi_2+\alpha)=T_{mn}(\varphi_1,\ \theta,\ \varphi_2)+\alpha\frac{\partial T_{mn}}{\partial \varphi_2}+\dots$$

and the transformation A_3 is the differential operator

$$A_3 = \frac{\partial}{\partial \varphi_2}. \tag{5}$$

In the general case the expansion of $T_{mn}(gg_1) = T_{mn}(\varphi_1', \theta', \varphi_2')$ takes the form

$$T_{mn}(\varphi_1', \theta', \varphi_2') = T_{mn}(\varphi_1, \theta, \varphi_3) + \\ + \alpha \left[\frac{\partial T_{mn}}{\partial \varphi_1} \frac{d\varphi_1'}{d\alpha} + \frac{\partial T_{mn}}{\partial \theta} \frac{d\theta'}{d\alpha} + \frac{\partial T_{mn}}{\partial \varphi_2} \frac{d\varphi_2'}{d\alpha} \right]_{\alpha=0} + \dots \tag{6}$$

We determine $\left.\frac{d\varphi_1'}{d\alpha}\right|_{\alpha=0}$, $\left.\frac{d\theta'}{d\alpha}\right|_{\alpha=0}$ and $\left.\frac{d\varphi_2'}{d\alpha}\right|_{\alpha=0}$ for the case when the rotation g_1 takes place about Ox through an angle a.

To do this we regard the matrix of the rotation g as a function of the Euler angles. As was shown in Sect.1, it has the form

$$g(\varphi_1, \theta, \varphi_2) = \| g_{ik}(\varphi_1, \theta, \varphi_2) \| = \\ = \left\| \begin{matrix} \cos\varphi_1 \cos\varphi_2 - \cos\theta \sin\varphi_1 \sin\varphi_2, & -\cos\varphi_1 \sin\varphi_2 - \cos\theta \sin\varphi_1 \cos\varphi_2, & \sin\varphi_1 \sin\theta \\ \sin\varphi_1 \cos\varphi_2 + \cos\theta \cos\varphi_1 \sin\varphi_2, & -\sin\varphi_1 \sin\varphi_2 + \cos\theta \cos\varphi_1 \cos\varphi_2, & -\cos\varphi_1 \sin\theta \\ \sin\varphi_1 \sin\theta, & \cos\varphi_2 \sin\theta, & \cos\theta \end{matrix} \right\|. \tag{7}$$

The matrices of the rotation gg_1 correspond to various values of the parameter: φ_1', θ', φ_2', which depend on the angle of rotation a and reduce to φ_1, θ, φ_2 when $\alpha = 0$. Expanding the matrix gg_1 in terms of a we obtain

$$gg_1 = \| g_{ik}(\varphi_1, \theta, \varphi_2) \| + \alpha \left\| \frac{\partial g_{ik}}{\partial \varphi_1} \frac{d\varphi_1'}{d\alpha} + \frac{\partial g_{ik}}{\partial \theta} \frac{d\theta'}{d\alpha} + \frac{\partial g_{ik}}{\partial \varphi_2} \frac{d\varphi_2'}{d\alpha} \right\|_{\alpha=0} + \dots \tag{8}$$

On the other hand, since g_1 is a rotation through an angle a about the axis Ox its matrix is

$$g_1 = \left\| \begin{matrix} 1 & 0 & 0 \\ 0 & \cos\alpha & -\sin\alpha \\ 0 & \sin\alpha & \cos\alpha \end{matrix} \right\| = \left\| \begin{matrix} 1 & 0 & 0 \\ 0 & 1 & 0 \\ 0 & 0 & 1 \end{matrix} \right\| + \alpha \left\| \begin{matrix} 0 & 0 & 0 \\ 0 & 0 & -1 \\ 0 & 1 & 0 \end{matrix} \right\| + \dots$$

and consequently,

$$gg_1 = \| g_{ik}(\varphi_1, \theta, \varphi_2) \| + \alpha \left\| \begin{matrix} 0 & g_{13} & -g_{12} \\ 0 & g_{23} & -g_{22} \\ 0 & g_{33} & -g_{32} \end{matrix} \right\| + \dots \tag{9}$$

Equating the coefficients of a in the expressions (8) and (9) for the matrix

gg_1, we obtain equations from which $\left.\frac{d\varphi_1'}{d\alpha}\right|_{\alpha=0}$, $\left.\frac{d\theta'}{d\alpha}\right|_{\alpha=0}$ and $\left.\frac{d\varphi_2'}{d\alpha}\right|_{\alpha=0}$ may be determined.

We need to take the three simplest elements in the matrices multiplied by a in equations (8) and (9), that is to say the elements in the top right and bottom right and left corners. Differentiating the expressions for g_{ik} from equation (7) we get

$$-\sin\theta\left.\frac{d\theta'}{d\alpha}\right|_{\alpha=0}=-\cos\varphi_2\sin\theta,$$

$$\cos\varphi_2\sin\theta\left.\frac{d\varphi_2'}{d\alpha}\right|_{\alpha=0}+\sin\varphi_1\cos\theta\left.\frac{d\theta'}{d\alpha}\right|_{\alpha=0}=0,$$

$$\cos\varphi_2\sin\theta\left.\frac{d\varphi_1'}{d\alpha}\right|_{\alpha=0}+\sin\varphi_1\cos\theta\left.\frac{d\theta'}{d\alpha}\right|_{\alpha=0}=\cos\varphi_1\sin\varphi_2+\cos\theta\sin\varphi_1\cos\varphi_2,$$

from which we find that

$$\left.\frac{d\theta'}{d\alpha}\right|_{\alpha=0}=\cos\varphi_2,\quad \left.\frac{d\varphi_2'}{d\alpha}\right|_{\alpha=0}=-\sin\varphi_2\cot\theta,\quad \left.\frac{d\varphi_1'}{d\alpha}\right|_{\alpha=0}=\frac{\sin\varphi_2}{\sin\theta}.$$

By substituting these expressions in equation (6) we can determine the differential operator corresponding to an infinitesimal rotation about the axis Ox

$$A_1=-\cot\theta\sin\varphi_2\frac{\partial}{\partial\varphi_2}+\frac{\sin\varphi_2}{\sin\theta}\frac{\partial}{\partial\varphi_2}+\cos\varphi_2\frac{\partial}{\partial\theta}. \tag{10}$$

The operator A_2 may be obtained in a similar way and has the form

$$A_2=-\cot\theta\cos\varphi_2\frac{\partial}{\partial\varphi_2}+\frac{\cos\varphi_2}{\sin\theta}\frac{\partial}{\partial\varphi_1}-\sin\varphi_2\frac{\partial}{\partial\theta}. \tag{11}$$

Before we proceed further it will be convenient to write down the expressions for the operators H_+, H_-, H_3

$$\left.\begin{aligned}
H_+&=H_1+iH_2=iA_1-A_2=e^{-i\varphi_2}\left(\cot\theta\frac{\partial}{\partial\varphi_2}-\frac{1}{\sin\theta}\frac{\partial}{\partial\varphi_1}+i\frac{\partial}{\partial\theta}\right),\\
H_-&=H_1-iH_2=iA_1+A_2=e^{i\varphi_2}\left(-\cot\theta\frac{\partial}{\partial\varphi_2}+\frac{1}{\sin\theta}\frac{\partial}{\partial\varphi_1}+i\frac{\partial}{\partial\theta}\right),\\
H_3&=iA_3=i\frac{\partial}{\partial\varphi_2}.
\end{aligned}\right\} \tag{12}$$

The operator $H^2=H_1^2+H_2^2+H_3^2$ can now be found. Consequently we can set down the differential equation satisfied by all of the functions $T_{mn}(\varphi_1,\theta,\varphi_2)$ just as was done in Sect. 3 for spherical functions:

$$\frac{\partial^2U}{\partial\theta^2}+\cot\theta\frac{\partial U}{\partial\theta}+\\
+\frac{1}{\sin^2\theta}\left(\frac{\partial^2U}{\partial\varphi_1^2}-2\cos\theta\frac{\partial^2U}{\partial\varphi_1\partial\varphi_2}+\frac{\partial^2U}{\partial\varphi_2^2}\right)+l(l+1)U=0. \tag{13}$$

We will write down the solutions of this equation in explicit form just as we did for the spherical functions $Y_l^m(\varphi, \vartheta)$. in Sect. 3.

We will, however, make one change in the notation used in Sect.3, in that we will use the letter ϑ for the polar angle while retaining θ for the second Euler angle.

§ 3. The Dependence of the Matrix Elements on the Euler Angles φ_1, φ_2.

The functions $T_{mn}(\varphi_1, \theta, \varphi_2)$ depend on the values of φ_1 and φ_2. In fact we know that an arbitrary rotation g may be represented as a product of three rotations: a rotation through an angle φ_1 about Oz, then a rotation through an angle θ about Ox and finally a rotation about Oz through an angle φ_2. Denoting the matrices corresponding (in the representation) to each of these rotations by T_{φ_1}, T_θ and T_{φ_2} respectively we have

$$T_g = T_{\varphi_1} T_\theta T_{\varphi_2} \tag{14}$$

(recall that the matrix of the product of two rotations is equal to the product of their matrices in reverse order). But the matrix corresponding to a rotation through an angle φ about the axis Oz in an irreducible representation of weight l has already been found. From the results of Sect.2 it has the form

$$T_\varphi = \begin{Vmatrix} e^{il\varphi} & 0 & 0 \ldots & 0 \\ 0 & e^{i(l-1)\varphi} & 0 \ldots & 0 \\ \ldots & \ldots & \ldots & \ldots \\ 0 & 0 & 0 \ldots & e^{-il\varphi} \end{Vmatrix} \tag{14'}$$

(see the equation on p.32). We note that the row and column indices of the matrix vary from $-l$ to $+l$, so that at the intersection of the m-th row and the m-th column we have $e^{-im\varphi}$.)

Substituting for T_{φ_2} and T_{φ_1} from equation (14') in (14), and multiplying out we find:

$$T_{mn}(\varphi_1, \theta, \varphi_2) = e^{-im\varphi_1} u_{mn}(\theta) e^{-in\varphi_2}, \tag{15}$$

where the elements of the matrix T_θ are denoted by $u_{mn}(\theta)$. It remains to determine the $u_{mn}(\theta)$.

By substituting $T_{mn}(\varphi_1, \theta\ \varphi_2)$ from equation (15) in the differential equation (13), which this function must satisfy, we obtain the ordinary differential equation satisfied by the function $u_{mn}(\theta)$ This has the form

$$\frac{d^2u}{d\theta^2} + \cot\theta \frac{du}{d\theta} + \left[l(l+1) - \frac{n^2 - 2mn\cos\theta + m^2}{\sin^2\theta}\right] u = 0.$$

By effecting the substitutions

$$\tau = \sin^2\frac{\theta}{2},$$

$$u(\theta) = \tau^{\frac{|m-n|}{2}} (1-\tau)^{\frac{|m+n|}{2}} v(\tau),$$

this equation may be put in the hypergeometric form

$$\tau(1-\tau)\frac{d^2v}{d\tau^2}+[c-(a+b+1)\tau]\frac{dv}{d\tau}-abv(\tau)=0,$$

where

$$a=l+1+\frac{1}{2}(|m-n|+|m+n|),$$

$$b=-l+\frac{1}{2}(|m-n|+|m+n|),$$

$$c=|m-n|+1.$$

In the following section we will find a simple expression for the functions $u_{mn}(\theta)$.

§ 4. The Generalized Spherical Functions

The functions $T_{mn}(\varphi_1, \theta, \varphi_2)$ for any m are the eigenfunctions of the operator H_3 corresponding to the eigenvalue n of this operator. In particular $T_{ml}(\varphi_1, \theta, \varphi_2)$ correspond to the greatest eigenvalue l. It follows (see Sect.2, paragraph 3) that they must satisfy the equation $H_+T_{ml}(\varphi_1, \theta, \varphi_2)=0$ or

$$e^{-i\varphi_1}\left(\cot\theta\frac{\partial T_{ml}}{\partial\varphi_1}-\frac{1}{\sin\theta}\frac{\partial T_{ml}}{\partial\varphi_2}+i\frac{\partial T_{ml}}{\partial\theta}\right)=0.$$

Putting $T_{ml}(\varphi_1, \theta, \varphi_2)=e^{-im\varphi_2}u_{ml}(\theta)e^{-il\varphi_1}$ in this equation and cancelling $e^{-im\varphi_2}e^{-i(l+1)\varphi_1}$ we find an ordinary first order differential equation from which $u_{ml}(\theta)$ may be determined.

$$\frac{du_{ml}(\theta)}{d\theta}+\frac{m-l\cos\theta}{\sin\theta}u_{ml}(\theta)=0. \tag{16}$$

The general solution of this equation has the form

$$u_{ml}(\theta)=C_m\frac{\sin^l\theta}{\operatorname{tg}^m\frac{\theta}{2}}, \tag{17}$$

or, writing $\mu=\cos\theta$, $u_{ml}(\arccos\theta)=P_{ml}(\mu)$, the form

$$P_{ml}(\mu)=C_m(1-\mu)^{\frac{l-m}{2}}(1+\mu)^{\frac{l+m}{2}}. \tag{18}$$

All the elements of the last column of the matrix T_g, namely the $T_{ml}(\varphi_1, \theta, \varphi_2)$ are determined by this equation, to within an arbitrary constant multiplier C_m. Leaving these constants undetermined, we will find the remaining elements $T_{mn}(\varphi_1, \theta, \varphi_2)$. To do this we apply the operator H_- to the functions T_{ml} which have already been determined and make use of the fact that $H_-T_{mn}=\alpha_nT_{m,n-1}$, where $\alpha_n=\sqrt{(l+n)(l-n+1)}$ (see Sect. 2, paragraph 3). Having substituted the operator H_- in this equation, from (12), and the functions $T_{mn}(\varphi_1, \theta, \varphi_2)$ from (15), we obtain a relation from which the function $u_{mn}(\theta)$ may be determined, namely

$$\frac{du_{mn}}{d\theta} - \frac{m - n\cos\theta}{\sin\theta} u_{mn} = -i\alpha_n u_{m,\, n-1},$$

After the substitution $\mu = \cos\theta$, $u_{mn}(\theta) = P_{mn}(\mu)$, this takes the form

$$(1-\mu^2)^{\frac{1}{2}} \frac{dP_{mn}}{d\mu} + \frac{m - n\mu}{(1-\mu^2)^{\frac{1}{2}}} P_{mn}(\mu) = i\alpha_n P_{m,\, n-1}(\mu). \tag{19}$$

Making the substitution

$$P_{mn}(\mu) = (1-\mu)^{-\frac{n-m}{2}} (1+\mu)^{-\frac{n+m}{2}} v_{mn}(\mu). \tag{20}$$

in equation (19) we obtain a simple equation for the $v_{mn}(\mu)$

$$\frac{dv_{mn}}{d\mu} = i\alpha_n v_{m,\, n-1}(\mu). \tag{21}$$

Writing the previously-determined functions $P_{ml}(\mu)$ (see (18)), in the form (20),

$$P_{ml}(\mu) = C_m (1-\mu)^{-\frac{l-m}{2}} (1+\mu)^{-\frac{l+m}{2}} v_{ml}(\mu),$$

we see that

$$v_{ml}(\mu) = C_m (1-\mu)^{l-m} (1+\mu)^{l+m}.$$

It follows immediately from this and from equation (21) that

$$v_{mn}(\mu) = (-i)^{l-n} \frac{C_m}{\alpha_l \alpha_{l-1} \cdots \alpha_{n+1}} \frac{d^{l-n}}{d\mu^{l-n}} [(1-\mu)^{l-m}(1+\mu)^{l+m}]$$

and consequently that the functions $P_{mn}(\mu)$, which we will denote by $P^l_{mn}(\mu)$, have the form

$$P^l_{mn}(\mu) = (-i)^{l-n} \frac{C_m}{\alpha_l \alpha_{l-1} \cdots \alpha_{n+1}} (1-\mu)^{-\frac{n-m}{2}} (1+\mu)^{-\frac{n+m}{2}} \times$$

$$\times \frac{d^{l-n}}{d\mu^{l-n}} [(1-\mu)^{l-m}(1+\mu)^{l+m}].$$

Writing $u_{mn}(\theta) = P^l_{mn}(\cos\theta)$ in equation (15) we obtain the functions $T_{mn}(\varphi_1, \theta, \varphi_2)$ for any value of the indices m and n.

The expressions obtained for the $T_{mn}(\varphi_1, \theta, \varphi_2)$, however, contain the $(2l + 1)$ undetermined constants C_m. These may be determined by the conditio that to the rotation $g_0 = g(0, 0, 0)$ with zero Euler angles there corresponds the unit matrix E. This means that $T_{mm}(0, 0, 0) = 1$, i.e.

$$P^l_{mm}(1) = (-i)^{l-m} \frac{C_m}{\alpha_l \alpha_{l-1} \cdots \alpha_{m+1}} 2^{-m} (-1)^{l-m} (l-m)!\ 2^{l+m} = 1.$$

Hence, substituting for $\alpha_l, \alpha_{l-1}, \ldots, \alpha_{m+1}$, we obtain

$$C_m = \frac{(-1)^{l-m} i^{l-m}}{2^l (l-m)!} \sqrt{\frac{(2l)!\ (l-m)!}{(l+m)!}},$$

and by substituting for C_m and $\alpha_l, \alpha_{l-1}, \ldots, \alpha_{n+1}$ in the expression for $P^l_{m,n}(\mu)$ we have finally the equation

$$P^l_{m,n}(\mu) = \frac{(-1)^{l-m} i^{n-m}}{2^l (l-m)!} \sqrt{\frac{(l-m)!\ (l+n)!}{(l+m)!\ (l-n)!}} \times$$

$$\times (1-\mu)^{-\frac{n-m}{2}} (1+\mu)^{-\frac{n+m}{2}} \frac{d^{l-n}}{d\mu^{l-n}} [(1-\mu)^{l-m} (1+\mu)^{l+m}]. \tag{22}$$

Thus we have shown that for an irreducible representation of weight l, the matrix corresponding to an arbitrary rotation g with Euler angles $\varphi_1, \theta, \varphi_2$ has, in the canonical basis, the form

$$T^l_g = \| T^l_{mn}(\varphi_1, \theta, \varphi_2) \| \quad (m, n = -l, -l+1, \ldots, l)$$

where

$$T^l_{mn}(\varphi_1, \theta, \varphi_2) = e^{-im\varphi_2} P^l_{mn}(\cos\theta)\, e^{-in\varphi_1}$$

and

$$P^l_{mn}(\mu) = A (1-\mu)^{-\frac{n-m}{2}} (1+\mu)^{-\frac{n+m}{2}} \frac{d^{l-n}}{d\mu^{l-n}} [(1-\mu)^{l-m} (1+\mu)^{l+m}]. \tag{23}$$

The constant A in this is $\frac{(-1)^{l-m} i^{n-m}}{2^l (l-m)!} \sqrt{\frac{(l-m)!\ (l+n)!}{(l+m)!\ (l-n)!}}$. The functions $T^l_{mn}(\varphi_1, \theta, \varphi_2)$ will in future be referred to briefly as the generalized spherical functions of the order l.

As examples of the generalized spherical functions we obtain those of order $\frac{1}{2}$, 1 and 2. Since the dependence of these functions on the arguments φ_1 and φ_2 is known to us, the discussion will be shortened if we write down the matrices of the functions $P^{\frac{1}{2}}_{mn}(\cos\theta)$, $P^1_{mn}(\cos\theta)$ and $P^2_{mn}(\cos\theta)$. They have the form:

$$l = \frac{1}{2}$$

$$\left\| \begin{matrix} \frac{1}{\sqrt{2}}(1+\cos\theta)^{\frac{1}{2}} & \frac{i}{\sqrt{2}}(1-\cos\theta)^{\frac{1}{2}} \\ \frac{i}{\sqrt{2}}(1-\cos\theta)^{\frac{1}{2}} & \frac{1}{\sqrt{2}}(1+\cos\theta)^{\frac{1}{2}} \end{matrix} \right\| = \left\| \begin{matrix} \cos\frac{\theta}{2} & i\sin\frac{\theta}{2} \\ i\sin\frac{\theta}{2} & \cos\frac{\theta}{2} \end{matrix} \right\|,$$

$$l = 1$$

$$\left\| \begin{matrix} \frac{1}{2}(1+\cos\theta) & -\frac{i}{\sqrt{2}}\sin\theta & \frac{1}{2}(\cos\theta-1) \\ \frac{i}{\sqrt{2}}\sin\theta & \cos\theta & \frac{i}{\sqrt{2}}\sin\theta \\ \frac{1}{2}(\cos\theta-1) & \frac{i}{\sqrt{2}}\sin\theta & \frac{1}{2}(1+\cos\theta) \end{matrix} \right\|,$$

$$l = 2$$

$$\left\| \begin{matrix} \frac{1}{4}(\cos\theta+1)^2 & \frac{i}{2}\sin\theta(\cos\theta+1) & -\frac{1}{2}\sqrt{\frac{3}{2}}(1-\cos^2\theta) & \frac{i}{2}\sin\theta(\cos\theta-1) & \frac{1}{4}(\cos\theta-1)^2 \\ \frac{i}{2}\sin\theta(\cos\theta+1) & \frac{1}{2}(2\cos^2\theta+\cos\theta-1) & \sqrt{\frac{3}{2}}\,i\sin\theta\cos\theta & \frac{1}{2}(2\cos^2\theta-\cos\theta-1) & \frac{i}{2}\sin\theta(\cos\theta-1) \\ -\frac{1}{2}\sqrt{\frac{3}{2}}(1-\cos^2\theta) & \sqrt{\frac{3}{2}}\,i\sin\theta\cos\theta & \frac{1}{2}(3\cos^2\theta-1) & \sqrt{\frac{3}{2}}\,i\sin\theta\cos\theta & -\frac{1}{2}\sqrt{\frac{3}{2}}(1-\cos^2\theta) \\ \frac{i}{2}\sin\theta(\cos\theta-1) & \frac{1}{2}(2\cos^2\theta-\cos\theta-1) & \sqrt{\frac{3}{2}}\,i\sin\theta\cos\theta & \frac{1}{2}(2\cos^2\theta+\cos\theta-1) & \frac{i}{2}\sin\theta(\cos\theta+1) \\ \frac{1}{4}(\cos\theta-1)^2 & \frac{i}{2}\sin\theta(\cos\theta-1) & -\frac{1}{2}\sqrt{\frac{3}{2}}(1-\cos^2\theta) & \frac{i}{2}\sin\theta(\cos\theta+1) & \frac{1}{4}(\cos\theta+1)^2 \end{matrix} \right\|.$$

In all cases considered the rows are numbered from top to bottom and the columns from left to right with the numbers $-l, -l+1, \ldots, l$. To obtain the generalized spherical function $T^l_{mn}(\varphi_1, \theta, \varphi_2)$, it is necessary to multiply the (m, n)-th element of the corresponding matrix by $e^{-im\varphi_1}e^{-in\varphi_2}$.

The functions $T^l_{mn}(\varphi_1, \theta, \varphi_2)$ for integral l and $m=0$ take the form

$$T^l_{0n}(\varphi_1, \theta, \varphi_2)=$$

$$= e^{-in\varphi_2}(-1)^l i^n \frac{1}{2^l \cdot l!}\sqrt{\frac{(l+n)!}{(l-n)!}}(1-\mu^2)^{-\frac{n}{2}} \frac{d^{l-n}}{d\mu^{l-n}}(1-\mu^2)^l.$$

By comparing this with the corresponding spherical function $Y^n_l(\varphi, \theta)$ of the l-th order (see equation (14), Sect.3) we see that

$$T^l_{0n}(\varphi_1, \theta, \varphi_2)=\sqrt{\frac{2}{2l+1}}\, Y^n_l\left(\frac{\pi}{2}-\varphi_1, \theta\right). \tag{24}$$

i.e. the elements of the row with index zero (the central row) of the matrix T_g (this row exists only in the case when the matrix is of odd order giving a representation with integral weight l) coincide with the spherical functions of the l-th order*, to within multiplication by $\sqrt{\frac{2}{2l+1}}$ and replacement of φ by $\frac{\pi}{2}-\varphi$. In particular $P^l_{00}(\mu)$ is the same as the Legendre polynomial (not normalized by the condition $\int_{-1}^{1} P^2_l(\mu)\, d\mu = 1$)

For arbitrary m and n the functions $P^l_{mn}(\mu)$ are closely connected with other polynomials met with in analysis - the polynomials of Jacobi. The corresponding equation will be given later.

Let us consider the properties which the functions $P^l_{mn}(\mu)$ possess in virtue of the fact that the matrix T_g is unitary. Let the rotation g have Euler angles $\varphi_1, \theta, \varphi_2$. Then the rotation g^{-1} is given by the Euler angles $\pi-\varphi_2, \theta, \pi-\varphi_1$ (see Sect.1) and the fact that the matrix is unitary gives

$$T^*_g = T^{-1}_g = T_{g^{-1}}$$

so that the elements of the matrix T_g must satisfy the condition

$$\overline{T_{nm}(\varphi_1, \theta, \varphi_2)} = T_{mn}(\pi-\varphi_2, \theta, \pi-\varphi_1).$$

Dividing throughout by $e^{im\varphi_1}e^{in\varphi_2}$ we have:

$$\overline{u_{nm}(\theta)} = u_{mn}(\theta)\cdot(-1)^{m+n}, \tag{25}$$

or, putting $\cos\theta=\mu$,

$$P^l_{mn}(\mu) = \overline{P^l_{nm}(\mu)}\cdot(-1)^{m+n}. \tag{25'}$$

---

* Equation (24) may be obtained from the fact that the elements of the "null" row are independent of φ_2 and may be regarded as functions on the surface of a sphere (cf. Sect. 3).

Since from equation (23) $\overline{P^l_{nm}(\mu)}$ contains the factor i^{n-m} then $\overline{P^l_{nm}(\mu)} = (-1)^{n-m}P^l_{nm}(\mu)$. Thus, finally,

$$P^l_{mn}(\mu) = P^l_{nm}(\mu), \tag{26}$$

i.e. the matrix of the functions $P^l_{mn}(\mu)$ is symmetrical with respect to the leading diagonal.

Hence it follows that, in virtue of equation (23), we may write the functions $P^l_{mn}(\mu)$ in the following form:

$$\left.\begin{aligned} P^l_{mn}(\mu) &= A'(1-\mu)^{-\frac{m-n}{2}}(1+\mu)^{-\frac{m+n}{2}}\frac{d^{l-m}}{d\mu^{l-m}}[(1-\mu)^{l-n}(1+\mu)^{l+n}], \\ \text{where}\quad A' &= (-1)^{l-n}\frac{i^{m-n}}{2^l(l-n)!}\sqrt{\frac{(l-n)!(l+m)!}{(l+n)!(l-m)!}}. \end{aligned}\right\} \tag{23'}$$

Another symmetric property possessed by the functions $P^l_{mn}(\mu)$ is obtained from the following observation. Consider the rotation g_0 with Euler angles $(0, \pi, 0)$ (a rotation through 180° about Ox). For this rotation $\mu = \cos\pi$ i.e. $\mu + 1 = 0$ and the elements of the matrix T_g corresponding to it are zero for $m + n \neq 0$ (see equation (23)). For $n = -m$ we have

$$T_{m,-m} = P_{m,-m}(-1) = (-1)^l.$$

Now let g be an arbitrary rotation with Euler angles $\varphi_1, \theta, \varphi_2$ and $T_{mn}(\varphi_1, \theta, \varphi_2)$ be the elements of the matrix corresponding to it. Consider the rotation $\tilde{g} = g_0 g g_0^{-1}$. To this corresponds the matrix $T_{\tilde{g}} = T_{g_0}T_g T_{g_0^{-1}}$. Having multiplied the matrices we easily see that the element in the m-th row and n-th column is $T_{-m,-n}(\varphi_1, \theta, \varphi_2)$. On the other hand the rotation g interchanges the directions of the axes Ox and Oy. Therefore if $g = g(\varphi_1, \theta, \varphi_2)$ then $g_0 g g_0^{-1} = g(-\varphi_1, \theta, -\varphi_2)$.* Consequently, $T_{\tilde{g}} = T(-\varphi_1, \theta, -\varphi_2)$, and we have the equation

$$T_{-m,-n}(\varphi_1, \theta, \varphi_2) = T_{mn}(-\varphi_1, \theta, -\varphi_2).$$

Dividing throughout by $e^{im\varphi_1}e^{in\varphi_2}$ and putting $\cos\theta = \mu$ we obtain:

$$P^l_{-m,-n}(\mu) = P^l_{mn}(\mu). \tag{27}$$

* The matrix $\tilde{g} = g_0 g g_0^{-1}$ may be regarded as the matrix of the same rotation g in a new system of coordinates obtained from the old one by the rotation g_0. This system is obtained from the old one by interchanging the directions of the axes Oy and Oz.

From the symmetrical properties of the functions $P^l_{mn}(\mu)$ it is immediately evident that these functions depend not on the indices m and n but on the value of $|m+n|$ and $|m-n|$.

We now discuss the connexion between the functions $P^l_{mn}(\mu)$ with the Jacobi polynomials mentioned above. The Jacobi polynomials are given by the equation

$$P_s^{\alpha\beta}(\mu) = \frac{(-1)^s}{2^s \cdot s!}(1-\mu)^{-\alpha}(1+\mu)^{-\beta}\frac{d}{d\mu^s}[(1-\mu)^{s+\alpha}(1+\mu)^{s+\beta}].$$

Obviously if $s = l - \frac{1}{2}(|m+n| + |m-n|)$, $\alpha = |n-m|$ $\beta = |n+m|$, then the functions $P^l_{mn}(\mu)$ will be expressed in terms of the Jacobi polynomals by the equation

$$P^l_{mn}(\mu) = K(1-\mu)^{\frac{\alpha}{2}}(1+\mu)^{\frac{\beta}{2}} P_s^{\alpha\beta}(\mu),$$

where K is a constant.

We note that, since the matrix $T^l_{mn}(g)$ is unitary, it follows that

$$\sum_{n=-l}^{+l} |P^l_{mn}(\cos\theta)|^2 = 1.$$

§ 5. The Addition Formula for Matrix Elements

Equation (3) of this section, $T_{mn}(g'g'') = \sum_{s=-l}^{l} T_{ms}(g')T_{sn}(g'')$, provides a rule by which the generalized spherical functions are added. It contains as a particular case the addition rule for the Legendre polynomials. We shall derive a simple form for the addition formula. Let the rotation g' be defined by the Euler angles 0, θ' φ_2', and the rotation g'' by the angles φ_1'', θ'', 0 and finally, the rotation $g'g''$ by the angles φ_1, θ, φ_2. Then

$$T_{mn}(\varphi_1, \theta, \varphi_2) = \sum_{s=-l}^{s=l} T_{ms}(0, \theta', \varphi_2')T_{sn}(\varphi_1'', \theta'', 0).$$

Replacing $T_{mn}(\varphi_1, \theta, \varphi_2)$ by $e^{-im\varphi_1}P^l_{mn}(\cos\theta)e^{-in\varphi_2}$ and denoting the sum $\varphi_1' + \varphi_2''$ by φ, we obtain:

$$e^{-im\varphi_1}P^l_{mn}(\cos\theta)e^{-in\varphi_2} = \sum_{s=-l}^{s=l} e^{-is\varphi}P^l_{ms}(\cos\theta')P^l_{sn}(\cos\theta''). \qquad (28)$$

It remains to express φ_1, θ, φ_2 in terms of φ, θ', θ''. With this in view, we consider the matrices $T_{g'}$, $T_{g''}$ and T_g for $l = \frac{1}{2}$. From the equation

$T_{g'}T_{g''}=T_g$, i.e.

$$\left\|\begin{matrix}\cos\frac{\theta'}{2}e^{i\frac{\varphi_2'}{2}} & -i\sin\frac{\theta'}{2}e^{-i\frac{\varphi_2'}{2}}\\ -i\sin\frac{\theta'}{2}e^{i\frac{\varphi_2'}{2}} & \cos\frac{\theta'}{2}e^{-i\frac{\varphi_2'}{2}}\end{matrix}\right\|\cdot\left\|\begin{matrix}\cos\frac{\theta''}{2}e^{i\frac{\varphi_1''}{2}} & -i\sin\frac{\theta''}{2}e^{i\frac{\varphi_1''}{2}}\\ -i\sin\frac{\theta''}{2}e^{-i\frac{\varphi_1''}{2}} & \cos\frac{\theta''}{2}e^{-i\frac{\varphi_1''}{2}}\end{matrix}\right\|=$$

$$=\left\|\begin{matrix}\cos\frac{\theta}{2}e^{i\frac{\varphi_1+\varphi_2}{2}} & -i\sin\frac{\theta}{2}e^{-i\frac{\varphi_2-\varphi_1}{2}}\\ -i\sin\frac{\theta}{2}e^{i\frac{\varphi_2-\varphi_1}{2}} & \cos\frac{\theta}{2}e^{-i\frac{\varphi_1+\varphi_2}{2}}\end{matrix}\right\|,$$

we obtain two complex equations for θ', θ'' and $\varphi=\varphi_1'+\varphi_2''$:

$$\cos\frac{\theta}{2}e^{i\frac{\varphi_1+\varphi_2}{2}}=\cos\frac{\theta'}{2}\cos\frac{\theta''}{2}e^{i\frac{\varphi}{2}}-\sin\frac{\theta'}{2}\sin\frac{\theta''}{2}e^{-i\frac{\varphi}{2}},$$

$$\sin\frac{\theta}{2}e^{i\frac{\varphi_2-\varphi_1}{2}}=\cos\frac{\theta'}{2}\sin\frac{\theta''}{2}e^{i\frac{\varphi}{2}}+\sin\frac{\theta'}{2}\cos\frac{\theta''}{2}e^{-i\frac{\varphi}{2}}.$$

Equating the moduli and arguments of the right- and left-hand sides we obtain:

$$\left.\begin{aligned}\cos\theta&=\cos\theta'\cos\theta''-\sin\theta'\sin\theta''\cos\varphi,\\ \tan\varphi_1&=\frac{\sin\varphi\,\sin\theta''}{\cos\theta'\sin\theta''\cos\varphi+\cos\theta''\sin\theta'},\\ \tan\varphi_2&=\frac{\sin\varphi\,\sin\theta'}{\sin\theta'\cos\theta''\cos\varphi+\cos\theta'\sin\theta''}.\end{aligned}\right\}\qquad(29)$$

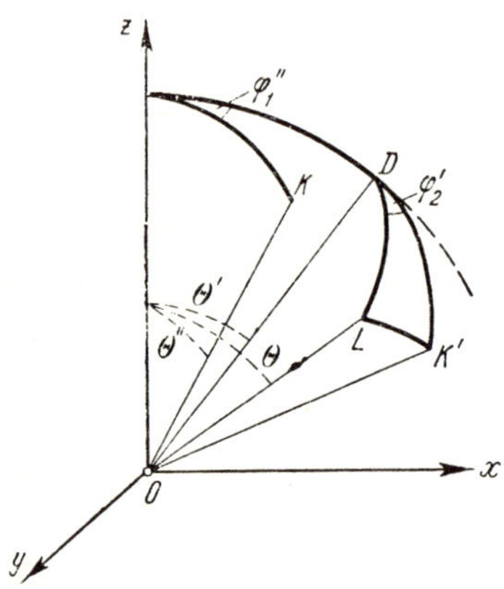

Fig. 5.

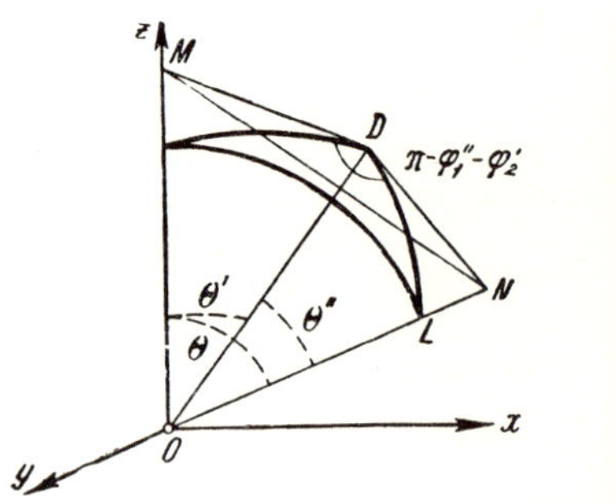

Fig. 6.

Equations (29) may be obtained by purely geometric methods. As an example we will derive the first of them. Consider the rotation g''. The axis Oz is transformed into the line OK (Fig.5). Now apply the rotation g'. Upon a a rotation about Ox through an angle θ' the ray OK is carried into the ray OK', and the axis Oz into the straight line OD. Then, after a rotation through an angle φ_2' about OD, the line OK' takes the position OL. We have to determine the angle θ between the axis Oz and the line OL. It is obvious that the axis Oz may be carried into the position of the straight line OL in the following manner: (see Fig.6) in the plane xOy carry out a rotation through an angle θ', and then a rotation in the plane DOL towards the plane xOy through an angle $\varphi = \varphi_1' + \varphi_2'$ the ray OD being turned through an angle θ''; after this it takes up the position OL. From Fig.6 it is clear that the problem of determining the angle θ involves finding the side of the spherical triangle whose other sides θ' and θ'' contain the angle $\pi - \varphi$ between them. To do this we construct the straight lines MD and DN, which touch the rays ZD and DL. From Fig.6 we see that:

$$OM = \sec\theta', \qquad MD = \tan\theta', \qquad ON = \sec\theta'', \qquad DN = \tan\theta''.$$

From the triangles MDN and MNO we have:

$$MN^2 = MD^2 + DN^2 - 2MD\,DN\cos(\pi - \varphi) = \\ = \tan^2\theta' + \tan^2\theta'' + 2\tan\theta'\tan\theta''\cos\varphi.$$

$$MN^2 = MO^2 + ON^2 - 2MO\,NO\cos\theta = \\ = \sec^2\theta'' + \sec^2\theta' - 2\sin\theta'\sec\theta''\cos\theta.$$

By equating the right-hand sides of these equations we obtain the first of equations (29). In a similar manner we may obtain the other two.

Thus we have obtained the following result. If φ, θ', θ'' are any angles such that $(0 \leqslant \varphi < 2\pi,\ 0 \leqslant \theta',\ \theta'' \leqslant \pi)$, and θ, φ_1, φ_2 are given in terms of φ_1, θ', θ'' by equations (29) then the generalized functions obey the following relation:

$$e^{-im\varphi_1} P^l_{mn}(\cos\theta)\, e^{-in\varphi_2} = \sum_{s=-l}^{l} e^{-is\varphi} P^l_{ms}(\cos\theta') P^l_{sn}(\cos\theta''). \qquad (28')$$

In particular, if $m = n = 0$ then equation (28') reduces to the usual addition formula for Legendre polynomials, expressing the Legendre polynomial for $\cos\theta = \cos\theta'\cos\theta'' - \sin\theta'\sin\theta''\cos\varphi$ in terms of the associated functions of θ' and θ''.

Putting $m = 0$, we obtain the usual addition formula for spherical functions $P^l_{0n}(\cos\theta)\, e^{-in\varphi_2} = Y^n_l\left(\frac{\pi}{2} - \varphi_2, \theta\right)$:

$$P^l_{0n}(\cos\theta) = \sum_{s=-l}^{l} e^{-is\varphi} P^l_{0s}(\cos\theta') P^l_{sn}(\cos\theta'').$$

If $\varphi = 0$ the equation takes a particularly simple form; for in consequence we get $\varphi_1 = \varphi_2 = 0$. In this case $\theta = \theta' + \theta''$, and we have:

$$P^l_{mn}[\cos(\theta' + \theta'')] = \sum_{s=-l}^{l} P^l_{ms}(\cos\theta')\, P^l_{sn}(\cos\theta'') \tag{30}$$

Remembering the connexion between the $P^l_{mn}(\mu)$ and the Jacobi polynomials (see paragraph 4) we may interpret this relation as the sum formula for Jacobi polynomials. Since

$$P^l_{0n}[\cos(\theta' + \theta'')] = \sum_{s=-l}^{l} P^l_{0s}(\cos\theta')\, P^l_{sn}(\cos\theta''),$$

where the $P^l_{0n}[\cos(\theta' + \theta'')]$ are the associated Legendre functions, it will be seen that the generalized spherical functions may be expressed uniquely by expanding $P^l_{0n}[\cos(\theta' + \theta'')]$ as a series of associated functions of θ'. These same functions, as follows from equations (28') and (30), form a system closed with respect to addition. The sum formula makes it possible to express Legendre polynomials of a sum of several angles in terms of functions each of which depends on only one of the angles. Thus, for example,

$$P_l[\cos(\theta' + \theta'' + \theta''')] =$$

$$= \sum_{s_1=-l}^{l} \sum_{s_2=-l}^{l} P^l_{0s_1}(\cos\theta')\, P^l_{s_1 s_2}(\cos\theta')\, P^l_{s_2 0}(\cos\theta''').$$

§ 6. The Expansion of Functions Defined on the Rotation Group in Series of Generalized Spherical Functions

To determine the matrix elements $T_{mn}(g)$ we construct in the first part of this paragraph the irreducible representation consisting of the transformations U_{g_1}, which act in the finite-dimensional space R^m of functions on the rotation group.

Let us now consider all functions of g:

$$f(g) = f(\varphi_1, \theta, \varphi_2),$$

for which the following integral exists

$$\int |f(g)|^2\, dg = \frac{1}{8\pi^2} \int_0^{2\pi} \int_0^{\pi} \int_0^{2\pi} |f(\varphi_1, \theta, \varphi_2)|^2 \sin\theta\, d\varphi_1\, d\theta\, d\varphi_2. \tag{31}$$

If we define the scalar product of $f_1(g)$ and $f_2(g)$ by the equation

$$(f_1, f_2) = \frac{1}{8\pi^2} \int_0^{2\pi} \int_0^{\pi} \int_0^{2\pi} f_1(\varphi_1, \theta, \varphi_2)\, \overline{f_2(\varphi_1, \theta, \varphi_2)} \sin\theta\, d\varphi_1\, d\theta\, d\varphi_2, \tag{32}$$

then these functions form a Hilbert space of functions on the group. The transformation

$$U_{g_1} f(g) = f(g g_1)$$

forms an infinite unitary representation in this space.* It is called the regular representation of the rotation group. The irreducible representations into which it may be resolved are known to us: they are the representations in the subspace of generalized spherical functions $T^l_{mn}(\varphi_1, \theta, \varphi_2)$ for a fixed l and m. It is not difficult to show that this contains all irreducible representations of the transformations U_{g_1}. For in each subspace in which there acts an irreducible representation of weight l there must, in agreement with the general theory, exist an element satisfying the equations $H_3 f = lf$, $H_+ f = 0$, i.e. (see equation (12)) the equations $i\frac{\partial f}{\partial \varphi_1} = lf$ and $\cot\theta \frac{\partial f}{\partial \varphi_1} - \frac{1}{\sin\theta}\frac{\partial f}{\partial \varphi_2} + i\frac{\partial f}{\partial \theta} = 0$. The solution of the first has the form $f = F(\theta, \varphi_2)\, e^{-il\varphi_1}$, and the second after substitution of this function takes the form

$$\frac{\partial F}{\partial \theta} + \frac{i}{\sin\theta}\frac{\partial F}{\partial \varphi_2} - l\cot\theta F = 0.$$

The solutions of this equation which are periodic in φ_2 may be written in the form

$$F(\theta, \varphi_2) = \sum_m u_m(\theta)\, e^{-im\varphi_2},$$

where the $u_m(\theta)$ satisfy the equation

$$\frac{du}{d\theta} + \frac{m - l\cos\theta}{\sin\theta}\, u = 0.$$

Hence (see equation (17) of this section)

$$u_m(\theta) = C\frac{\sin^l\theta}{\tan^m\frac{\theta}{2}}.$$

The possible values of m are determined by the condition that the function $u_m(\theta)\, e^{-im\varphi_2} e^{-il\varphi_1}$ must belong to the Hilbert space i.e. from the condition that the following integral must exist

$$\int_0^\pi |u_m(\theta)|^2 \sin\theta\, d\theta.$$

Putting $\cos\theta = \mu$, it is easy to show that the integral exists provided that $-l \leqslant m \leqslant l$. But in this case the solutions are linear combinations of the generalized spherical functions.

Thus the resolution of the regular representation into irreducible representations means that every function on the rotation group the square of whose modulus is integrable over the group may be expanded as a series in the generalized spherical functions $T^l_{mn}(\varphi_1, \theta, \varphi_2)$. The functions $T^l_{mn}(\varphi_1, \theta, \varphi_2)$ and

---

* This representation is unitary because the scalar product introduced is invariant with respect to multiplication from the right by the element g_0.

$T^{l'}_{m'n'}(\varphi_1, \theta, \varphi_2)$ are mutually orthogonal for $l' \neq l$ since they belong to invariant orthogonal subspaces in which distinct irreducible representations of the rotation group act. For $l = l'$ and $m \neq m'$ they are again orthogonal since $\int_0^{2\pi} e^{-im\varphi_2} e^{im'\varphi_2} d\varphi_2 = 0$. Similarly they are orthogonal for $l' = l$ and $n' \neq n$. We have for the integral over the group of the squared modulus of each of the functions $T^l_{mn}(\varphi_1, \theta, \varphi_2)$:

$$\frac{1}{8\pi^2}\int_0^{2\pi}\int_0^{\pi}\int_0^{2\pi} | T^l_{mn}(\varphi_1, \theta, \varphi_2)|^2 \sin\theta \, d\varphi_1 \, d\theta \, d\varphi_2 = \frac{2}{2l+1}.$$

We may now state the following theorem: <u>the set of generalized spherical functions $T^l_{mn}(\varphi_1, \theta, \varphi_2)$ (l an integer) forms a complete orthogonal system in the space of functions $f(\varphi_1, \theta, \varphi_2)$ $(0 \leqslant \theta \leqslant \pi;\ 0 \leqslant \varphi_1, \varphi_2 < 2\pi)$, if the scalar product is given by equation (32).</u>

Appendix to Section 7. Recurrence Relations between the Generalized Spherical Functions

There is a whole series of recurrence relations between the generalized spherical functions T^l_{mn}. Some of these connect generalized spherical functions of the same order (with one and the same l), others connect functions of different orders. In paragraph 4, Sect. 7, we met the recurrence relations connecting functions of a given order. They derived from the equations $H_- T^l_{mn} = \alpha_n T^l_{m,n-1}$ and $H_+ T^l_{mn} = \alpha_{n+1} T^l_{m,n+1}$ by substituting for the operators H_- and H_+ from equation (12), Sect. 7. We put T^l_{mn} equal to $e^{-im\varphi_1} u_{mn}(\theta) e^{-i\ldots}$ so that the relations take the form*

$$\left.\begin{aligned} \frac{du_{mn}}{d\theta} - \frac{m - n\cos\theta}{\sin\theta} u_{m,n} &= -i\alpha_n u_{m,n-1}, \\ \frac{du_{mn}}{d\theta} + \frac{m - n\cos\theta}{\sin\theta} u_{m,n} &= -i\alpha_{n+1} u_{m,n+1}. \end{aligned}\right\} \quad (1)$$

After the substitution $\mu = \cos\theta$ these equations become

$$\left.\begin{aligned} \sqrt{1-\mu^2}\frac{dP^l_{mn}(\mu)}{d\mu} + \frac{m - n\mu}{\sqrt{1-\mu^2}} P^l_{mn}(\mu) &= i\alpha_n P^l_{m,n-1}, \\ \sqrt{1-\mu^2}\frac{dP^l_{mn}(\mu)}{d\mu} - \frac{m - n\mu}{\sqrt{1-\mu^2}} P^l_{mn}(\mu) &= i\alpha_{n+1} P^l_{m,n+1}, \end{aligned}\right\} \quad (1')$$

where

$$\alpha_n = \sqrt{(l+n)(l-n+1)}.$$

---

* For brevity we write $u_{mn}(\theta)$ for $P^l_{mn}(\cos\theta)$.

As was shown in Sect.7, (equation 26, paragraph 4) $u_{mn}(\theta)=u_{nm}(\theta)$. Putting $u_{nm}(\theta)$ for u_{mn} in equations (1) and interchanging m and n we obtain

$$\left.\begin{aligned} \frac{du_{mn}}{d\theta}-\frac{n-m\cos\theta}{\sin\theta}u_{mn} &= -i\alpha_m u_{m-1,\,n},\\ \frac{du_{mn}}{d\theta}+\frac{n-m\cos\theta}{\sin\theta}u_{mn} &= -i\alpha_{m+1}u_{m+1,\,n}. \end{aligned}\right\} \tag{2}$$

From each pair of equations (1) and (2) a relation between any three consecutive elements of a row (or column) may be obtained without introducing derivatives. They have the form

$$\alpha_{n+1}u_{m,\,n+1}-\alpha_n u_{m,\,n-1}=2i\frac{m-n\cos\theta}{\sin\theta}u_{mn}, \tag{3}$$

$$\alpha_{m+1}u_{m+1,\,n}-\alpha_m u_{m-1,\,n}=2i\frac{n-m\cos\theta}{\sin\theta}u_{mn}, \tag{3'}$$

where again $u_{mn}(\theta)$ denotes $P^l_{mn}(\cos\theta)$.

Equations of another type may be obtained if we use a property of the matrix of a representation introduced in Sect.2. In paragraph 2, Sect.2 we showed that if $\eta=(\eta_1,\ \eta_2,\ \eta_3)$ is any vector and $A_\eta=A_1\eta_1+A_2\eta_2+A_3\eta_3$, where the A_k are the matrices corresponding to infinitesimal rotations about the coordinate axes then for any rotation g

$$T_gA_\eta T_{g^{-1}}=A_{\tilde\eta}, \tag{4}$$

where $\tilde\eta=g\eta$ and $A_{\tilde\eta}=A_1\tilde\eta_1+A_2\tilde\eta_2+A_3\tilde\eta_3$. Let us now multiply equation (4) on the right by the matrix T_g and put $\eta=(1,\ 0,\ 0)$. Then the components of the vector $\tilde\eta$ will be the elements of the first column of the matrix $\|g_{ik}\|$. In Sect.1 we wrote down the matrix $\|g_{ik}\|$ as a function of the Euler angles (see equation (9), Sect.1). Having substituted the elements g_{11}, g_{21}, g_{31} of this matrix in the relation $A_1T_g=g_{11}T_gA_1+g_{21}T_gA_2+g_{31}T_gA_3$, we obtain the equation

$$\begin{aligned} T_gA_1 &= (\cos\varphi_1\cos\varphi_2-\cos\theta\sin\varphi_1\sin\varphi_2)A_1T_g+\\ &+(\sin\varphi_1\cos\varphi_2+\cos\theta\cos\varphi_1\sin\varphi_2)A_2T_g+\sin\varphi_2\sin\theta\, A_3T_g. \end{aligned}$$

It will be more convenient for us later on if we now express the A_k in terms of H_+, H_- and H_3 by the equations

$$A_1=-\frac{i}{2}(H_++H_-),\quad A_2=\frac{1}{2}(H_--H_+),\quad A_3=-iH_3.$$

Substituting and collecting together the coefficients of H_+ and H_- we obtain:

$$\begin{aligned} T_g(H_++H_-) &= (\cos\varphi_2-i\cos\theta\sin\varphi_2)e^{-i\varphi_1}H_+T_g+\\ &+(\cos\varphi_2+i\cos\theta\sin\varphi_2)e^{i\varphi_1}H_-T_g-2\sin\varphi_2\sin\theta\, H_3T_g. \end{aligned}$$

Let us now apply the transformations on both sides of this equation to the vector f_n of the canonical basis and equate the coefficients of f_m in the expressions obtained. Since $T_g f_n = \sum_m T_{mn} f_m$, $H_+ f_n = \alpha_{n+1} f_{n+1}$, $H_- f_n = \alpha_n f_{n-1}$, $H_3 f_n = n f_n$, we find the following relation between the functions T_{mn}:

$$\alpha_{n+1} T_{m,\,n+1} + \alpha_n T_{m,\,n-1} = (\cos\varphi_2 - i\cos\theta\sin\varphi_2)\, e^{-i\varphi_1} \alpha_m T_{m-1,\,n} + (\cos\varphi_2 + i\cos\theta\sin\varphi_2)\, e^{i\varphi_1} \alpha_{m+1} T_{m+1,\,n} - 2m\sin\varphi_2\sin\theta\, T_{mn}$$

Having substituted $T_{m,\,n} = e^{-im\varphi_1} u_{mn}(\theta)\, e^{-in\varphi_2}$ in this expression and multiplied by $e^{im\varphi_1 + in\varphi_2}$ we obtain a relation between the functions $u_{mn}(\theta)$, which holds for any values of φ_1:

$$\alpha_{n+1} u_{m,\,n+1} e^{-i\varphi_2} + \alpha_n u_{m,\,n-1} e^{i\varphi_2} = (\cos\varphi_2 - i\cos\theta\sin\varphi_2)\, \alpha_m u_{m-1,\,n} + (\cos\varphi_2 + i\cos\theta\sin\varphi_2)\, \alpha_{m+1} u_{m+1,\,n} - 2m\sin\varphi_2\sin\theta\, u_{mn}$$

By putting $\cos\varphi_2 = \dfrac{e^{i\varphi_2} + e^{-i\varphi_2}}{2}$ and $\sin\varphi_2 = \dfrac{e^{i\varphi_2} - e^{-i\varphi_2}}{2i}$ and equating the coefficients of $e^{i\varphi_2}$ and $e^{-i\varphi_2}$ on the left- and right-hand sides we obtain finally the equations

$$\alpha_{n+1} u_{m,\,n+1} = \frac{1}{2}(1+\cos\theta)\,\alpha_m u_{m-1,\,n} + \frac{1}{2}(1-\cos\theta)\,\alpha_{m+1} u_{m+1,\,n} - im\sin\theta\, u_{mn}, \tag{5}$$

$$\alpha_n u_{m,\,n-1} = \frac{1}{2}(1-\cos\theta)\,\alpha_m u_{m-1,\,n} + \frac{1}{2}(1+\cos\theta)\,\alpha_{m+1} u_{m+1,\,n} + im\sin\theta\, u_{mn}. \tag{5'}$$

These equations connect three neighbouring elements of the n-th column with an element of the $n+1$-th or the $n-1$-th column. Here $u_{mn}(\theta)$ denotes $P^l_{mn}(\cos\theta)$. By using the fact that $u_{mn}(\theta) = u_{nm}(\theta)$ we may, as above, obtain similar equations connecting three neighbouring elements of a row with an element of a higher or lower row:

$$\alpha_{m+1} u_{m+1,\,n} = \frac{1}{2}(1+\cos\theta)\,\alpha_n u_{m,\,n-1} + \frac{1}{2}(1-\cos\theta)\,\alpha_{n+1} u_{m,\,n+1} - in\sin\theta\, u_{mn}, \tag{6}$$

$$\alpha_m u_{m-1,\,n} = \frac{1}{2}(1-\cos\theta)\,\alpha_n u_{m,\,n-1} + \frac{1}{2}(1+\cos\theta)\,\alpha_{n+1} u_{m,\,n+1} + in\sin\theta\, u_{mn}. \tag{6'}$$

If, in equation (4), we take $\eta = (0,\ 1,\ 0)$, we again obtain precisely equations (5), (5'). If we put $\eta = (0,\ 0,\ 1)$ we obtain the equation

$$\alpha_{m+1} u_{m+1,\,n} - \alpha_m u_{m-1,\,n} = -2i\frac{n - m\cos\theta}{\sin\theta} u_{mn},$$

by the method introduced above.

We turn now to the derivation of equations connecting the elements T^l_{mn} of the matrix T_g corresponding to different values of l (associated with the different irreducible representations). To do this we use the results of paragraph 4, Sect.4 concerning the resolution of the product of two irreducible representations into irreducible representations. We saw then that the product $g \to T_g$ of an irreducible representation of weight 1 and an irreducible representation of weight l could be resolved into three irreducible representations: $g \to T^{l+1}_g$ of weight $l+1$, $g \to T^l_g$ of weight l and $g \to T^{l-1}_g$ of weight $l-1$. The basis $e_k f_{m-k}$ in the space $R_1 \times R_2$ in which the representation acts is connected with the canonical bases $\{g^{l+1}_m\}$, $\{g^l_m\}$ $\{g^{l-1}_m\}$ in the spaces of the irreducible representations $g \to T^{l+1}_g$, $g \to T^l_g$, $g \to T^{l-1}_g$ by the equations

$$\left.\begin{aligned} e_{-1}f_{m+1} &= \overset{m}{c}_{11}g^{l+1}_m + \overset{m}{c}_{12}g^l_m + \overset{m}{c}_{13}g^{l-1}_m, \\ e_0 f_m &= \overset{m}{c}_{21}g^{l+1}_m + \overset{m}{c}_{22}g^l_m + \overset{m}{c}_{23}g^{l-1}_m, \\ e_1 f_{m-1} &= \overset{m}{c}_{31}g^{l+1}_m + \overset{m}{c}_{32}g^l_m + \overset{m}{c}_{33}g^{l-1}_m. \end{aligned}\right\} \tag{7}$$

The values of the coefficients $\overset{m}{c}_{ik}$ were obtained on p.57 (equations (9), Sect.4).

Let us now denote by T^l_g the matrix of the irreducible representation of weight l in the canonical basis and apply the transformation T_g to the left- and right-hand sides of equations (7). On the left, using the definition of a product of representations, we obtain:

$$T_g e_k f_{m-k} = T^1_g e_k T^l_g f_{m-k} =$$
$$= \overset{m}{c}_{k+2,1} T^{l+1}_g g^{l+1}_m + \overset{m}{c}_{k+2,2} T^l_g g^l_m + \overset{m}{c}_{k+2,3} T^{l-1}_g g^{l-1}_m$$
$$(k = -1,\ 0,\ 1).$$

Denoting the elements of the matrix T^l_g by T^l_{mn} we find from this relation, that

$$\left(T^1_{-1k}e_{-1} + T^1_{0\,k}e_0 + T^1_{1k}e_1\right)\sum T^l_{j,\,m-k}f_j =$$
$$= \sum \overset{m}{c}_{k+2}T^{l+1}_{jm}g^{l+1}_j + \overset{m}{c}_{k+2,2}T^l_{jm}g^l_j + \overset{m}{c}_{k+2,3}T^{l-1}_{jm}g^{l-1}_j.$$

Let us now substitute on the right-hand side for the vectors g^{l+1}_j, g^l_j, g^{l-1}_j the vectors $e_{-1}f_{j+1}$, $e_0 f_j$ and $e_1 f_{j-1}$ as given by equations (7)* and then equate the coefficients of the vectors $e_{-1}f_{j+1}$, $e_0 f_j$ and $e_1 f_{j-1}$ on the left- and right-hand sides of the equation thus obtained. We get three equations depending on k. Giving k its three possible values -1, 0, 1 and substituting

* Remembering that since the matrix is orthogonal its inverse equals its transpose.

for the functions T^1_{mn} $(\varphi_1, \theta, \varphi_2)$ see the second matrix of Sect. 7 (on p.86) we get the following nine recurrence formulae:

$$\left.\begin{aligned}
\overset{m}{c}_{11}T^{l+1}_{jm}\overset{j}{c}_{11}+\overset{m}{c}_{12}T^{l}_{jm}\overset{j}{c}_{12}+\overset{m}{c}_{13}T^{l-1}_{jm}\overset{j}{c}_{13}&=\frac{1}{2}(1+\cos\theta)\,e^{i\varphi_1+i\varphi_2}\,T^l_{j+1,\,m+1},\\
\overset{m}{c}_{11}T^{l+1}_{jm}\overset{j}{c}_{21}+\overset{m}{c}_{12}T^{l}_{jm}\overset{j}{c}_{22}+\overset{m}{c}_{13}T^{l-1}_{jm}\overset{j}{c}_{23}&=\frac{-i}{\sqrt{2}}\sin\theta\,e^{i\varphi_1}T^l_{j,\,m+1},\\
\overset{m}{c}_{11}T^{l+1}_{jm}\overset{j}{c}_{31}+\overset{m}{c}_{12}T^{l}_{jm}\overset{j}{c}_{32}+\overset{m}{c}_{13}T^{l-1}_{jm}\overset{j}{c}_{33}&=\frac{1}{2}(\cos\theta-1)\,e^{i\varphi_1-i\varphi_2}\,T^l_{j-1,\,m+1},\\
\overset{m}{c}_{21}T^{l+1}_{jm}\overset{j}{c}_{11}+\overset{m}{c}_{22}T^{l}_{jm}\overset{j}{c}_{12}+\overset{m}{c}_{23}T^{l-1}_{jm}\overset{j}{c}_{13}&=\frac{-i}{\sqrt{2}}\sin\theta\,e^{i\varphi_2}T^l_{j+1,\,m},\\
\overset{m}{c}_{21}T^{l+1}_{jm}\overset{j}{c}_{21}+\overset{m}{c}_{22}T^{l}_{jm}\overset{j}{c}_{22}+\overset{m}{c}_{23}T^{l-1}_{jm}\overset{j}{c}_{23}&=\cos\theta\,T^l_{j,\,m},\\
\overset{m}{c}_{21}T^{l+1}_{jm}\overset{j}{c}_{31}+\overset{m}{c}_{22}T^{l}_{jm}\overset{j}{c}_{32}+\overset{m}{c}_{23}T^{l-1}_{jm}\overset{j}{c}_{33}&=\frac{-i}{\sqrt{2}}\sin\theta\,e^{-i\varphi_2}\,T^l_{j-1,\,m},\\
\overset{m}{c}_{31}T^{l+1}_{jm}\overset{j}{c}_{11}+\overset{m}{c}_{32}T^{l}_{jm}\overset{j}{c}_{12}+\overset{m}{c}_{33}T^{l-1}_{jm}\overset{j}{c}_{13}&=\frac{1}{2}(\cos\theta-1)\,e^{-i\varphi_1+i\varphi_2}T^l_{j+1,\,m-1},\\
\overset{m}{c}_{31}T^{l+1}_{jm}\overset{j}{c}_{21}+\overset{m}{c}_{32}T^{l}_{jm}\overset{j}{c}_{22}+\overset{m}{c}_{33}T^{l-1}_{jm}\overset{j}{c}_{23}&=\frac{-i}{\sqrt{2}}\sin\theta\,e^{-i\varphi_1}\,T^l_{j,\,m-1},\\
\overset{m}{c}_{31}T^{l+1}_{jm}\overset{j}{c}_{31}+\overset{m}{c}_{32}T^{l}_{jm}\overset{j}{c}_{32}+\overset{m}{c}_{33}T^{l-1}_{jm}\overset{j}{c}_{33}&=\frac{1}{2}(1+\cos\theta)\,e^{-i\varphi_1-i\varphi_2}T^l_{j-1,\,m-1}.
\end{aligned}\right\}\tag{8}$$

In these equations the $\overset{m}{c}_{ik}$ are known constants, to wit, the elements of the matrix $C^{(m)}$ given by equation (9) of Sect.4 on p.57.

If we do not substitute all the values of these in equations (8) restricting ourselves only to those which give the relation between the functions T^{l+1}_{mn}, T^l_{mn} and T^{l-1}_{mn} standing in the same place in the corresponding matrices we obtain the equations

$$\frac{\sqrt{(l+m+1)(l-m+1)(l+j+1)(l-j+1)}}{(2l+1)(l+1)}\,T^{l+1}_{jm}+\frac{mj}{l(l+1)}\,T^l_{jm}+$$
$$+\frac{\sqrt{(l+m)(l-m)(l+j)(l-j)}}{l(2l+1)}\,T^{l-1}_{jm}=\cos\theta\,T^l_{jm},\tag{9}$$

or, after multiplying by $e^{-im\varphi_1-ij\varphi_2}$ the equation

$$\frac{\sqrt{(l+m+1)(l-m+1)(l+j+1)(l-j+1)}}{(2l+1)(l+1)}\,P^{l+1}_{jm}(\mu)+$$
$$+\frac{mj}{l(l+1)}P^l_{jm}(\mu)+\frac{\sqrt{(l+m)(l-m)(l+j)(l-j)}}{l(2l+1)}\,P^{l-1}_{jm}(\mu)=\mu P^l_{jm}(\mu).\tag{9'}$$

If $j=m=0$ this equation reduces to the recurrence relation between the Legendre polynomials.

Section 8. Expansion of Vector and Tensor Fields

In Sect.3 we expanded the functions defined on a sphere in terms of the spherical functions (for their definition see the end of paragraph 2, Sect.7). Such expansions are normally used in the solution of problems which involve spherical symmetry about a point (problems which are invariant under rotation).

In this section we will concern ourselves with an analogous resolution for functions which are not scalars but vectors, tensors or some other quantities. Thus the contents of this section will be a development of the results of Sect.3, where such a resolution was carried out for scalar functions. As in Sect. 3 the resolution of these functions, is achieved by finding the resolution of the representation arising from transformations of these functions, into irreducible representations.

The special functions in terms of which the resolution is carried out are, as we have seen, the generalized spherical functions of which the ordinary spherical functions are a special case.

For the sake of greater clarity we will first, in paragraph 1 give the solution for functions taking vector values, and then for functions taking any magnitude.

Similar resolutions of various magnitudes are met with in many problems of physics.

Thus the solution of the Laplace equation in spherical coordinates involves the resoltuion of a function into a series of spherical functions while the solution of Maxwell equations in spherical coordinates necessitates the expansion in series of vector-valued functions.

Similarly the solution of Dirac equations involves the expansion in terms of functions on the sphere of spinor-valued functions. The functions used in this case are the spherical spinor functions introduced by V.A.Fock.

In the solution of equations in the theory of elasticity G.I.Petrashen' introduced the so-called spherical vectors and successfully applied them to the solution of problems.

Finally, through the work of A.Z.Dolginov, V.B.Beresteskii and K.A.Ter-Martirosyan, there was introduced the expansion of functions of l-vectors (those magnitudes transformed by an irreducible representation of weight l). This expansion is carried out in terms of the so-called spherical (l, L)-vector functions, the components of which are linear combinations of the ordinary spherical functions. The coefficients of these linear combinations coincide with the coefficients c_{ik}^{m} determined in paragraph 4, Sect.4. In the general case they are quite difficult to obtain.

As indicated above, we will carry out the expansion of components of magnitudes in terms of the generalized spherical functions. The connexion between these functions and the ordinary spherical functions may be obtained from the recurrence formulae of Sect. 7.

§ 1. Expansion of Vector Functions

Let us consider the function $a(x)$ where x is a point in three-dimensional space and a is a vector, i.e. let us consider a vector field.

Clearly such functions are transformed under rotation. Let us apply to the vector field $a(x)$ an arbitrary rotation g_0. As a result we obtain a new vector field $a'(x)$. We will express $a'(x)$ in terms of $a(x)$. After the rotation g_0 the point $g_0^{-1}x$ becomes the point x. Evidently the vector $a(g_0^{-1}x)$ is transformed by the rotation and, together with the entire vector field undergoes the rotation g_0.

Consequently as a result of the rotation g_0 the vector field $a(x)$ is transformed into the vector field $a'(x) = g_0 a(g_0^{-1}x)$. Thus to every rotation g_0 there corresponds a transformation T_{g_0} of the vector function $a(x)$ which is given by the equation

$$T_{g_0} a(x) = g_0 a(g_0^{-1}x). \tag{1}$$

It is clear that the transformation is linear.

Furthermore, it follows from the very definition of T_g that the transformation $T_{g_0 g_1}$ corresponding to the product of the rotations g_0 and g_1, is the same as the product of the transformations T_{g_0} and T_{g_1}:

$$T_{g_0 g_1} = T_{g_0} T_{g_1}. \tag{2}$$

The points of any sphere with centre the origin remain on the sphere after the rotation g_0. Hence, for the study of the transformations we will confine ourselves to functions defined on the surface of the unit sphere, i.e. to functions $a(P) = a(\vartheta, \varphi)$, where a is a vector depending on the point P of the sphere with spherical coordinates ϑ and φ. By equation (2), the transformations T_g, constitute a representation of the rotation group in the space of vector-valued functions on the surface of the sphere*.

We will separate out from the space of vector-valued functions on the sphere those subspaces in which there act irreducible representations. To do this we first determine those finite systems of vector-valued functions which are transformed into linear combinations of themselves under the action of the transformation T_g.

For a field of scalar functions, such systems abound; the spherical functions of a given order l are examples.

---

* This representation will be unitary if we define the scalar product of two vector functions $a(P)$ and $b(P)$ by the equation

$$(a(P), b(P)) = \int_0^{2\pi}\int_0^{\pi} \{a_1(\vartheta, \varphi)\overline{b_1(\vartheta, \varphi)} + a_2(\vartheta, \varphi)\overline{b_2(\vartheta, \varphi)} + a_3(\vartheta, \varphi)\overline{b_3(\vartheta, \varphi)}\} \sin\vartheta \, d\vartheta \, d\varphi.$$

where ϑ, φ are the spherical coordinates of P and a_k, b_k the components of a, b in some system of coordinates.

It is also possible in the case of a vector field to expand each component of the vector $a(P)$ in terms of the spherical functions. Such a resolution, however, is not very convenient as each component of a vector under rotation is transformed into a combination of all three components.

There does exist a more convenient method of dealing with a vector field. We may, for example, take the component $a_r(P)$ of the point P which is normal to the sphere. Since the normal component $a_r(P)$ at the point P is transformed by a rotation into the normal component of the vector at the point $g_0^{-1}P$, i.e.

$$T_{g_0}a_r(P) = a_r(g_0^{-1}P),$$

the function $a_r(P)$ is transformed under rotation as a scalar function (see Sect.3) The problem of the resolution of this representation into irreducible representations becomes that of the resolution of the functions $a_r(\vartheta, \varphi)$ in a series of spherical functions. We shall in future, therefore, consider the vector component $a_r(P)$.

We will now give a method of representing a vector field by using three such functions each of which, under rotation, is transformed independently of the other two.

A function on the surface of the sphere is a function of two variables ϑ and φ. A new approach, which consists in transforming this function into one of three variables φ_1, θ, φ_2 (functions of the rotation g), provides an easy solution to the problem by making it possible to apply the results of the preceding section.

With this end in view, let us consider a normalized orthogonal triad e_1, e_2, and e_3, at the point P, the third vector of which is normal to the surface of the sphere. Associated with each such triad is a rotation g which transforms the triad at the 'North Pole' into the given triad.

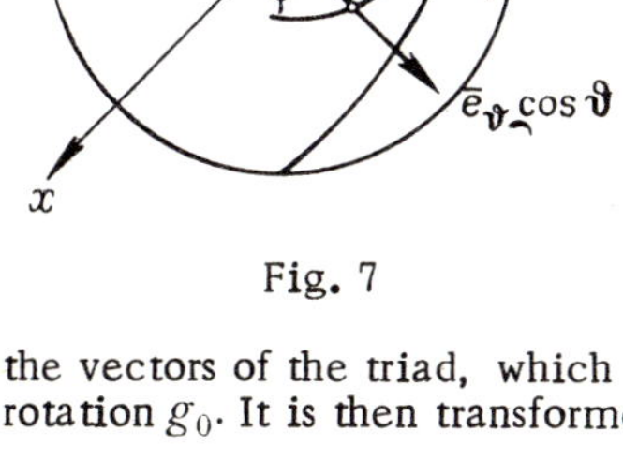

Fig. 7

We will subject all the vectors of the triad, which corresponds to the rotation g, to an arbitrary rotation g_0. It is then transformed into another triad.

Since the first triad is obtained from the normalized rotation g and the second from the first by the rotation g_0 the second triad is obtained by the rotation $g_0 g$, i.e. it is defined by the rotation $g_0 g$.

Every triad, as shown above, is determined by a rotation g, i.e. by the Euler angles φ_1, θ, φ_2. It is obvious how to represent a triad by using the angles φ_1, θ, φ_2.

We note that the element g of the group determines the triad and the point P with which it is associated. The spherical coordinates ϑ, φ of the point P of the sphere, to which the 'North Pole' of the sphere is transformed by the rotation g, are connected with the Euler angles of the rotation g by the relations

$$\vartheta = \theta, \qquad \varphi = \varphi_2 - \frac{\pi}{2}, \tag{3}$$

i.e. this point is independent of the first Euler angle, φ_1.*

The third vector of the triad, which is normal to the sphere, is completely determined by the position of the point on the sphere and hence is also independent of the angle φ_1.

The vectors e_1 and e_2 lie in the tangent plane at P and depend on φ_1. To find this relation consider the two rotations $g = g(\varphi_1, \theta, \varphi_2)$ and $g_1 = g(\varphi_1 + \varphi^*, \theta, \varphi_2)$ with different values of the first Euler angle. Obviously these rotations transform the normal triad into two triads with the same origin and consequently with the same value of the vector e_3. This second triad (determined by the rotation g_1) is obtained from the first by a rotation through an angle φ^* in a positive direction about e_3. If e_1 and e_2 are the vectors of the first triad and e_1', e_2' the corresponding vectors of the second then we have according to the usual formulae:

$$e_1' = e_1 \cos\varphi^* + e_2 \sin\varphi^*,$$
$$e_2' = -e_1 \sin\varphi^* + e_2 \cos\varphi^*.$$

Let us put $\varphi_1 = \pi$ and denote φ^* by φ_1 and $e_k(\pi, \theta, \varphi_2)$, $k = 1, 2, 3$ by e_k^0. We then obtain:

$$\left.\begin{aligned} e_1(\varphi_1 + \pi, \theta, \varphi_2) &= e_1^0 \cos\varphi_1 + e_1^0 \sin\varphi_1, \\ e_2(\varphi_1 + \pi, \theta, \varphi_2) &= -e_1^0 \sin\varphi_1 + e_2^0 \cos\varphi_1. \end{aligned}\right\}$$

The vectors $e_1^0 = e_1(\pi, \theta, \varphi_2)$ $e_2^0 = e_2(\pi, \theta, \varphi_2)$ have a simple geometrical significance. They are the unit vectors directed along the tangents to the parallel of latitude and line of longitude at the point P. By putting $\varphi_1 = \pi$,

* The Cartesian coordinates of the point into which the 'North Pole' is transformed by the rotation are the elements of the third column of the matrix $\| g_{ik} \|$, i.e. $\sin\varphi_2 \sin\theta$, $-\cos\varphi_2 \sin\theta$, $\cos\theta$ (see paragraph 2, Sect.1). By equating these expressions with the spherical coordinates of the point $\cos\varphi \sin\vartheta$, $\sin\varphi \sin\vartheta$, $\cos\vartheta$, we obtain equation (3).

in the matrix $\|g_{ik}\|$ we obtain expressions for the Cartesian coordinates of these vectors: $e_1 = (-\cos\varphi_2, -\sin\varphi_2, 0)$, $e_2 = (\cos\theta \sin\varphi_2, -\cos\theta\cos\varphi_2, -\sin\theta)$. On the other hand from Fig.7 it is clear that the unit vector e_φ directed along the parallel of latitude in the direction of φ increasing has the components $(-\sin\varphi, \cos\varphi, 0)$ and e_ϑ, the vector along the line of longitude in the direction of ϑ increasing, has the components $(\cos\vartheta\cos\varphi, \cos\vartheta\sin\varphi, -\sin\vartheta)$. Remembering that the spherical coordinates ϑ and φ of the point P are connected with the Euler angles θ and φ_2 by equations (3) we find that $e_1^0 = -e_\varphi$, $e_2^0 = e_\vartheta$.

Thus, we have finally

$$\left.\begin{aligned} e_1(\varphi_1+\pi, \theta, \varphi_2) &= -e_\varphi\cos\varphi_1 + e_\vartheta\sin\varphi_1, \\ e_2(\varphi_1+\pi, \theta, \varphi_2) &= e_\varphi\sin\varphi_1 + e_\vartheta\cos\varphi_1, \\ e_3(\varphi_1+\pi, \theta, \varphi_2) &= e_r, \end{aligned}\right\} \qquad (4)$$

where e_r, e_φ, e_ϑ are unit vectors directed along the normal to the sphere and parallel to the tangents to the parallel of latitude and line of longitude at the point P with spherical coordinates $\varphi = \varphi_2 - \frac{\pi}{2}$ and $\vartheta = \theta$.

Let us now consider a vector function $a(P)$ (a vector field on the sphere). If we take a triad at the point P and resolve $a(P)$ along the vectors of this triad we obtain three quantities a_1, a_2, a_3 which are the components of a in terms of the vectors of the triad. These quantities, as are the vectors of the triad themselves, functions of the rotation g, i.e. of the Euler angles φ_1, θ, φ_2. Hence, since e_3 is independent of φ_1, the component a_3 is also independent of φ_1. It is the function $a_r(P)$ at the point P with spherical coordinates $\varphi_2 - \frac{\pi}{2}$ and θ, which we have already considered above.

From equation (4) it is clear that the components a_1 and a_2 depend on φ_1. Forming the scalar product of $a(P)$ and equations (4) we obtain

$$\left.\begin{aligned} a_1(\varphi_1+\pi, \theta, \varphi_2) &= -a_\varphi\cos\varphi_1 + a_\vartheta\sin\varphi_1, \\ a_2(\varphi_1+\pi, \theta, \varphi_2) &= a_\varphi\sin\varphi_1 + a_\vartheta\cos\varphi_1, \\ a_3(\varphi_1+\pi, \theta, \varphi_2) &= a_r. \end{aligned}\right\} \qquad (5)$$

We recall how to express the components a_φ and a_ϑ, and also a_r in terms of the usual components of a, i.e. in terms of a_x, a_y and a_z. To do this we need to take the scalar product of the vector $a = (a_x, a_y, a_z)$ and the vectors $e_\varphi = (-\sin\varphi, \cos\varphi, 0)$, $e_\vartheta = (\cos\vartheta\cos\varphi, \cos\vartheta\sin\varphi, -\sin\vartheta)$ and $e_r = (\sin\vartheta\cos\varphi, \sin\vartheta\sin\varphi, \cos\vartheta)$.

$$\left.\begin{aligned} a_\varphi(\varphi, \vartheta) &= -a_x\sin\varphi + a_y\cos\varphi, \\ a_\vartheta(\varphi, \vartheta) &= a_x\cos\vartheta\cos\varphi + a_y\cos\vartheta\sin\varphi - a_z\sin\vartheta, \\ a_r(\varphi, \vartheta) &= a_x\sin\vartheta\cos\varphi + a_y\sin\vartheta\sin\varphi + a_z\cos\vartheta. \end{aligned}\right\} \qquad (6)$$

We now introduce the complex components of the vector

$$a_+ = a_1 - ia_2,$$

$$a_- = a_1 + ia_2.$$

From (5) the expressions for a_+ and a_- given below may be obtained

$$\left.\begin{aligned} a_+(\varphi_1, \theta, \varphi_2) &= a_+(\pi, \theta, \varphi_2)\, e^{i\varphi_1} = (-a_\varphi - ia_\vartheta)\, e^{i\varphi_1}, \\ a_-(\varphi_1, \theta, \varphi_2) &= a_-(\pi, \theta, \varphi_2)\, e^{-i\varphi_1} = (-a_\varphi + ia_\vartheta)\, e^{-i\varphi_1}, \end{aligned}\right\} \qquad (7)$$

a_φ and a_ϑ refer to the point with coordinates $\varphi = \varphi_2 - \frac{\pi}{2}$, $\vartheta = \theta$.

We now consider what happens to the functions a_1, a_2 and a_3 (or, what is the same thing, to the functions a_+, a_- and a_3) under rotation. If both the vectors of the field $a(P)$ and the vectors of the triad are subjected to the same rotation then it is clear that the components of the transformed vector a' in terms of the new triad are the same as the components of the old vector in terms of the old triad. The old triad is given by the rotation g, and the new, as we have seen above, by the rotation $g_0 g$. Thus we have:

$$a'_k(g_0 g) = a_k(g)$$

or, replacing $g_0 g$ by g,

$$a'_k(g) = a_k(g_0^{-1} g). \qquad (8)$$

Hence we see that the functions $a_1(g)$, $a_2(g)$ and $a_3(g)$ and consequently the functions $a_+(g)$, $a_-(g)$ and $a_3(g)$ are transformed under rotation independently of one another, i.e. the transformation of each of these components into itself gives rise to a representation of the rotation group.

Having resolved each of these representations into irreducible representations we obtain for the components of the vector $a(P)$ resolutions which are invariant with respect to rotations, and are independent of each other.

In Sect.7 we solved the problem of resolving a representation of transformed functions of rotations (the regular representation) into irreducible representations. There, however, for the sake of convenience, we considered that transformation which consists in multiplying g by g_0 from the right rather than multiplication by g_0^{-1} from the left. It is possible to pass from one of these transformations to the other without difficulty. For this it will be sufficient to consider instead of the functions $a_k(g)$ the functions $\tilde{a}_k(g) = a_k(g^{-1})$, which may be obtained by changing the arguments from φ_1, θ, φ_2 to $\pi - \varphi_2$, θ, $\pi - \varphi_1$. After making this substitution the transformation (8) may be written in the form

$$\tilde{a}'_k(g) = a'_k(g^{-1}) = a_k(g_0^{-1} g^{-1}) = a_k((g, g_0)^{-1}) = \tilde{a}_k(g g_0),$$

i.e.

$$\tilde{a}'_k(g) = \tilde{a}_k(g g_0). \qquad (8')$$

In future we will consider the functions a_1, $\tilde{a}_2$, $\tilde{a}_3$ of g instead of a_1, a_2 and a_3. Since we know that the old functions a_1, a_2 and a_3 depend on the arguments φ_1, θ, φ_2 we may substitute $\pi - \varphi_2$, θ, $\pi - \varphi_1$, for these and obtain the result that a_3 is independent of φ_2 and that

$$\tilde{a}_+(\varphi_1, \theta, \varphi_2) = -e^{-i\varphi_2}\tilde{a}_+(\varphi_1, \theta, 0) = e^{-i\varphi_2}(a_\varphi + ia_\vartheta),$$
$$\tilde{a}_-(\varphi_1, \theta, \varphi_2) = -e^{i\varphi_2}\tilde{a}_-(\varphi_1, \theta, 0) = e^{i\varphi_2}(a_\varphi - ia_\vartheta).$$

Now, the point P which is given by the vector a with these components has the spherical coordinates $\frac{\pi}{2} - \varphi_1$ and θ. The irreducible representations obtained from $\tilde{a}_+$, $\tilde{a}_-$ and $\tilde{a}_r$ are now clear. Since the functions a_r are independent of φ_2, they are resolved into the functions occurring in the null (central) rows of all the matrices T_g^l $(l = 0, 1, 2, \ldots)$ i.e. as found earlier for the spherical functions of $\frac{\pi}{2} - \varphi_1$ and θ. The functions $\tilde{a}_+(\varphi_1, \theta, \varphi_2)$ contain the factor $e^{-i\varphi_2}$ as do the elements of the first rows of the matrices T_g^l — the generalized spherical functions $T_{1n}^l(\varphi_1, \theta, \varphi_2)$ and consequently they may be expanded in a series of these functions.

Similarly the functions $\tilde{a}$ $(\varphi_1, \theta, \varphi_2)$ may be expanded in terms of the generalized spherical functions $T_{-1,n}^l(\varphi_1, \theta, \varphi_2)$.

Thus we have proved the following. Let there be given a vector function $a(P)$ on the surface of a sphere. The expansion of this function in a series which is invariant with respect to rotations is obtained in the following manner. At the point P, with spherical coordinates φ and ϑ take a vector with components a_φ, a_ϑ and a_r along the parallel of latitude, the line of longitude, and the radius, and consider the functions $\tilde{a}_+$, $\tilde{a}_-$ and $\tilde{a}_r$ given by the following expressions

$$\left.\begin{aligned} \tilde{a}_+(\varphi_1, \theta, \varphi_2) &= e^{-i\varphi_2}[a_\varphi(\varphi, \theta) + ia_\vartheta(\varphi, \theta)], \\ \tilde{a}_-(\varphi_1, \theta, \varphi_2) &= e^{i\varphi_2}[a_\varphi(\varphi, \theta) - ia_\vartheta(\varphi, \theta)], \\ \tilde{a}_r(\varphi_1, \theta, \varphi_2) &= a_r(\varphi, \theta), \text{where } \varphi = \frac{\pi}{2} - \varphi_1. \end{aligned}\right\} \tag{9}$$

Each of these functions may be expanded in a series of generalized spherical functions; the functions $\tilde{a}_+(\varphi_1, \theta, \varphi_2)$ may be expanded in a series of the functions $T_{1n}^l(\varphi_1, \theta, \varphi_2)$ $(l = 0, 1, 2, \ldots; -l \leqslant n \leqslant l)$, $\tilde{a}_-(\varphi_1, \theta, \varphi_2)$ in terms of the functions $T_{-1n}^l(\varphi_1, \theta, \varphi_2)$ and the functions $\tilde{a}_r(\varphi_1, \theta, \varphi_2)$ in terms of the functions $T_{0n}^l(\varphi_1, \theta, \varphi_2) = \sqrt{\frac{2}{2l+1}}\, Y_n^l\left(\frac{\pi}{2} - \varphi_1, \theta\right)$. By cancelling the factor $e^{\pm i\varphi_2}$ common to the first two series and making the substitution $\frac{\pi}{2} - \varphi_1$ for φ we obtain the expansions of the components of the vectors $a_\varphi \pm ia_\vartheta$ and a_r, which correspond to the expansions of the scalar functions given in Sect. 3.

§ 2. Expansion of Arbitrary Quantities

In paragraph 1, we studied the problem of the expansion of functions whose values at each point are vectors in three-dimensional space. Let us turn now to the case when, in the space of the given function $f(x)$, to each point there is associated a magnitude transformed according to an irreducible representation of weight l_0. This means that at each point x of the space we are given a set of $2l_0+1$ numbers c_m $(-l_0 \leqslant m \leqslant l_0)$ - the components of the magnitude f which are subjected by the rotation g_0 to the linear transformation

$$c'_m = \sum_{n=-l_0}^{l_0} a_{mn}(g_0)\, c_n. \tag{10}$$

As usual, we will write the transformation (10) as

$$f' = U_{g_0} f. \tag{10'}$$

A field of quantities is transformed by the rotation g_0 in the following manner; in the first place, as a result of the rotation g_0 of the space, the point x now has associated with it the magnitude previously associated with the point $g_0^{-1}x$, and, in the second place, this magnitude undergoes the transformation U_{g_0}. Thus for the rotation g_0 the magnitude $f(x)$ is changed to $f'(x) = U_{g_0} f(g_0^{-1}x)$ i.e. the transformation T_{g_0} of the field of magnitudes is given by the equation

$$T_{g_0} f(x) = U_{g_0} f(g_0^{-1}x). \tag{11}$$

(In the case of a vector field, the vectors have undergone the same rotation as the space itself, and thus we arrived at equation (1) $T_{g_0} a(x) = g_0 a(g_0^{-1}x)$ which is a particular case of equation (11).)

As in the case of a vector field it is obvious that these transformations form a representation of the rotation group.

As in paragraph 1 we limit ourselves to the consideration of points lying on the unit sphere $f(P) = f(\vartheta, \varphi)$ and, as in the case of the vector field, we consider the problem of the expansion of the field which is invariant under rotation. As before, instead of the usual components we introduce $2l_0+1$ functions of the three variables $\varphi_1, \theta, \varphi_2$, i.e. of the rotation g, which define the magnitude f and, for the rotation g_0, are transformed independently of each other. We then obtain $2l_0+1$ different representations each of which is easily resolved into irreducible representations.

To define the components of the magnitude, which depend on the rotation g, we make use of the relation established in paragraph 1 between rotations g and the triads with origin at some arbitrary point of the sphere, the third vector being directed along the normal to the sphere. The value of the magnitude at a given point of the sphere will be given by the components corresponding to the different coordinate systems (the different triads) in space. If c_m denotes the components of the magnitude and g the rotation which carries the coordinate vectors into the triad e_1, e_2, e_3, then the components corresponding to the triad e_1, e_2, e_3, will be $c'_m = \sum_{n=-l_0}^{l_0} a_{mn}(g)\, c_m$ (see equation (10)). The magnitude at the point P of the sphere will be given by its components relative to the triad with origin at the point P and which depends on the rotation which carries the 'North Pole' into P. The components of the magnitude f at the point P with respect to such a triad will be the functions $c_m(g)$ $(-l_0 \leqslant m \leqslant l_0)$, obtained from the original components of the magnitude $f(P)$ by the transformation U_g.

It is not difficult to see how the functions $c_m(g)$ are transformed by an arbitrary rotation g_0. The triad, which depends on g, is altered by the rotation g_0. As we saw in paragraph 1 it goes into a triad which depends on $g_0 g$. On the other hand, under the rotation g_0 the magnitude f and the triads are transferred from point to point and are transformed (rotated) accordingly, and therefore the components of the magnitude f in the corresponding basis will be unchanged after rotation. Hence we see that

$$c'_m(g_0 g) = c_m(g)$$

or, substituting $g_0 g$ for g,

$$c'_m(g) = c_m(g_0^{-1} g). \tag{12}$$

We see, then, that the functions $c_m(g)$ defining the magnitude are transformed independently of each other by the rotation g_0.

Having introduced, as in paragraph 1, the equivalent function $\tilde{c}_m(g) = c_m(g^{-1})$, instead of $c_m(g)$, we obtain from (12) the transformation for $\tilde{c}_m(g)$ corresponding to the rotation g_0:

$$T_{g_0}\tilde{c}_m(g) = \tilde{c}_m(g g_0), \tag{13}$$

i.e. the transformation for the function $\tilde{c}_m(g)$ for any m is the same as the transformation studied in Sect. 7.

We need to resolve each of these representations into irreducible representations. This is most easily done when the functions $\tilde{c}_m(g)$ are related to the first angle φ_1 in the particular form

$$\tilde{c}_m(g) = \tilde{c}_m(\varphi_1, \theta, \varphi_2) = e^{-im\varphi_1}\tilde{c}_m(0, \theta, \varphi_2). \tag{14}$$

Functions of this form may be expanded in terms of the elements of the m-th row of all the matrices T_g^l i.e. in terms of the functions $T_{mn}^l(\varphi_1, \theta, \varphi_2)$ with a fixed value of m.

If the components of the magnitude f have the form (14) then by rotating the space about the axis e_3, through an angle φ^*, i.e. by adding φ^* to φ_1 we multiply each function $\tilde{c}_m$ $(\varphi_1, \theta, \varphi_2)$ by $e^{-im\varphi^*}$. This means that the matrix corresponding to a rotation about the axis Oz is diagonal i.e. the basis consists of the eigenvectors of this matrix.*

Thus finally we have obtained the following result. For a correct resolution of the field of magnitudes on the surface of the sphere, transformed under the rotation g_0 by the equation

$$T_{g_0}f(P) = U_{g_0}f\left(g_0^{-1}P\right), \tag{15}$$

we must proceed in the following manner. First express the magnitude $f(P)$ at each point P with spherical coordinates φ and ϑ in terms of its components $c_m^0(\varphi, \vartheta)$ in some basis consisting of the eigenvectors of the transformation which corresponds to a rotation about the axis Oz. Then transform to the components $c_m(g)$, depending on the rotation $g = g(\varphi_1, \theta, \varphi_2)$, where $\varphi_2 = \frac{\pi}{2} - \varphi$, $\theta = \vartheta$, and subject the magnitude $\left\{c_m^0\left(\frac{\pi}{2} - \varphi_1, \theta\right)\right\}$ to the transformation U_g which corresponds to this rotation. Now expand each function $\tilde{c}_m(g) = \tilde{c}_m(\varphi_1, \theta, \varphi_2)$ in a series in the elements of the m-th rows of the matrices T_g^l for all $l \geqslant m$, i.e. put

$$\tilde{c}_m(\varphi_1, \theta, \varphi_2) = \sum_{l=m}^{\infty} \sum_{n=-l}^{l} a_{mn}^l T_{mn}^l(\varphi_1, \theta, \varphi_2). \tag{16}$$

In conclusion we consider the case when the magnitude is transformed according to an arbitrary representation of the rotation group (not necessarily an irreducible one, as in the case just considered). In this case, of course, it is possible to resolve the magnitude into components each of which is transformed according to an irreducible representation, and then, with each of these, to carry out the process described above.

It is, however, sufficient to give the magnitude in terms of the components which, for a rotation about the axis Oz through an angle φ are multiplied by $e^{-im\varphi}$, where m is an integer, or half an odd integer, then transform it to spherical coordinates (basis - e_φ, e_ϑ, e_r) and finally to expand the m-th component in terms of the functions $T_{mn}^l(\varphi_1, \theta, \varphi_2)$ $(l \geqslant m, \; -l \leqslant n \leqslant l)$. In the case of a reducible representation as distinct from an irreducible representation, however, the same value of m will be met with more than once, i.e. different components may be broken down into the same functions.

§ 3. Example. A Field of Tensors of the Second Rank

To illustrate what has been said in paragraph 2, we will take the simple

---

* It follows from the results of Sect.2 that this basis, if normalized, is the canonical basis which was introduced in that section.

example of tensors of the second rank $a = a_{ij}$. This tensor is a magnitude which is transformed according to a reducible representation of dimensionality 9, the resolution of which into irreducible representations was obtained in Sect.5

We find first the components of the tensor a_{ij}, which transform in the required manner by a rotation about the axis Oz. For this we utilize the fact that we know the components of the vector a_i after a rotation about Oz (see paragraph 1 of this section). The components $a_x - ia_y = a_1 - ia_2$ (multiplication by $e^{-i\varphi}$), $a_z = a_3$ (multiplication by 1) $a_x + ia_y = a_1 + ia_2$ (multiplication by $e^{i\varphi}$). The components of a tensor of the second rank are transformed by a rotation in the same way as a product of the components of two vectors.

First we distinguish three groups of components of this tensor according to the action of a rotation on the first index $a_{1j} - ia_{2j}$, a_{3j}, $a_{1j} + ia_{2j}$ $(j = 1, 2, 3)$. We then obtain the nine complex components of a tensor of the second rank, that is;

$$a_{11} - ia_{21} - i(a_{12} - ia_{22}) = a_{11} - a_{22} - i(a_{21} + a_{12}),$$
$$a_{13} - ia_{23},$$
$$a_{11} - ia_{21} + i(a_{12} - ia_{22}) = a_{11} + a_{22} - i(a_{21} - a_{12}),$$
$$a_{31} - ia_{32},$$
$$a_{33},$$
$$a_{31} + ia_{32},$$
$$a_{11} + ia_{21} - i(a_{21} + ia_{22}) = a_{11} + a_{22} + i(a_{21} - a_{12}),$$
$$a_{13} + ia_{23},$$
$$a_{11} + ia_{21} + i(a_{12} + ia_{22}) = a_{11} - a_{22} + i(a_{21} + a_{12}).$$

To expand in terms of the generalized spherical functions we must operate on the tensor for each point P with spherical coordinates φ, ϑ with the rotation $g = g\left(\frac{\pi}{2} - \varphi_1, \vartheta, 0\right)$, which carries this point to the 'North Pole' of the sphere.* As a result of this rotation the basic vectors e_x, e_y and e_z become $-e_\varphi$, e_ϑ and e_r and we obtain the following nine complex components of the tensor, depending on φ and ϑ:

a_{rr}, $a_{\varphi\varphi} + a_{\vartheta\vartheta} \pm i(a_{\varphi\vartheta} - a_{\vartheta\varphi})$ - the components expanded in terms of the central row of the matrices T_g^l (the usual spherical functions $Y_l^n\left(\frac{\pi}{2} - \varphi, \vartheta\right)$);

$-a_{\varphi r} - ia_{\vartheta r}$, $-a_{r\varphi} - ia_{r\vartheta}$ - the components expanded in terms of the functions $T_{1n}^l\left(\frac{\pi}{2} - \varphi, \vartheta, 0\right)$;

---

* The inverse rotation g^{-1} must carry the 'North Pole' into P (see paragraph 1 of the present section)

$-a_{\varphi r}+ia_{\vartheta r}$, $-a_{r\varphi}+ia_{r\vartheta}$ - the components expanded in terms of the functions $T^l_{-1n}\left(\frac{\pi}{2}-\varphi,\ \vartheta,\ 0\right)$;

$a_{\varphi\varphi}-a_{\vartheta\vartheta}+i(a_{\vartheta\varphi}+a_{\varphi\vartheta})$ - the components expanded in terms of the functions $T^l_{-2,\,n}\left(\frac{\pi}{2}-\varphi,\ \vartheta,\ 0\right)$.

§ 4. Solution of Maxwell's Equations

As an example of the application of the expansions obtained we consider the solution of Maxwell's equations.

We examine the case when these equations may be written in the form

$$\ddot{\tilde{A}}-\Delta\tilde{A}=0, \qquad \operatorname{div}\tilde{A}=0,$$

where $\tilde{A}$ is a vector function of x, y, z, and t (the so-called vector-potential and $\Delta\tilde{A}$ is the vector with components $\Delta\tilde{A}_x$, $\Delta\tilde{A}_y$, $\Delta\tilde{A}_z$ where

$$\Delta=\frac{\partial^2}{\partial x^2}+\frac{\partial^2}{\partial y^2}+\frac{\partial^2}{\partial z^2}.$$

Putting $\tilde{A}=A(x,\ y,\ z)\,e^{ikt}$ we obtain the system of equations

$$\Delta A+k^2A=0,$$
$$\operatorname{div}A=0.$$

Since these equations are invariant under rotation, to determine the solution it will be best to search for a series which is also invariant under rotation, i.e. a series of generalized spherical functions. With this end in view we transform to spherical coordinates and instead of components A_x, A_y, A_z the vector A will be determined by the components A_r, A_φ, A_ϑ, given by equations (6) of the present section.* Having made the required transformation of the independent variables and the functions to be determined, we obtain the following four equations:

$$\left.\begin{aligned}
&\frac{\partial^2A_r}{\partial r^2}+\frac{2}{r}\frac{\partial A_r}{\partial r}+\frac{1}{r^2}\frac{d^2A_r}{\partial\vartheta^2}+\frac{1}{r^2\sin^2\vartheta}\frac{\partial^2A_r}{\partial\varphi^2}+\frac{\cot\vartheta}{r^2}\frac{\partial Ar}{\partial\vartheta}-\\
&\qquad-\frac{2}{r^2}\frac{\partial A_\vartheta}{\partial\vartheta}-\frac{2}{r^2\sin\vartheta}\frac{\partial A_\varphi}{\partial\varphi}-\frac{2A_r}{r^2}-\frac{2\cot\vartheta}{r^2}A_\vartheta+k^2A_r=0,\\
&\frac{\partial^2A_\varphi}{\partial r^2}+\frac{2}{r}\frac{\partial A_\varphi}{\partial r}+\frac{1}{r^2}\frac{\partial^2A_\varphi}{\partial\vartheta^2}+\frac{1}{r^2\sin^2\vartheta}\frac{\partial^2A_\varphi}{\partial\varphi^2}+\frac{\cot\vartheta}{r^2}\frac{\partial A_\varphi}{\partial\vartheta}+\\
&\qquad+\frac{2\cos\vartheta}{r^2\sin^2\vartheta}\frac{\partial A_\vartheta}{\partial\varphi}+\frac{2}{r^2\sin\vartheta}\frac{\partial A_r}{\partial\varphi}-\frac{A_\varphi}{r^2\sin^2\vartheta}+k^2A_\varphi=0,
\end{aligned}\right\}\quad(17)$$

---

* Clearly the letter a must be replaced by A throughout formulae (6).

$$\left.\begin{aligned}&\frac{\partial^2 A_\vartheta}{\partial r^2}+\frac{2}{r}\frac{\partial A_\vartheta}{\partial r}+\frac{1}{r^2}\frac{\partial A_\vartheta}{\partial \vartheta^2}+\frac{1}{r^2\sin^2\vartheta}\frac{\partial^2 A_\vartheta}{\partial \varphi^2}+\frac{\cot\vartheta}{r^2}\frac{\partial A_\vartheta}{\partial \vartheta}-\\&\qquad-\frac{2\cos\vartheta}{r^2\sin^2\vartheta}\frac{\partial A_\varphi}{\partial \varphi}+\frac{2}{r^2}\frac{\partial A_r}{\partial \vartheta}-\frac{A_\vartheta}{r^2\sin^2\vartheta}+k^2A_\vartheta=0,\\&\frac{\partial A_r}{\partial r}+\frac{1}{r}\frac{\partial A_\vartheta}{\partial \vartheta}+\frac{1}{r\sin\vartheta}\frac{\partial A_\varphi}{\partial \varphi}+\frac{2}{r}A_r+\frac{\cot\vartheta}{r}A_\vartheta=0.\end{aligned}\right\}\tag{17}$$

We now introduce the following combinations of components:

$$A_+=-\frac{1}{\sqrt{2}}(A_\varphi+iA_\vartheta),$$

$$A_0=A_r,$$

$$A_-=\frac{1}{\sqrt{2}}(A_\varphi-iA_\vartheta).$$

In terms of these components equations (17) take the form

$$\left.\begin{aligned}&\frac{\partial^2 A_0}{\partial r^2}+\frac{2}{r}\frac{\partial A_0}{\partial r}+\frac{1}{r^2}\frac{\partial^2 A_0}{\partial \vartheta^2}+\frac{1}{r^2}\cot\vartheta\frac{\partial A_0}{\partial \vartheta}+\frac{1}{r^2\sin^2\vartheta}\frac{\partial^2 A_0}{\partial \varphi^2}-\frac{2}{r^2}A_0+k^2A_0-\\&\qquad-\frac{i\sqrt{2}}{r^2}\left[\frac{\partial A_+}{\partial \vartheta}+\frac{i}{\sin\vartheta}\frac{\partial A_+}{\partial \varphi}+\cot A_+\right]-\\&\qquad\qquad-\frac{i\sqrt{2}}{r^2}\left[\frac{\partial A_-}{\partial \vartheta}-\frac{i}{\sin\vartheta}\frac{\partial A_-}{\partial \varphi}+\cot\vartheta A_-\right]=0,\\&\frac{\partial^2 A_-}{\partial r^2}+\frac{2}{r}\frac{\partial A_-}{\partial r}+\frac{1}{r^2}\frac{\partial^2 A_-}{\partial \vartheta^2}+\frac{1}{r^2}\cot\vartheta\frac{\partial A_-}{\partial \vartheta}+\frac{1}{r^2\sin^2\vartheta}\frac{\partial^2 A_-}{\partial \varphi^2}+\frac{2i\cos\vartheta}{r^2\sin^2\vartheta}\frac{\partial A_-}{\partial \varphi}-\\&\qquad-\frac{1}{r^2\sin^2\vartheta}A_-+k^2A_--\frac{i\sqrt{2}}{r^2}\left(\frac{\partial A_0}{\partial \vartheta}+i\frac{\partial A_0}{\partial \varphi}\right)=0,\\&\frac{\partial^2 A_+}{\partial r^2}+\frac{2}{r}\frac{\partial A_+}{\partial r}+\frac{1}{r^2}\frac{\partial^2 A_+}{\partial \vartheta^2}+\frac{1}{r^2}\cot\vartheta\frac{\partial A_+}{\partial \vartheta}+\frac{1}{r^2\sin^2\vartheta}\frac{\partial^2 A_+}{\partial \varphi^2}-\frac{2i\cos\vartheta}{r^2\sin^2\vartheta}\frac{\partial A_+}{\partial \varphi}-\\&\qquad-\frac{1}{r^2\sin^2\vartheta}A_++k^2A_+-\frac{i\sqrt{2}}{r^2}\left(\frac{\partial A_0}{\partial \vartheta}-i\frac{\partial A_0}{\partial \varphi}\right)=0,\\&\frac{\partial A_0}{\partial r}+\frac{2}{r}A_0+\frac{i}{r\sqrt{2}}\left(\frac{\partial A_+}{\partial \vartheta}+\frac{i}{\sin\vartheta}\frac{\partial A_+}{\partial \varphi}+\cot\vartheta A_+\right)+\\&\qquad+\frac{i}{\sqrt{2}}\left(\frac{\partial A_-}{\partial \vartheta}-\frac{i}{\sin\vartheta}\frac{\partial A_-}{\partial \varphi}+\cot\vartheta A_-\right)=0.\end{aligned}\right\}\tag{17'}$$

We look for a solution of equations (17') in a series of generalized spherical functions of the form

$$A_0(r,\varphi,\vartheta)=\sum_{l=0}^{\infty}f_l^0(r)\sum_{n=-l}^{l}\alpha_{l,n}T_{0n}^l\left(\frac{\pi}{2}-\varphi,\vartheta,0\right),$$

$$A_+(r, \varphi, \vartheta) = \sum_{l=0}^{\infty} f_l^+(r) \sum_{n=-l}^{l} \beta_{l,n} T_{1n}^l \left(\frac{\pi}{2} - \varphi, \vartheta, 0\right),$$

$$A_-(r, \varphi, \vartheta) = \sum_{l=0}^{\infty} f_l^-(r) \sum_{n=-l}^{l} \gamma_{l,n} T_{-1n}^l \left(\frac{\pi}{2} - \varphi, \vartheta, 0\right).$$

The best method of finding a solution is by separating the variables and thus reducing the problem to that of solving ordinary equations for functions of r.

Since the equations are linear we may substitute the terms of the series in the system of equations one by one, i.e. substitute the functions

$$\left.\begin{aligned} A_0(r, \varphi, \vartheta) &= f_l^0(r)\, T_{0n}^l\left(\frac{\pi}{2} - \varphi, \vartheta, 0\right), \\ A_+(r, \varphi, \vartheta) &= f_l^+(r)\, T_{1n}^l\left(\frac{\pi}{2} - \varphi, \vartheta, 0\right), \\ A_-(r, \varphi, \vartheta) &= f_l^-(r)\, T_{-1n}^l\left(\frac{\pi}{2} - \varphi, \vartheta, 0\right). \end{aligned}\right\} \tag{18}$$

The first of the equations (18) then acquires the form

$$\begin{aligned} &\left[\frac{d^2 f_l^0}{dr^2} + \frac{2}{r}\frac{df_l^0}{dr} + k^2 f_l^0(r)\right] T_{0n}^l\left(\frac{\pi}{2} - \varphi, \vartheta, 0\right) + \\ &\quad + \frac{1}{r^2} f^0(r)\left[\frac{\partial^2 T_{0n}^l}{\partial\vartheta^2} + \cot\vartheta\,\frac{\partial T_{0n}}{\partial\vartheta} + \frac{1}{\sin^2\vartheta}\frac{\partial^2 T_{0n}}{\partial\varphi^2} - \frac{2}{r^2} T_{0n}\right] - \\ &\quad - \frac{i\sqrt{2}}{r^2} f_l^+(r)\left[\frac{\partial T_{1n}}{\partial\vartheta} + \frac{i}{\sin\vartheta}\frac{\partial T_{1n}}{\partial\varphi} + \cot\vartheta\, T_{1n}\right] - \\ &\quad - \frac{i\sqrt{2}}{r^2} f_l^-(r)\left[\frac{\partial T_{-1n}}{\partial\vartheta} - \frac{i}{\sin\vartheta}\frac{\partial T_{-1n}}{\partial\varphi} + \cot\vartheta\, T_{-1n}\right] = 0. \end{aligned} \tag{19}$$

Remembering that $T_{mn}\left(\frac{\pi}{2} - \varphi, \vartheta, 0\right) = e^{-in\left(\frac{\pi}{2} - \varphi\right)} u_{mn}(\vartheta)$, and using the differential equation for $u_{mn}(\vartheta)$ (see Sect.7)

$$\frac{d^2 u_{0n}}{d\vartheta^2} + \operatorname{ctg}\vartheta\,\frac{du_{0n}}{d\vartheta} + \left[l(l+1) - \frac{n^2}{\sin^2\vartheta}\right] u_{0n}(\vartheta) = 0$$

and the recurrence formulæ (see appendix to Sect.7)

$$\frac{du_{1n}}{d\vartheta} - \frac{n - \cos\vartheta}{\sin\vartheta} u_{1n} = -i\sqrt{l(l+1)}\, u_{0n},$$

$$\frac{du_{-1n}}{d\vartheta} + \frac{n + \cos\vartheta}{\sin\vartheta} u_{-1n} = -i\sqrt{l(l+1)}\, u_{0n},$$

we may, with the help of the relations, eliminate from equation (19)

$$\frac{d^2u_{0n}}{d\vartheta^2}, \quad \frac{du_{0n}}{d\vartheta}, \quad \frac{du_{1n}}{d\vartheta}, \quad u_{1n}, \quad \frac{du_{-1n}}{d\vartheta}, \quad u_{-1n}$$

and cancel $e^{-in\left(\frac{\pi}{2}-\varphi\right)} u_{0n}(\vartheta)$, from the equation obtained. Thus we obtain an equation which contains functions of r only:

$$\frac{d^2f_l^0}{dr^2}+\frac{2}{r}\frac{df_l^0}{dr}+\left[k^2-\frac{2+l(l+1)}{r^2}\right]f_l^0(r)-$$
$$-\frac{\sqrt{2}\sqrt{l(l+1)}}{r^2}\left[f_l^+(r)+f_l^-(r)\right]=0. \tag{20}$$

We see that in fact our substitution did succeed in separating the variables in equation (19). Analogous substitution and transformation of the two other equations of (18) gives the familiar equations

$$\left.\begin{aligned}\frac{d^2f_l^+}{dr^2}+\frac{2}{r}\frac{df_l^+}{dr}+\left[k^2-\frac{l(l+1)}{r^2}\right]f_l^+(r)-\frac{\sqrt{2}\sqrt{l(l+1)}}{r^2}f_l^0(r)=0,\\ \frac{d^2f_l^-}{dr^2}+\frac{2}{r}\frac{df_l^-}{dr}+\left[k^2-\frac{l(l+1)}{r^2}\right]f_l^-(r)-\frac{\sqrt{2}\sqrt{l(l+1)}}{r^2}f_l^0(r)=0.\end{aligned}\right\} \tag{21}$$

The equation $\operatorname{div} \boldsymbol{A}=0$ after corresponding transformations reduces to

$$\frac{df_l^0}{dr}+\frac{2}{r}f_l^0(r)+\frac{\sqrt{l(l+1)}}{r\sqrt{2}}\left[f_l^-(r)+f_l^+(r)\right]=0. \tag{22}$$

It remains for us to solve equations (20), (21), (22). To do this we find $f_l^-(r)+f_l^+(r)$ from equations (20) and (22). We then obtain the following equation of the second order for $f_l^0(r)$:

$$\frac{d^2f_l^0}{dr^2}+\frac{4}{r}\frac{df_l^0}{dr}+\left[k^2+\frac{2-l(l+1)}{r^2}\right]f_l^0(r)=0. \tag{23}$$

By transforming the function to be determined this becomes the Bessel equation of order $l+\frac{1}{2}$. Thus the solution of equation (23) has the form

$$f_l^0(r)=C_1\frac{J_{l+\frac{1}{2}}(kr)}{r^{\frac{3}{2}}}. \tag{24}$$

From this and equation (22) we obtain:

$$f_l^+(r)+f_l^-(r)=-\frac{C_1\sqrt{2}}{\sqrt{l(l+1)}}\left[\frac{1}{2}\frac{J_{l+\frac{1}{2}}(kr)}{r^{\frac{3}{2}}}+\frac{kJ'_{l+\frac{1}{2}}(kr)}{r^{\frac{1}{2}}}\right]. \tag{25}$$

Subtracting equations (21) and denoting $f_l^+(r)-f_l^-(r)$ by $\varphi(r)$, we obtain the equation for $\varphi(r)$

$$\frac{d^2\varphi}{dr^2}+\frac{2}{r}\frac{d\varphi}{dr}+\left[k^2-\frac{l(l+1)}{r^2}\right]\varphi(r)=0,$$

from which

$$\varphi(r)=f_l^+(r)-f_l^-(r)=C_2\frac{J_{l+\frac{1}{2}}(r)}{r^{\frac{1}{2}}}. \tag{26}$$

Solving equations (25) and (26) we obtain:

$$f_l^+(r)=\frac{1}{2}\left\{-\frac{C_1}{\sqrt{2}\sqrt{l(l+1)}}\frac{J_{l+\frac{1}{2}}(kr)}{r^{\frac{3}{2}}}-kC_1\frac{\sqrt{2}}{\sqrt{l(l+1)}}\frac{J'_{l+\frac{1}{2}}(kr)}{r^{\frac{1}{2}}}+C_2\frac{J_{l+\frac{1}{2}}(kr)}{r^{\frac{1}{2}}}\right\},$$

$$f_l^-(r)=\frac{1}{2}\left\{-\frac{C_1}{\sqrt{2}\sqrt{l(l+1)}}\frac{J_{l+\frac{1}{2}}(kr)}{r^{\frac{3}{2}}}-kC_1\frac{\sqrt{2}}{\sqrt{l(l+1)}}\frac{J'_{l+\frac{1}{2}}(kr)}{r^{\frac{1}{2}}}-C_2\frac{J_{l+\frac{1}{2}}(kr)}{r^{\frac{1}{2}}}\right\}.$$

Substituting $J'_{l+\frac{1}{2}}(kr)\ \frac{l+\frac{1}{2}}{kr}J_{l+\frac{1}{2}}(kr)$ for $J_{l+\frac{3}{2}}(kr)$ from the recurrence formula we may write these equations in the form of a combination of Bessel functions

$$f_l^+(r)=\frac{1}{2}\left\{\frac{kC_1\sqrt{2}}{\sqrt{l(l+1)}}\frac{J_{l+\frac{2}{3}}(kr)}{r^{\frac{1}{2}}}-\frac{C_1\sqrt{2}\,(l+1)}{\sqrt{l(l+1)}\ r^{\frac{3}{2}}}J_{l+\frac{1}{2}}(kr)+C_2\frac{J_{l+\frac{1}{2}}(kr)}{r^{\frac{1}{2}}}\right\},$$

$$f_l^-(r)=\frac{1}{2}\left\{\frac{kC_1\sqrt{2}}{\sqrt{l(l+1)}}\frac{J_{l+\frac{3}{2}}(kr)}{r^{\frac{1}{2}}}-\frac{C_1\sqrt{2}\,(l+1)}{\sqrt{l(l+1)}\ r^{\frac{3}{2}}}J_{l+\frac{1}{2}}(kr)-C_2\frac{J_{l+\frac{1}{2}}(kr)}{r^{\frac{1}{2}}}\right\},$$

$$f_l^0(r)=C_1\frac{J_{l+\frac{1}{2}}(kr)}{r^{\frac{3}{2}}}.$$

Thus for every l and n we obtain a solution depending on the two arbitrary constants C_1 and C_2.

Having substituted these functions in equation (18) we obtain particular solutions of Maxwell's equations. A general solution may be expressed in the form of a series of such particular solutions.

By putting $C_1=1,\ C_2=0$, and then $C_1=0,\ C_2=1$, we obtain two solutions with different physical significance. They are distinguished from each other by their behaviour under reflection in the origin of coordinates.* The first of them (for which $C_2 = 0$) is multiplied by $(-1)^l$ on reflection and is called the vector potential of an electrical multipole of order l. The second solution (for which $C_1 = 0$) is multiplied by $(-1)^{l+1}$ on reflection and is called the vector potential of a magnetic multipole of order l.

By using the recurrence formulae expressing T^l_{1n} and $T^l_{-1,n}$ in terms of T^{l-1}_{0n}, T^l_{0n} and T^{l+1}_{0n} (see equation (8) of the appendix to Sect.7), it is possible to express the solution which has been determined in terms of the usual spherical functions of orders $l-1$, l and $l+1$.

Section 9. Equations Invariant with Respect to Rotation

In this section we shall examine systems of partial differential equations that are invariant with respect to rotation of the frame of reference.

Let $\psi_1(x_1, x_2, x_3)$, $\psi_2(x_1, x_2, x_3)$, ..., $\psi_N(x_1, x_2, x_3)$ be the unknown functions. We shall denote the column whose components are these functions by $\psi(x_1, x_2, x_3)$. A system of differential equations may then be written in the form

$$L_1 \frac{\partial\psi}{\partial x_1} + L_2 \frac{\partial\psi}{\partial x_2} + L_3 \frac{\partial\psi}{\partial x_3} + \varkappa\psi = 0, \tag{1}$$

where L_1, L_2, L_3 are N-th order matrices and $\varkappa$ is a number **

In order that speaking of the invariance of equation (1) under rotation be meaningful, we should describe the law according to which $\psi_1, \psi_2, \ldots, \psi_N$ are transformed upon rotation. Since, as a result of a rotation a set of conditions must be satisfied for the corresponding transformations upon the functions

---

* To find out what happens on reflection to A_+, A_- and A_0 we note that the point r, φ, ϑ is transformed on reflection into the point with coordinates $r, \varphi+\pi, \pi-\vartheta$ the components $A_r, A_\varphi, A_\vartheta$ being transformed into $A_r, A_\varphi, -A_\vartheta$ and

$$e^{-in\left(\frac{\pi}{2}-\varphi+\pi\right)} = (-1)^n e^{-in\left(\frac{\pi}{2}-\varphi\right)},$$

and $u^l_{mn}(\pi-\vartheta) = (-1)^{m+n+l} u^l_{mn}(\vartheta)$.

** The general form for a system of equations of the first order is

$$A_1 \frac{\partial\psi}{\partial x_1} + A_2 \frac{\partial\psi}{\partial x_2} + A_3 \frac{\partial\psi}{\partial x_3} + B\psi = 0.$$

If the matrix B is non-singular, then multiplying both sides by B^{-1} we obtain a system of the form (1). In the same way, multiplying the system by the matrix CB^{-1} we make the matrix applied to the function ψ equal to C.

We note that the application to the system of some non-singular matrix does not alter the system, because it is equivalent to replacing the given equations by linear combinations of themselves.

$\psi_1, \ldots, \psi_N$ the quantity ψ should be transformed according to a certain representation of the rotation group (which is generally speaking reducible). In this way, under the rotation g the quantity ψ transforms to $\psi' = T_g\psi$.

The system of equations (1) is said to be invariant with respect to rotation if it remains unchanged under the transformation $x' = gx$ of the independent variables (where g is an arbitrary rotation) and under the corresponding transformation $\psi' = T_g\psi$ of the unknown functions.

In this section we shall find the general system of first order equation invariant under rotation; furthermore with the aid of an analysis by generalized spherical functions we shall reduce this system to a system of ordinary first order equations capable of solution in terms of cylindrical functions.

§ 1. Definition of Invariant Equations

In order to find the form of the invariant equations we shall first of all write the conditions for the invariance of system (1).

We shall subject the frame of reference to the rotation $g: x' = g^{-1}x$, i.e. effect a change of the independent variables $x'_i = \sum_{k=1}^{3} g_{ki}x_k$. In place of ψ we should now substitute the quantity $\psi' = T_g\psi$. Replacing ψ by $T_g^{-1}\psi'$, and a differentiation by x_k by a differentiation by x'_i according to the formula

$$\frac{\partial}{\partial x_k} = \sum g_{ik}\frac{\partial}{\partial x'_i},$$

we obtain the system

$$\sum_{i=1}^{3}\left[g_{i1}L_1\frac{\partial(T_g^{-1}\psi')}{\partial x'_i} + g_{i2}L_2\frac{\partial(T_g^{-1}\psi')}{\partial x'_i} + g_{i3}L_3\frac{\partial(T_g^{-1}\psi')}{\partial x'_i}\right] + \varkappa T_g^{-1}\psi' = 0$$

or, because T_g is a constant matrix, the system

$$\sum_i\left[g_{i1}L_1T_g^{-1}\frac{\partial\psi'}{\partial x'_i} + g_{i2}L_2T_g^{-1}\frac{\partial\psi'}{\partial x'_i} + g_{i3}L_3T_g^{-1}\frac{\partial\psi'}{\partial x'_i}\right] + \varkappa T_g^{-1}\psi' = 0.$$

In order to obtain the condition of coincidence of this system of equations with system (1) we firstly make the coefficient of ψ' equal to $\varkappa$ by applying the transformation T_g. We obtain the system

$$\sum_k\sum_i g_{ik}T_gL_kT_g^{-1}\frac{\partial\psi'}{\partial x_i} + \varkappa\psi' = 0. \tag{2}$$

The stipulation of invariance means, consequently, that for any rotation g there should hold the following relation between the matrices L_k:

$$\sum_k g_{ik}T_gL_kT_g^{-1} = L_i. \tag{3}$$

We shall rewrite the condition for invariance in another, sometimes more convenient form. To this end, we consider the matrix $\sum L_i p_i$ where p_1, p_2, p_3 are the components of some vector. Since a rotation carries the matrix L_i to $\sum g_{ik} L_k$ it follows that $\sum L_i p_i$ transforms to $\sum L_k p'_k$ where $p'_k = \sum g_{ik} p_i$, i.e.the numbers p_i transform under rotation in the same way as the differential operators $\frac{\partial}{\partial x_i}$.

From formula (3) there result the equations

$$\sum_{i=1}^{3} L_i p_i = \sum_{k=1}^{3} \sum_{i=1}^{3} g_{ik} p_i T_g L_k T_g^{-1} = \sum_{k=1}^{3} T_g L_k p'_k T_g^{-1} = T_g \left(\sum L_k p'_k\right) T_g^{-1},$$

i.e.

$$\sum_{i=1}^{3} L_i p_i = T_g \left(\sum_{k=1}^{3} L_k p'_k \right) T_g^{-1}. \tag{4}$$

Equations (4) and (3) represent forms of writing the condition for system (1) to be invariant under rotation. With the aid of formula (4) it is possible to find directly the characteristic polynomial of an arbitrary invariant system*.

Taking determinants on both sides of equality (4), we see that, since $\det T_g = \frac{1}{\det T_g^{-1}}$,

$$\det \sum L_i p_i = \det \sum L_k p'_k,$$

i.e. the characteristic polynomial is not altered by rotation. Since any two vectors of the same length may be interchanged one for the other by some rotation, such a function is constant on the surface of each sphere with centre at the origin of the coordinate axes, i.e. depends only upon

$$r = \sqrt{p_1^2 + p_2^2 + p_3^2}.$$

But $\det \sum L_i p_i$ is a homogeneous function of order N, where N is the number of equations and unknown functions in the system. Consequently, the characteristic polynomial of system (1) is equal to

$$C\left(p_1^2 + p_2^2 + p_3^2\right)^{N/2}.$$

Since $\det \sum L_i p_i$ is clearly a rational function of p_1, p_2, p_3, then in order that it be different from zero, $N/2$ must be an integer and therefore the number of equations and unknown functions must be even.

* The expression $\det \sum L_i p_i$ in the p_i, is known as the characteristic polynomial of system (1).

§ 2. Reformulation of the Conditions for Invariance

Condition (3) represents, in fact, an infinite number of equations insofar as the rotation g occurring in it is arbitrary. We shall now replace these equations by a finite number of algebraic relations. In order to obtain these relations we alter the rotation g by a small rotational displacement about each of the coordinate axes.

First of all we consider the rotation g equal to $e + a_1\xi + \ldots$ where a_1 is an infinitesimally small displacement about the axis Ox (see Sect.2). The matrix of such a rotation to within a small quantity of higher order than ξ has the form:

$$\begin{Vmatrix} 1 & 0 & 0 \\ 0 & 1 & -\xi \\ 0 & \xi & 1 \end{Vmatrix}.$$

The corresponding transformation T_g is $E + \xi A_1 + \ldots$, where A_1 is the transformation representing the infinitesimally small displacement a_1 (this transformation was determined in Sect.2). The reciprocal transformation T_g^{-1} is to within a small quantity of higher order $E - \xi A_1$. Substituting $e + a_1\xi + \ldots$, $E + A_1\xi + \ldots$, $E - A_1\xi + \ldots$ in system (3) we obtain three equations correct to within terms of the second order:

$$(E + A_1\xi)\, L_1\, (E - A_1\xi) = L_1,$$
$$(E + A_1\xi)\,(L_2 - \xi L_3)\,(E - A_1\xi) = L_2,$$
$$(E + A_1\xi)\,(\xi L_2 + L_3)\,(E - A_1\xi) = L_3.$$

Expanding the brackets, we see that the terms not containing ξ cancel out as expected, and those terms containing the first power of ξ give us the equations

$$A_1L_1 - L_1A_1 = 0,$$
$$A_1L_2 - L_2A_1 - L_3 = 0,$$
$$A_1L_3 - L_3A_1 + L_2 = 0,$$

which may be written more concisely as

$$[A_1, L_1] = 0,$$
$$[A_1, L_2] = L_3,$$
$$[A_1, L_3] = -L_2.$$

In this way, taking $g = e + a_1\xi + \ldots$, we find the commutators of all three matrices L_1, L_2, L_3 with the known matrix A_1. Similarly, taking $g = e + a_2\xi + \ldots$ and $g = e + a_3\xi + \ldots$, we find the commutators of the matrices L_1, L_2, L_3 with A_2 and A_3. The results are given in the following table:

$$[A_1, L_1] = 0, \qquad [A_1, L_2] = L_3, \qquad [A_1, L_3] = -L_2,$$
$$[A_2, L_1] = -L_3, \qquad [A_2, L_2] = 0, \qquad [A_2, L_3] = L_1,$$

$$[A_3, L_1] = L_2, \qquad [A_3, L_2] = -L_1, \quad [A_3, L_3] = 0. \tag{5}$$

From these equations we find the possible forms of the matrices L_1, L_2, L_3.

We shall not prove now that equations (3) or (4) follow from equation (5). This proof, that the validity of certain facts for infinitesimal rotations implies their validity for all finite rotations, is entirely analogous to the procedure in Sect.2, whereby it was proved that the matrices corresponding to infinitesimal rotations completely determine the transformations T_g for all g.

In order to find L_1, L_2, L_3 we first of all eliminate the matrices L_1 and L_2 from system (5). To this end we utilize the transformations introduced in Sect. 2:

$$H_+ = H_1 + iH_2 = iA_1 - A_2,$$
$$H_- = H_1 - iH_2 = iA_1 + A_2$$

and we compute the commutator $[[L_3, H_-], H_+]$. Firstly we find the commutator of L_3 and H_-. Using equation (5), we have:

$$[L_3, H_-] = i[L_3, A_1] + [L_3, A_2] = iL_2 - L_1.$$

We compute the commutator of the resulting operator with H_+; then

$$[iL_2 - L_1, H_+] = [iL_2 - L_1, iA_1 - A_2] = -[L_2, A_1] + [L_1, A_2] = 2L_3.$$

In this way we obtain two equations

$$\left.\begin{aligned} &[L_3, H_3] = 0, \\ &[[L_3, H_-], H_+] = 2L_3, \end{aligned}\right\} \tag{6}$$

which the matrix L_3 must satisfy.

It can be shown that if we have a matrix L_3 which satisfies these equations and if further we define L_1 and L_2 by the formulae $L_1 = [A_2, L_3]$, $L_2 = -[A_1, L_3]$, then the matrices L_1, L_2, L_3 obtained satisfy the complete system (5).

§ 3. Determination of the Matrices L_1, L_2, L_3

In this paragraph we determine explicity the form of a matrix L_3 which satisfies the conditions (6), and then find L_1 and L_2. The quantity ψ transforms according to some representation which we shall assume to be resolved into irreducible representations. We shall number the components of the quantity ψ with the (possibly repeated) indices l and m, where l is the weight of the irreducible representation and m is the number of componėnts in the representation of weight l. If there are representations with one and the same weight l occurring more than once upon resolving the representation of ψ into irreducible components, then, in this case, in order to distinguish between these representations we shall add a further index τ, indicating the number of the representation of weight l. In this fashion, using these components the quantity

ψ will be written as

$$\psi(x_1, x_2, x_3) = \{\psi^{\tau}_{lm}(x_1, x_2, x_3)\}.$$

Denoting by $\xi^{\tau_1}_{l_0 m_0}$ the quantity for which $\psi^{\tau_1}_{l_0 m_0} = 1$ and the rest of the components are zero, we may write

$$\psi(x_1, x_2, x_3) = \sum_{l, m, \tau} \psi^{\tau}_{lm}(x_1, x_2, x_3)\, \xi^{\tau}_{lm}\ *$$

Because the quantity ξ^{τ}_{lm} is determined by three indices, a transformation of this quantity, in particular the matrix L_3, involves six indices. The transformation L_3 of the vectors ξ^{τ}_{lm} consequently has the form

$$L_3 \xi^{\tau}_{lm} = \sum c^{\tau'\tau}_{l'l,\, m'm}\, \xi^{\tau'}_{l'm'}.$$

In order to find the numbers $c^{\tau'\tau}_{l'l,\, m'm}$ we make use of system (6).

Recalling (see Sect.2) that:

$$H_3 \xi^{\tau}_{lm} = m \xi^{\tau}_{lm},$$
$$H_+ \xi^{\tau}_{lm} = \alpha^{l}_{m+1}\, \xi^{\tau}_{l, m+1},$$
$$H_- \xi^{\tau}_{lm} = \alpha^{l}_{m}\, \xi^{\tau}_{l, m-1},$$

where $(\alpha^{l}_{m})^2 = (l + m)(l - m + 1)$ we have:

$$L_3 H_3 \xi^{\tau}_{lm} = m L_3 \xi^{\tau}_{lm} = m \sum_{l', \tau', m'} c^{\tau'\tau}_{l'l,\, m'm}\, \xi^{\tau'}_{l'm'},$$

$$H_3 L_3 \xi^{\tau}_{lm} = H_3 \sum c^{\tau'\tau}_{l'l,\, m'm}\, \xi^{\tau'}_{l'm'} = \sum_{l' m' \tau'} m' c^{\tau'\tau}_{l'l,\, m'm}\, \xi^{\tau'}_{l'm'}.$$

Hence, because of the first of equations (6) $(L_3 H_3 - H_3 L_3)\, \xi^{\tau}_{lm} = 0$ we have:

$$\sum_{l', m', \tau'} (m - m')\, c^{\tau'\tau}_{l'l,\, m'm}\, \xi^{\tau'}_{l'm'} = 0.$$

Equating the coefficients of $\xi^{\tau'}_{l'm'}$ to zero we conclude that the coefficients $c^{\tau'\tau}_{l'l,\, m'm}$ can be different from zero only if $m' = m$.

For brevity, we shall denote $c^{\tau'\tau}_{l'l,\, mm}$ simply by $c^{\tau'\tau}_{l'l,\, m}$.

Now making use of the second equation of system (6), we have

$$L_3 H_- \xi^{\tau}_{lm} = L_3 \alpha^{l}_{m}\, \xi^{\tau}_{l, m-1} = \alpha^{l}_{m} \sum_{l'\tau'} c^{\tau'\tau}_{l'l,\, m-1}\, \xi^{\tau'}_{l', m-1},$$

$$H_- L_3 \xi^{\tau}_{lm} = H_- \sum_{l'\tau'} c^{\tau'\tau}_{l'l,\, m}\, \xi^{\tau'}_{l', m} = \sum_{l'\tau'} \alpha^{l'}_{m}\, c^{\tau'\tau}_{l'l,\, m}\, \xi^{\tau'}_{l', m-1},$$

$$[L_3, H_-]\, \xi^{\tau}_{lm} = \sum_{l'\tau'} [\alpha^{l}_{m}\, c^{\tau'\tau}_{l'l,\, m-1} - \alpha^{l'}_{m}\, c^{\tau'\tau}_{l'l,\, m}]\, \xi^{\tau'}_{l', m-1}.$$

---

* The quantities ξ^{τ}_{lm} form, in this manner, a basis in the space in which the representation $g \to T_g$ acts, and ψ^{τ}_{lm} are the component quantities in this basis.

Further

$$[L_3, \; H_-]\, H_+ \xi^{\tau}_{lm} = [L_3, \; H_-]\, \alpha^{l}_{m+1}\, \xi^{\tau}_{l,\, m+1} =$$

$$= \alpha^{l}_{m+1} \sum_{l'\tau'} [\alpha^{l}_{m+1}\, c^{\tau'\tau}_{l'l,\, m} - \alpha^{l'}_{m+1}\, c^{\tau'\tau}_{l'l,\, m+1}]\, \xi^{\tau'}_{l'm},$$

$$H_+ [L_3, \; H_-]\, \xi^{\tau}_{lm} = H_+ \sum_{l'\tau'} [\alpha^{l}_{m}\, c^{\tau'\tau}_{l'l,\, m-1} - \alpha^{l'}_{m}\, c^{\tau'\tau}_{l'l,\, m}]\, \xi^{\tau'}_{l',\, m-1} =$$

$$= \sum_{l'\tau'} \alpha^{l'}_{m}\, [\alpha^{l}_{m}\, c^{\tau'\tau}_{l'l,\, m-1} - \alpha^{l'}_{m}\, c^{\tau'\tau}_{l'l,\, m}]\, \xi^{\tau'}_{l'm},$$

$$[\,[L_3, \; H_-], \; H_+]\, \xi^{\tau}_{lm} =$$

$$= \sum_{l'\tau'} \{[\,(\alpha^{l}_{m+1})^2 + (\alpha^{l'}_{m})^2]\, c^{\tau'\tau}_{l'l,\, m} - \alpha^{l'}_{m}\, \alpha^{l}_{m}\, c^{\tau'\tau}_{l'l,\, m-1} -$$

$$- \alpha^{l'}_{m+1}\, \alpha^{l}_{m+1}\, c^{\tau'\tau}_{l'l,\, m+1}\}\, \xi^{\tau'}_{l',\, m}.$$

The second of the equations (6) gives, in this manner, a system of equations for determining $c^{\tau'\tau}_{l'l,\, m}$:

$$2 c^{\tau'\tau}_{l'l,\, m} = [\,(\alpha^{l}_{m+1})^2 + (\alpha^{l}_{m})\,]\, c^{\tau'\tau}_{l'l,\, m} -$$

$$- \alpha^{'}_{m}\, \alpha^{l}_{m}\, c^{\tau'\tau}_{l'l,\, m-1} - \alpha^{l'}_{m+1}\, \alpha^{l}_{m+1}\, c^{\tau'\tau}_{l'l,\, m+1},$$

or substituting the values of α^{l}_{m}, we have the system

$$2 c^{\tau'\tau}_{l'l,\, m} = [(l + m + 1)(l - m) + (l' + m)(l' - m + 1)\, c^{\tau'\tau}_{l'l,\, m} -$$

$$- \sqrt{(l' + m)(l' - m + 1)(l + m)(l - m + 1)}]\, c^{\tau'\tau}_{l'l,\, m-1} -$$

$$- \sqrt{(l' + m + 1)(l' - m)(l + m + 1)(l - m)}\, c^{\tau'\tau}_{l'l,\, m+1}. \tag{7}$$

This system can be solved with fixed indices l', l, τ' and τ, which are then allowed to take all possible values. We fix upon some values of l', l, τ', τ and for the time being denote $c^{\tau'\tau}_{l'l,\, m}$ by c_m.

We obtain a system of homogeneous equations for c_m where $-\min(l', l) \leqslant m \leqslant \min(l', l)$ * and the number of equations is equal to the number of unknowns. These equations are most conveniently solved by successive elimination of the unknowns. When m assumes its maximum possible value $m_0 = \min(l', l)$, we obtain an equation containing two unknowns, c_{m_0} and c_{m_0-1}, by which c_{m_0-1} is determined from c_{m_0}.

Now giving m the value $m_0 - 1$ we obtain an equation connecting c_{m_0-2}, c_{m_0-1}, c_{m_0}, by which we can define c_{m_0-2} once again through c_m. Continuing this process, we arrive at the smallest possible value of m for

* Since $-l' \leqslant m' \leqslant l'$, $-l \leqslant m \leqslant l$ and $c^{\tau'\tau}_{l'l,\, m'm} \neq 0$ only for $m' = m$, therefore m varies from $-\min(l', l)$ to $\min(l', l)$. We note, by the way, that when m assumes its maximum permitted value m_0, we may (still) formally use the equation since the coefficient of $c^{\tau'\tau}_{l'l,\, m_0+1}$ is equal to zero.

which the equation again contains only two unknowns with the minimum value of m and with the next largest value. Since both these unknowns are already defined by a previous equation, this relationship either will be a consequence of the previous one or it will follow from it that c_{m_0}, and thus all the unknowns are equal to zero. This procedure shows that $c^{\tau'\tau}_{l'l,\,m}$ can differ from zero only when $|l'-l|\leqslant 1$ i.e. if $l'=l,\ l'=l-1$ and $l'=l+1$.

We shall find $c^{\tau'\tau}_{l'l,\,m}$ in these cases by the method indicated.

The proof of the fact that $c^{\tau'\tau}_{l'l,\,m}$ for other values of l' is equal to zero proceeds in an entirely analogous fashion and we leave it for the reader to establish.

We consider first of all, $l'=l$, τ' and τ, arbitrary. Then equation (7) takes the form

$$[2-(l+m+1)(l-m)-(l+m)(l-m+1)]\,c^{\tau'\tau}_{ll,\,m}+$$
$$+(l+m)(l-m+1)\,c^{\tau'\tau}_{ll,\,m-1}+(l+m+1)(l-m)\,c^{\tau'\tau}_{ll,\,m+1}=0.$$

Assuming $m=l$ we find that $(1-l)\,c^{\tau'\tau}_{ll,\,l}+l c^{\tau'\tau}_{ll,\,l-1}=0$, whence $c^{\tau'\tau}_{ll,\,l}=$ $=c^{\tau'\tau}_{ll}\cdot l;\ c^{\tau'\tau}_{ll,\,l-1}=c^{\tau'\tau}_{ll}(l-1)$, where $c^{\tau'\tau}_{ll}$ is an arbitrary constant not depending on m. Taking $m=l-1$ we find similarly that $c^{\tau'\tau}_{ll,\,l-2}=$ $=c^{\tau'\tau}_{ll}(l-2)$. By substitution one is easily convinced that the regularity observed is general, i.e. for all m

$$c^{\tau'\tau}_{ll,\,m}=c^{\tau'\tau}_{ll}\cdot m.$$

We now let $l'=l-1$. Equations (7) take the form:

$$[2-(l+m+1)(l-m)-(l+m-1)(l-m)]\,c^{\tau'\tau}_{l-1,\,l,\,m}+$$
$$+\sqrt{(l+m-1)(l-m)(l+m)(l-m+1)}\,c^{\tau'\tau}_{l-1,\,l,\,m-1}+$$
$$+\sqrt{(l+m)(l-m-1)(l+m+1)(l-m)}\,c^{\tau'\tau}_{l-1,\,l,\,m+1}=0.$$

In these equations we perform the substitution

$$c^{\tau'\tau}_{l-1,\,l,\,m}=\tilde{c}^{\tau'\tau}_{l-1,\,l,\,m}\sqrt{(l+m)(l-m)}.$$

After this substitution and the division of the m-th equation by $\sqrt{(l+m)(l-m)}$ we obtain the system

$$2\,[1-l^2+m^2]\,\tilde{c}^{\tau'\tau}_{l-1,\,l,\,m}+[l^2-(m-1)^2]\,\tilde{c}^{\tau'\tau}_{l-1,\,l,\,m-1}+$$
$$+[l^2-(m+1)^2]\,\tilde{c}^{\tau'\tau}_{l-1,\,l,\,m+1}=0.$$

It is easily verified that this system is satisfied by $\tilde{c}^{\tau'\tau}_{l-1,\,l,\,m}$, being independent of m. We therefore allow $\tilde{c}^{\tau'\tau}_{l-1,\,l,\,m}$ to be equal to $c^{\tau'\tau}_{l-1,\,l}$. Hence reverting to the old unknowns, we find

$$c^{\tau'\tau}_{l-1,\,l,\,m}=c^{\tau'\tau}_{l-1,\,l}\sqrt{l^2-m^2}.$$

Finally we let $l' = l + 1$ In the resulting equations

$$[2 - (l+m+1)(l-m) - (l+m+1)(l-m+2)]\, c^{\tau'\tau}_{l+1,\, l,\, m} +$$
$$+ \sqrt{(l+m+1)(l-m+2)(l+m)(l-m+1)}\, c^{\tau'\tau}_{l+1,\, l,\, m-1} +$$
$$+ \sqrt{(l+m+2)(l-m+1)(l+m+1)(l-m)}\, c^{\tau'\tau}_{l+1,\, l,\, m+1} = 0$$

We make the substitution

$$c^{\tau'\tau}_{l+1,\, l,\, m} = \sqrt{(l+m+1)(l-m+1)}\, \tilde{c}^{\tau'\tau}_{l+1,\, l,\, m}.$$

After this substitution and division by $\sqrt{(l+m+1)(l-m+1)}$ we arrive at the equation

$$2\,[m^2 - l^2 - 2l]\, \tilde{c}^{\tau'\tau}_{l+1,\, l,\, m} + [l^2 - m^2 + 2l + 2m]\, \tilde{c}^{\tau'\tau}_{l+1,\, l,\, m-1} +$$
$$+ [l^2 - m^2 + 2l - 2m]\, \tilde{c}^{\tau'\tau}_{l+1,\, l,\, m+1} = 0,$$

the solution of which is also independent of m. We denote it by $c^{\tau'\tau}_{l+1,\, l}$.

In this way we have found the following values of the elements of the matrix L_3:

$$\left.\begin{aligned} c^{\tau'\tau}_{l-1,\, l,\, m} &= c^{\tau'\tau}_{l-1,\, l} \sqrt{l^2 - m^2}, \\ c^{\tau'\tau}_{ll,\, m} &= c^{\tau'\tau}_{ll}\, m, \\ c^{\tau'\tau}_{l+1,\, l,\, m} &= c^{\tau'\tau}_{l+1,\, l} \sqrt{(l+1)^2 - m^2}. \end{aligned}\right\} \tag{8}$$

We shall now determine the matrices L_1, L_2. We denote the elements of the matrix L_1 by $a^{\tau'\tau}_{l'l,\, m'm}$, that is we let

$$L_1 \xi^{\tau}_{lm} = \sum_{l',\, m',\, \tau'} a^{\tau'\tau}_{l'l,\, m'm}\, \xi^{\tau'}_{l'm'}. \tag{9}$$

In order to find $a^{\tau'\tau}_{l'l,\, m'm}$ we use the fact that $L_1 = [A_2, L_3]$ and in place of A_2, L_3 substitute the known matrices. We then obtain

$$L_1 \xi^{\tau}_{lm} = A_2 L_3 \xi^{\tau}_{lm} - L_3 A_2 \xi^{\tau}_{lm} =$$
$$= A_2 \sum_{l',\, m',\, \tau'} c^{\tau'\tau}_{l'l,\, m'm}\, \xi^{\tau'}_{l'm'} - \frac{1}{2} L_3 (\alpha^l_m\, \xi^{\tau}_{l,\, m-1} - \alpha^l_{m+1}\, \xi^{\tau}_{lm+1}) =$$
$$= \frac{1}{2} \sum_{l,\, m',\, \tau'} c^{\tau'\tau}_{l'l,\, m'm} (\alpha^{l'}_{m'}\, \xi^{\tau'}_{l',\, m'-1} - \alpha^{l'}_{m'+1}\, \xi^{\tau'}_{l',\, m'+1}) -$$
$$- \frac{1}{2} \alpha^l_m \sum_{l',\, m'\, \tau} c^{\tau'\tau}_{l'l,\, m,\, m-1} \xi^{\tau'}_{l'm'} + \frac{1}{2} \alpha^l_{m+1} \sum_{l',\, m',\, \tau'} c^{\tau'\tau}_{l'l,\, m'+1}\, \xi^{\tau'}_{l'm'}.$$

Dividing the first summation into two and altering the corresponding index of summation in each of the resulting sums, we may write our result in the following manner:

$$L_1\xi^{\tau}_{lm}=\frac{1}{2}\sum_{l',m',\tau'}(\alpha^{l'}_{m'+1}c^{\tau'\tau}_{l'l,m'+1,m}-\alpha^{l'}_{m'}c^{\tau'\tau}_{l'l,m'-1,m}-\\-\alpha^{l}_{m}c^{\tau'\tau}_{l'l,m',m-1}+\alpha^{l}_{m+1}c^{\tau'\tau}_{l'l,m',m+1})\xi^{\tau'}_{l'm'}.$$

Consequently, the elements of the matrix L_1 have the form:

$$a^{\tau'\tau}_{l'l,m'm}=\frac{1}{2}(\alpha^{l'}_{m'+1}c^{\tau'\tau}_{l'l,m'+1,m}-\alpha^{l'}_{m'}c^{\tau'\tau}_{l'l,m'-1,m}-\\-\alpha^{l}_{m}c^{\tau'\tau}_{l'l,m',m-1}+\alpha^{l}_{m+1}c^{\tau'\tau}_{l'l,m',m+1}). \tag{10}$$

Since $c^{\tau'\tau}_{l'l,m'm}\neq 0$ only if $m'=m$ and $l'=l-1,\ l,\ l+1$, then for fixed $m,\ l,\ \tau,\ \tau'$ there prove to be six numbers $a^{\tau'\tau}_{l'l,m'm}$ different from zero, namely $a^{\tau'\tau}_{l-1,m-1,m}$, $a^{\tau'\tau}_{ll,m-1,m}$, $a^{\tau'\tau}_{l+1,l,m-1,m}$, $a^{\tau'\tau}_{l-1,l,m+1,m}$, $a^{\tau'\tau}_{ll,m+1,m}$, $a^{\tau'\tau}_{l+1,l,m+1,m}$.

Putting $\alpha^{l}_{m}=\sqrt{(l+m)(l-m+1)}$ and $c^{\tau'\tau}_{l'lm}$ from formulae (8) in formula (10) we find the following values of these coefficients

$$\left.\begin{aligned}
a^{\tau'\tau}_{l-1,l,m-1,m}&=-\frac{c_{l-1,l}}{2}\sqrt{(l+m)(l+m-1)},\\
a^{\tau'\tau}_{l,l,m-1,m}&=\frac{c_{ll}}{2}\sqrt{(l+m)(l-m+1)},\\
a^{\tau'\tau}_{l+1,l,m-1,m}&=\frac{c_{l+1,l}}{2}\sqrt{(l-m+1)(l-m+2)},\\
a^{\tau'\tau}_{l-1,l,m+1,m}&=\frac{c_{l-1,l}}{2}\sqrt{(l-m)(l-m-1)},\\
a^{\tau'\tau}_{l,l,m+1,m}&=\frac{c_{ll}}{2}\sqrt{(l+m+1)(l-m)},\\
a^{\tau'\tau}_{l+1,l,m+1,m}&=-\frac{c_{l+1,l}}{2}\sqrt{(l+m+1)(l+m+2)}.
\end{aligned}\right\} \tag{11}$$

The constants $c_{l-1,l}$, c_{ll}, $c_{l+1,l}$ are now those appearing in formula (8). Similarly it is possible to find the matrix $L_2=-[A_1,\ L_3]$.

Letting $L_2\xi^{\tau}_{lm}=\sum_{l',m',\tau'}b^{\tau'\tau}_{l'l,m'm}\xi^{\tau'}_{l'm'}$, we find that

$$\begin{aligned}
b^{\tau'\tau}_{l-1,l,m-1,m}&=-\frac{ic_{l-1,l}}{2}\sqrt{(l+m)(l+m-1)},\\
b^{\tau'\tau}_{l,l,m-1,m}&=\frac{ic_{ll}}{2}\sqrt{(l+m)(l-m+1)},\\
b^{\tau'\tau}_{l+1,l,m-1,m}&=\frac{ic_{l+1,l}}{2}\sqrt{(l-m+1)(l-m+2)},
\end{aligned}$$

$$\left.\begin{aligned} b^{\tau'\tau}_{l-1,l,m+1,m} &= -\frac{ic_{l-1,l}}{2}\sqrt{(l-m)(l-m-1)},\\ b^{\tau'\tau}_{l,l,m+1,m} &= -\frac{ic_{ll}}{2}\sqrt{(l+m+1)(l-m)},\\ b^{\tau'\tau}_{l+1,l,m+1,m} &= \frac{ic_{l+1,l}}{2}\sqrt{(l+m+1)(l+m+2)}, \end{aligned}\right\} \tag{12}$$

and all the remaining $b^{\tau'\tau}_{l'l,m'm} = 0$.

We have found in this manner the possible form of the matrices L_1, L_2, L_3 for systems of first order invariant equations. The constants $c^{\tau'\tau}_{l-1,l}$, $c^{\tau'\tau}_{ll}$ and $c^{\tau'\tau}_{l+1,l}$, may be selected arbitrarily.

By giving these constants specific values we obtain various invariant systems of equations.

§ 4. Solution of Invariant Equations

We show now that the solution of an invariant system of equations is conveniently sought in the form of series in the generalized spherical functions examined in the preceding section. Using this analysis we reduce the solution of an arbitrary invariant system of the form (1) to the solution of a system of ordinary differential equations, in similar fashion to that adopted in Sect. 8 for Maxwell's equations.

To this end, we transform equation (1) in the following manner. Firstly, we change over into spherical coordinates, allowing at each point that

$$\frac{\partial}{\partial x_1} = -\frac{\sin\varphi}{r\sin\vartheta}\frac{\partial}{\partial\varphi} + \frac{\cos\varphi\cos\vartheta}{r}\frac{\partial}{\partial\vartheta} + \cos\varphi\sin\vartheta\frac{\partial}{\partial r},$$

$$\frac{\partial}{\partial x_2} = \frac{\cos\varphi}{r\sin\vartheta}\frac{\partial}{\partial\varphi} + \frac{\sin\varphi\cos\vartheta}{r}\frac{\partial}{\partial\vartheta} + \sin\varphi\sin\vartheta\frac{\partial}{\partial r},$$

$$\frac{\partial}{\partial x_3} = -\frac{\sin\vartheta}{r}\frac{\partial}{\partial\vartheta} + \cos\vartheta\frac{\partial}{\partial r}.$$

It follows from the results of Sect.8 and can also be verified directly, that this change of independent variables has the same effect on system (1) as the following transformation. We subject space to the rotation $g = g\left(\frac{\pi}{2} - \varphi, \vartheta, \pi\right)$ (which is the inverse of $g\left(0, \vartheta, \frac{\pi}{2} + \varphi\right)$) at the point with spherical coordinates (φ, ϑ), and we set $\frac{\partial}{\partial x_1'} = \frac{1}{r\sin\vartheta}\frac{\partial}{\partial\varphi}$, $\frac{\partial}{\partial x_2'} = -\frac{1}{r}\frac{\partial}{\partial\vartheta}$, $\frac{\partial}{\partial x_3'} = \frac{\partial}{\partial r}$. Simultaneously we subject the unknown functions $\psi_1, \psi_2, \ldots, \ldots, \psi_N$ to the corresponding transformation $\psi' = T_g\psi$ where $T_g = T\left(\frac{\pi}{2} - \varphi, \vartheta, \pi\right)$. If we substitute in system (1) $\psi = T_{g^{-1}}\psi'$ and further, apply to the result the transformation T_g, then we obtain the system

$$T_g L_1\left[-\frac{\sin\varphi}{r\sin\vartheta}\frac{\partial\left(T_g^{-1}\psi'\right)}{\partial\varphi}+\frac{\cos\varphi\cos\vartheta}{r}\frac{\partial\left(T_g^{-1}\psi'\right)}{\partial\vartheta}+\cos\varphi\sin\vartheta\frac{\partial\left(T_g^{-1}\psi'\right)}{\partial r}\right]+$$
$$+T_g L_2\left[\frac{\cos\varphi}{r\sin\vartheta}\frac{\partial\left(T_g^{-1}\psi'\right)}{\partial\varphi}+\frac{\sin\varphi\cos\vartheta}{r}\frac{\partial\left(T_g^{-1}\psi'\right)}{\partial\vartheta}+\sin\varphi\sin\vartheta\frac{\partial\left(T_g^{-1}\psi'\right)}{\partial r}\right]+$$
$$+T_g L_3\left[-\frac{\sin\vartheta}{r}\frac{\partial\left(T_g^{-1}\psi'\right)}{\partial\vartheta}+\cos\vartheta\frac{\partial\left(T_g^{-1}\psi'\right)}{\partial r}\right]+\varkappa\psi'=0.$$

Because of the invariance of the system we obtain the equations (see formula (3) of this section):

$$T_g[-L_1\sin\varphi+L_2\cos\varphi]\,T_g^{-1}=L_1,$$
$$T_g[L_1\cos\varphi\cos\vartheta+L_2\sin\varphi\cos\vartheta-L_3\sin\vartheta]\,T_g^{-1}=L_2,$$
$$T_g[L_1\cos\varphi\sin\vartheta+L_2\sin\varphi\sin\vartheta+L_3\cos\vartheta]\,T_g^{-1}=L_3,$$

with the aid of which the system may be written in the form *

$$\frac{1}{r\sin\vartheta}L_1T_g\frac{\partial\left(T_g^{-1}\psi'\right)}{\partial\varphi}-\frac{1}{r}L_2T_g\frac{\partial\left(T_g^{-1}\psi'\right)}{\partial\vartheta}+L_3T_g\frac{\partial\left(T_g^{-1}\psi'\right)}{\partial r}+\varkappa\psi'=0. \tag{13}$$

The matrix T_g^{-1} depends on φ and ϑ. Therefore upon differentiation of $T_g^{-1}\psi'$ with respect to these variables we must differentiate both factors, after which the equation assumes the form

$$\frac{1}{r\sin\vartheta}L_1\frac{\partial\psi'}{\partial\varphi}-\frac{1}{r}\frac{\partial\psi'}{\partial\vartheta}+L_3\frac{\partial\psi'}{\partial\vartheta}+$$
$$+\left[\frac{1}{r\sin\vartheta}L_1T_g\frac{\partial T_g^{-1}}{\partial\varphi}-\frac{1}{r}L_2T_g\frac{\partial T_g^{-1}}{\partial\vartheta}+\varkappa E\right]\psi'=0. \tag{14}$$

The products $T_g\frac{\partial T_g^{-1}}{\partial\varphi}$ and $T_g\frac{\partial T_g^{-1}}{\partial\vartheta}$ entering into the coefficients for ψ', are in essence none other than a linear combination of the matrices A_k representing an infinitesimal (rotational) displacement about the coordinate axes. In fact, taking $g=g\left(\frac{\pi}{2}-\varphi,\ \vartheta,\ \pi\right)$ and $g^{-1}=g\left(0,\ \vartheta,\ \varphi+\frac{\pi}{2}\right)$ and writing out in parameters, we have, for example: for $T_g\frac{\partial T_g^{-1}}{\partial\varphi}$,

$$T_g\frac{\partial T_g^{-1}}{\partial\varphi}=\lim_{\Delta\varphi\to 0}\frac{1}{\Delta\varphi}T\left(\frac{\pi}{2}-\varphi,\ \vartheta,\ \pi\right)\left[T\left(0,\ \vartheta,\ \varphi+\Delta\varphi+\frac{\pi}{2}\right)-\right.$$

---

* In order to write the system in this manner, it is necessary to insert the product $T_g^{-1}T_g$ to the right of each matrix L_k.

$$-T\left(0,\,\vartheta,\,\varphi+\frac{\pi}{2}\right)\Big]=\lim_{\Delta\varphi\to 0}\frac{1}{\Delta\varphi}\left[T\left(\frac{\pi}{2}-\varphi,\,\vartheta,\,\pi\right)T\left(0,\,\vartheta,\,\varphi+\Delta\varphi+\frac{\pi}{2}\right)-E\right].$$

The first term in brackets is a matrix, which as $\Delta\varphi\to 0$ clearly tends to E. Therefore, the entire square bracket may be considered as a linear combination of the matrices A_k, multiplied by $\Delta\varphi+$ terms of higher order (see Sect. 2). The limit of the whole expression, i.e. $T_g\frac{\partial T_g^{-1}}{\partial\varphi}$ is simply a linear combination of A_1, A_2, A_3. In this case the coefficients in this linear combination depend upon the rotation $g\left(\frac{\pi}{2}-\varphi,\,\vartheta,\,\pi\right)g\left(0,\,\vartheta,\,\varphi+\Delta\varphi+\frac{\pi}{2}\right)$. Thus they will be the same for the product $T_g\frac{\partial T_g^{-1}}{\partial\varphi}$ and for the product $g\frac{\partial g^{-1}}{\partial\varphi}$. We compute directly, the following products

$$g=\begin{Vmatrix}-\sin\varphi & \cos\varphi & 0\\ -\cos\varphi\cos\vartheta & -\sin\varphi\cos\vartheta & \sin\vartheta\\ \cos\varphi\sin\vartheta & \sin\varphi\sin\vartheta & \cos\vartheta\end{Vmatrix},$$

$$g^{-1}=\begin{Vmatrix}-\sin\varphi & -\cos\varphi\cos\vartheta & \cos\varphi\sin\vartheta\\ \cos\varphi & -\sin\varphi\cos\vartheta & \sin\varphi\sin\vartheta\\ 0 & \sin\vartheta & \cos\vartheta\end{Vmatrix},$$

$$g\frac{\partial g^{-1}}{\partial\varphi}=\begin{Vmatrix}0 & -\cos\vartheta & \sin\vartheta\\ \cos\vartheta & 0 & 0\\ -\sin\vartheta & 0 & 0\end{Vmatrix}=a_2\sin\vartheta+a_3\cos\vartheta,$$

$$g\frac{\partial g^{-1}}{\partial\vartheta}=\begin{Vmatrix}0 & 0 & 0\\ 0 & 0 & -1\\ 0 & 1 & 0\end{Vmatrix}=a_1,$$

where a_1, a_2, a_3 are the matrices of an infinitesimal rotation in the fundamental (three-dimensional) representation. Therefore, for any representation $g\to T_g$ the following holds:

$$T_g\frac{\partial T_g^{-1}}{\partial\varphi}=A_2\sin\vartheta+A_3\cos\vartheta,\quad T_g\frac{\partial T_g^{-1}}{\partial\vartheta}=A_1.$$

Substituting these products in system (14) brings it to the final form:

$$\frac{1}{r\sin\vartheta}L_1\frac{\partial\psi}{\partial\varphi}-\frac{1}{r}L_2\frac{\partial\psi}{\partial\vartheta}+L_3\frac{\partial\psi}{\partial r}+$$
$$+\frac{1}{r}(L_1A_2-L_2A_1+\operatorname{ctg}\vartheta L_1A_3)\psi+\varkappa\psi=0. \tag{15}$$

We have performed on the quantity ψ at each point the same transformation which preceded (in Sect.8) the expansion of this quantity in generalized spherical functions. Recall that this expansion is invariant under rotations. Now expand each component ψ^{τ}_{lm} of the quantity ψ in a series of generalized spherical functions. Because of the invariance of the system, we may assume that each term of the series will satisfy the system. And this, as we shall convince ourselves below, leads to a separation of the variables, i.e., reduces our system to a system of ordinary equations. In order to perform the corresponding calculation we shall write system (15) in components. To begin with we shall calculate the matrix $D = L_1A_2 - L_2A_1$.

Assuming

$$D\xi^{\tau}_{l,m} = \sum_{l',m',\tau'} d^{\tau'\tau}_{l'l,m'm}\xi^{\tau'}_{l',m'}$$

and using formulae (8), (11) and (12) and also the expressions for A_1 and A_2 from Sect.2, we find that

$$\left.\begin{aligned} d^{\tau'\tau}_{l-1,l,mm} &= c^{\tau'\tau}_{l-1,l}(l-1)\sqrt{l^2-m^2},\\ d^{\tau'\tau}_{ll,mm} &= c^{\tau'\tau}_{ll}m,\\ d^{\tau'\tau}_{l+1,l,mm} &= -c^{\tau'\tau}_{l+1,l}\,l\sqrt{(l+1)^2-m^2} \end{aligned}\right\} \tag{16}$$

and the remaining $d^{\tau'\tau}_{l'l,m'm}$ are equal to zero. System (15) in components ψ^{τ}_{lm} has the form

$$\frac{1}{r\sin\varphi}\sum_{l',m',\tau'} a^{\tau\tau'}_{ll',mm'}\frac{\partial\psi^{\tau'}_{l'm'}}{\partial\varphi} - \frac{1}{r}\sum_{l',m',\tau'} b^{\tau\tau'}_{ll',mm'}\frac{\partial\psi^{\tau'}_{l'm'}}{\partial\vartheta} +$$

$$+\sum_{l',m',\tau'} c^{\tau\tau'}_{ll',mm'}\frac{\partial\psi^{\tau'}_{l'm'}}{\partial r} + \frac{1}{r}\sum_{l',m',\tau'} d^{\tau\tau'}_{ll',mm'}\psi^{\tau'}_{l'm'} -$$

$$-\frac{i}{r}\cot\vartheta\sum_{l',m',\tau'} m' a^{\tau\tau'}_{ll',mm'}\psi^{\tau'}_{l'm'} + \varkappa\psi^{\tau}_{lm} = 0,^{*} \tag{17}$$

where the coefficients $a^{\tau\tau'}_{ll',mm'}$, $b^{\tau\tau'}_{ll',mm'}$, $c^{\tau\tau'}_{ll',mm'}$, $d^{\tau\tau'}_{ll',mm'}$ are defined by formulae (8), (11), (12) and (16) of this Section.

Here in the 1st, 2nd and 5th summations they differ from zero in six components and in the 3rd and 4th, in three. In order to separate the variables, let us now write

$$\psi^{\tau}_{lm} = f^{l_0}_{lm\tau}(r)\,T^{l_0}_{mn}\left(\frac{\pi}{2}-\varphi,\ \vartheta,\ 0\right),$$

where $l_0 \geqslant l$ and $-l_0 \leqslant m$, $n \leqslant l_0$, and substitute these products in system (17). Now it happens that $T^{l_0}_{m-1,n}$ and $T^{l_0}_{m+1,n}$, which enter into the 1st,

* It should be noted that since in the action of the matrices L_k on the vectors ξ^{τ}_{lm} the summations are carried out over the first index of each pair, therefore, upon transformation of the functions ψ^{τ}_{lm} we must sum over the second index.

2nd and 5th summations may be expressed in terms of $T^{l_0}_{mn}$ by means of a recurrence formula; the effect of this is that each equation contracts to $T^{l_0}_{mn}$ and in it there remain only functions of r. At the same time we introduce the substituted values of the coefficients $a^{\tau\tau'}_{ll', mm'}$, $b^{\tau\tau'}_{ll', mm'}$, $c^{\tau\tau'}_{ll', mm'}$, $d^{\tau\tau'}_{ll', mm'}$ and collect terms with one and the same function of r. We then obtain the system

$$\sum_{\tau'} c^{\tau\tau'}_{l, l-1}\left\{\left[\sqrt{(l^2-m^2)}\frac{df^{l_0}_{l-1, m, \tau'}}{dr}-\right.\right.$$

$$\left.-\frac{(l-1)\sqrt{(l^2-m^2)}}{r}f^{l_0}_{l-1, m, \tau'}\right]T^{l_0}_{m, n}\left(\frac{\pi}{2}-\varphi, \vartheta, 0\right)-$$

$$-\frac{i}{2r}\sqrt{(l+m-1)(l+m)}f^{l_0}_{l-1, m-1, \tau'}\left[\frac{1}{\sin\vartheta}\frac{\partial T^{l_0}_{m-1, n}}{\partial\varphi}+\right.$$

$$\left.+\frac{\partial T^{l_0}_{m-1, n}}{\partial\vartheta}-\frac{(m-1)\cos\vartheta}{\sin\vartheta}T^{l_0}_{m-1, n}\right]-$$

$$-\frac{i}{2r}\sqrt{(l-m-1)(l-m)}f^{l_0}_{l-1, m+1, \tau'}\left[-\frac{1}{\sin\vartheta}\frac{\partial T^{l_0}_{m+1, n}}{\partial\varphi}+\right.$$

$$\left.\left.+\frac{\partial T^{l_0}_{m+1, n}}{\partial\vartheta}+\frac{(m+1)\cos\vartheta}{\sin\vartheta}T^{l_0}_{m+1, n}\right]\right\}+$$

$$+c^{\tau\tau'}_{ll}\left\{\left[m\frac{df^{l_0}_{lm\tau'}}{dr}+\frac{m}{r}f^{l_0}_{lm\tau}\right]T^{l_0}_{mn}\left(\frac{\pi}{2}-\varphi, \vartheta, 0\right)+\right.$$

$$+\frac{i}{2r}\sqrt{(l+m)(l-m+1)}f^{l_0}_{l, m-1, \tau'}\times$$

$$\times\left[\frac{1}{\sin\vartheta}\frac{\partial T^{l_0}_{m-1, n}}{\partial\varphi}+\frac{\partial T^{l_0}_{m-1, n}}{\partial\vartheta}-\frac{(m-1)\cos\vartheta}{\sin\vartheta}T^{l_0}_{m-1, n}\right]-$$

$$-\frac{i}{2r}\sqrt{(l+m+1)(l-m)}f^{l_0}_{l, m+1, \tau'}\times$$

$$\left.\times\left[-\frac{1}{\sin\vartheta}\frac{\partial T_{m-1, n}}{\partial\varphi}+\frac{\partial T_{m-1, n}}{\partial\vartheta}+\frac{(m+1)\cos\vartheta}{\sin\vartheta}T^{l_0}_{m+1, n}\right]\right\}+$$

$$+c^{\tau\tau'}_{l, l+1}\left\{\left[\sqrt{(l+1)^2-m^2}\frac{df^{l_0}_{l+1, m, \tau'}}{dr}+\right.\right.$$

$$\left.+\frac{l\sqrt{(l+1)^2-m^2}}{r}f^{l_0}_{l+1, m, \tau'}\right]T^{l_0}_{m, n}\left(\frac{\pi}{2}-\varphi, \vartheta, 0\right)+$$

$$+\frac{i}{2r}\sqrt{(l-m+2)(l-m+1)}f^{l_0}_{l+1,m-1,\tau}\times$$

$$\times\left[\frac{1}{\sin\vartheta}\frac{\partial T_{m-1,n}}{\partial\varphi}+\frac{\partial T_{m-1,n}}{\partial\vartheta}-\frac{(m-1)\cos\vartheta}{\sin\vartheta}T^{l_0}_{m-1,n}\right]+$$

$$+\frac{i}{2r}\sqrt{(l+m+2)(l+m+1)}f^{l_0}_{l+1,m+1,\tau'}\times$$

$$\times\left[-\frac{1}{\sin\vartheta}\frac{\partial T_{m+1,n}}{\partial\varphi}+\frac{\partial T^{l_0}_{m+1,n}}{\partial\vartheta}+\frac{(m+1)\cos\vartheta}{\sin\vartheta}T^{l_0}_{m+1,n}\right]\Bigg\}+$$

$$+i\varkappa f^{l_0}_{lm\tau}(r)\,T^{l_0}_{mn}\left(\frac{\pi}{2}-\varphi,\ \vartheta,\ 0\right)=0.$$

There appear in each equation of the system obtained three generalized spherical functions $T^{l_0}_{mn}$, $T^{l_0}_{m-1,n}$, $T^{l_0}_{m+1,n}$. Moreover, it is possible to apply recurrence formulae (see appendix to Sect.7) to the square brackets by means of which these brackets may also be expressed in terms of $T^{l_0}_{mn}$. At the same time, recalling that $T^{l_0}_{m-1,n}\left(\frac{\pi}{2}-\varphi,\ \vartheta,\ 0\right)=e^{-in\left(\frac{\pi}{2}-\varphi\right)}u^{l_0}_{mn}(\vartheta)$ and that, consequently, $\frac{\partial T_{m\pm1,n}}{\partial\varphi}=inT_{m\pm1,n}$, we may rewrite the corresponding square brackets in the form:

$$e^{in\left(\frac{\pi}{2}-\varphi\right)}\left(\frac{du_{m-1,n}}{d\vartheta}+\frac{n-(m-1)\cos\vartheta}{\sin\vartheta}u_{m-1,n}\right)$$

and

$$e^{in\left(\frac{\pi}{2}-\varphi\right)}\left(\frac{du_{m+1,n}}{d\vartheta}+\frac{n-(m+1)\cos\vartheta}{\sin\vartheta}u_{m+1,n}\right).$$

But according to formula (2) of the appendix to Sect.7 the first of these brackets should be equal to $-i\sqrt{(l_0+m)(l_0-m+1)}\,u^{l_0}_{mn}$ and the second bracket correspondingly equal to $-i\sqrt{(l_0+m+1)(l_0-m)}\,u^{l_0}_{mn}$

Having effected the substitution and denoting $e^{in\left(\frac{\pi}{2}-\varphi\right)}u^{l_0}_{mn}(\theta)$ by $T^{l_0}_{mn}$ anew, we can reduce all equations to $T^{l_0}_{mn}\left(\frac{\pi}{2}-\varphi,\ \vartheta,\ 0\right)$. As a consequence we obtain a system which contains only $f^{l_0}_{lm\tau}(r)$, this being

$$\sum_{\tau'}c^{\tau\tau'}_{l,l-1}\left[\sqrt{(l^2-m^2)}\frac{df^{l_0}_{l-1,m,\tau'}}{dr}-\frac{(l-1)\sqrt{(l^2-m^2)}}{r}f^{l_0}_{l-1,m,\tau'}-\right.$$

$$-\frac{1}{2r}\sqrt{(l+m-1)(l+m)}\sqrt{(l_0+m)(l_0-m+1)}f^{l_0}_{l-1,m-1,\tau'}-$$

$$\left.-\frac{1}{2r}\sqrt{(l-m-1)(l-m)}\sqrt{(l_0+m+1)(l_0-m)}f^{l_0}_{l-1,m+1,\tau'}\right]+$$

$$+c_{ll}^{\tau\tau'}\left[m\frac{df_{lm\tau'}^{l_0}}{dr}+mf_{lm\tau'}^{l_0}+\right.$$

$$+\frac{1}{2r}\sqrt{(l+m)(l-m+1)}\sqrt{(l_0+m)(l_0-m+1)}f_{l,m-1,\tau'}^{l_0}(r)-$$

$$\left.-\frac{1}{2r}\sqrt{(l+m+1)(l-m)}\sqrt{(l_0+m+1)(l_0-m)}f_{l,m+1,\tau'}^{l_0}(r)\right]+$$

$$+c_{l,l+1}^{\tau\tau'}\left[\sqrt{(l+1)^2-m^2}\frac{df_{l+1,m,\tau'}^{l_0}}{dr}+\frac{l\sqrt{(l+1)^2-m^2}}{r}f_{l+1,m,\tau'}^{l_0}(r)+\right.$$

$$+\frac{1}{2r}\sqrt{(l-m+2)(l-m+1)}\sqrt{(l_0+m)(l_0-m+1)}f_{l+1,m-1,\tau'}^{l_0}(r)+$$

$$\left.+\frac{1}{2r}\sqrt{(l+m+2)(l+m+1)}\sqrt{(l_0+m+1)(l_0-m)}f_{l+1,m+1,\tau}^{l_0}\right]+$$

$$+i\varkappa f_{lm\tau}^{l_0}(r)=0. \qquad (18)$$

§ 5. Solution of the Dirac Equations

A particular case of system (18) is the Dirac system of equations, which, after separation of the time dependent factor, may be written in the form:

$$\left(\frac{\partial}{\partial x}+i\frac{\partial}{\partial y}\right)u_4+\frac{\partial}{\partial z}u_3-\frac{i}{hc}[E-mc^2+V(r)]u_1=0,$$

$$\left(\frac{\partial}{\partial x}-i\frac{\partial}{\partial y}\right)u_3-\frac{\partial}{\partial z}u_4-\frac{i}{hc}[E-mc^2+V(r)]u_2=0,$$

$$\left(\frac{\partial}{\partial x}+i\frac{\partial}{\partial y}\right)u_2+\frac{\partial}{\partial z}u_1-\frac{i}{hc}[E+mc^2+V(r)]u_3=0,$$

$$\left(\frac{\partial}{\partial x}-i\frac{\partial}{\partial y}\right)u_1-\frac{\partial}{\partial z}u_2-\frac{i}{hc}[E+mc^2+V(r)]u_4=0.$$

Upon rotation the pair of functions $(u_1,\ u_2)$, and equally the pair $(u_3,\ u_4)$, transform according to the rotation group representation with weight $l=\frac{1}{2}$. An arbitrary invariant system concerning such unknown functions depends upon the selection of four constants $\overset{11}{c}_{1/2,\,1/2}$, $\overset{12}{c}_{1/2,\,1/2}$, $\overset{21}{c}_{1/2,\,1/2}$, $\overset{22}{c}_{1/2,\,1/2}$. The Dirac equations are obtained when $\overset{11}{c}_{1/2,\,1/2}=\overset{22}{c}_{1/2,\,1/2}=0$; $\overset{12}{c}_{1/2,\,1/2}=\frac{2hc}{E-mc^2+V(r)}$; $\overset{21}{c}_{1/2,\,1/2}=\frac{2hc}{E+mc^2+V(r)}$, $\varkappa=-1$.

The transformation carried out in paragraph 4 above, of a general invariant system gives, in our case, the following system of ordinary differential equations

$$\left.\begin{aligned}\frac{df_3^l}{dr}+\frac{1}{r}f_3^l(r)-\frac{i\left(l+\frac{1}{2}\right)}{r}f_4^l(r)-k_1f_1^l(r)=0,\\ -\frac{df_4^l}{dr}-\frac{1}{r}f_4^l(r)-\frac{i\left(l+\frac{1}{2}\right)}{r}f_3^l(r)-k_1f_2^l(r)=0,\end{aligned}\right\}$$

$$\left.\begin{aligned} \frac{df_1}{dr}+\frac{1}{r}f_1^l(r)-\frac{i\left(l+\frac{1}{2}\right)}{r}f_2^l(r)-k_2f_3^l(r)=0, \\ -\frac{df_2^l}{dr}-\frac{1}{r}f_2^l(r)-\frac{i\left(l+\frac{1}{2}\right)}{r}f_1^l(r)-k_2f_4^l(r)=0, \end{aligned}\right\} \qquad (19)$$

where

$$k_1=\frac{i}{hc}\,[E-mc^2+V(r)],\quad k_2=\frac{i}{hc}\,[E+mc^2+V(r)].$$

This system in four unknown functions can be reduced to a system of two second order equations. To see this, note that if it is assumed that $f_2^l(r)=$ $=\pm if_1^l(r)$ and $f_4^l(r)=\pm if_3^l(r)$ then the second equation is equivalent to the first, and the third to the fourth, and the system assumes the form

$$\left.\begin{aligned} \frac{df_3^l}{dr}+\frac{l+\frac{3}{2}}{r}f_3^l(r)-k_1f_1^l(r)=0, \\ \frac{df_1^l}{dr}-\frac{l-\frac{1}{2}}{r}f_1(r)-k_2f_3^l(r)=0. \end{aligned}\right\} \qquad (20)$$

From each solution $\left(f_1^l,\ f_3^l\right)$ of the system (20) it is possible to construct two linearly independent solutions $\left(f_1^l,\ -if_1^l,\ f_3^l,\ if_3^l\right)$ and $\left(f_1^l,\ if_1^l,\ f_3^l,\ -if_3^l\right)$ of the initial system (19).

In particular, if $V(r)\equiv 0$ then the solutions of system (20) are expressed in terms of cylindrical functions of half-integral order

$$f_1^l(r)=\frac{c_1}{\sqrt{r}}J_l\left(\sqrt{-k_1k_2}\cdot r\right)+\frac{c_2}{\sqrt{r}}J_{-l}\left(\sqrt{-k_1k_2}\cdot r\right),$$

$$f_3^l(r)=-\frac{c_1}{k_2\sqrt{r}}J_{l+1}\left(\sqrt{-k_1k_2}\cdot r\right)+\frac{c_2}{k_2\sqrt{r}}J_{-l-1}\left(\sqrt{-k_1k_2}\cdot r\right).$$

Consequently, the solutions that we have found for the Dirac equations have the form:

$$u_1(r,\varphi,\vartheta)=f_1^l(r)\,T^l_{\frac{1}{2},\,n}\left(\frac{\pi}{2}-\varphi,\vartheta,0\right),$$

$$u_2(r,\varphi,\vartheta)=\mp if_1^l(r)\,T^l_{-\frac{1}{2},\,n}\left(\frac{\pi}{2}-\varphi,\vartheta,0\right),$$

$$u_3(r,\varphi,\vartheta)=f_3^l(r)\,T^l_{\frac{1}{2},\,n}\left(\frac{\pi}{2}-\varphi,\vartheta,0\right),$$

$$u_4(r, \varphi, \vartheta) = \pm i f_3^l(r)\, T^l_{-\frac{1}{2}, n}\left(\frac{\pi}{2} - \varphi, \vartheta, 0\right)$$
$$\left(l = \frac{1}{2}, \frac{3}{2}, \frac{5}{2}, \ldots; \quad n = -l, -l+1, \ldots, l\right),$$

where

$$T^l_{\pm\frac{1}{2}, n}(\varphi_1, \vartheta, \varphi_2) = e^{\mp i \frac{\varphi_1}{2}} \cdot P^l_{\pm\frac{1}{2}, n}(\cos\vartheta) \cdot e^{-in\varphi_2},$$

$$P^l_{\pm\frac{1}{2}, n}(\mu) =$$
$$= A(1-\mu)^{-\frac{2n \mp 1}{4}}(1+\mu)^{-\frac{2n \pm 1}{4}} \frac{d^{l-n}}{d\mu^{l-n}}\left[(1-\mu)^{l \mp \frac{1}{2}}(1+\mu)^{l \pm \frac{1}{2}}\right],$$

$$A = \frac{(-1)^{l \mp \frac{1}{2}} i^{n \pm \frac{1}{2}}}{2^l \left(l \mp \frac{1}{2}\right)} \sqrt{\frac{\left(l \mp \frac{1}{2}\right)!\,(l+n)!}{\left(l \pm \frac{1}{2}\right)!\,(l-n)!}}.$$

The general solution may be analysed into a series of such particular solutions.

§ 6. The Matrices L_1, L_2, L_3 for the Case $\varkappa \neq 0$ (Further Conclusions

In this paragraph we shall show that the matrices L_1, L_2, L_3 in an invariant equation may be evaluated with the aid of the analysis of the product of two representations into irreducible components. We note, first of all, that relation (3) for the matrices L_i may, clearly, be written in the form

$$T_g L_i T_g^{-1} = \sum_{k=1,2,3} g_{ki} L_k, \tag{21}$$

where the T_g are the matrices of the representation $g \to T_g$ acting in the space R (the value $\psi(x)$ of the function belongs to the space R), and L_1, L_2, L_3 are the matrices in this space.

We shall consider all matrices acting on the space R. They also form a linear space. We denote this space by the letter S (the dimension of S is equal to n^2 if the dimension of R is equal to n.)

In the space S, we have a representation $g \to \tau_g$, acting according to the formula

$$\tau_g L = T_g L T_g^{-1} \tag{22}$$

(The reader can easily satisfy himself that this formula does in fact give a representation of the rotation group in the space S).

The representation is, generally speaking, reducible. Let us suppose that among its irreducible constituents there is a representation of weight $l = 1$. The representation with $l = 1$ is equivalent to the identical representation which acts in three-dimensional space. Therefore, if there occurs among the irreducible representations (22) of the subspaces in S, a space with $l = 1$, then we may choose in it the basis L_1, L_2, L_3 in which representation (22) acts according to the formula

$$\tau_g L_i = \sum_{k=1,2,3} g_{ki} L_k,$$

or

$$T_g L_i T_g^{-1} = \sum_{k=1,2,3} g_{ki} L_k. \tag{21'}$$

Thus, we see that in the three-dimensional subspace $R^{(3)}$ (irreducible with respect to the representation $g \to \tau_g$) of the matrix space S, there occurs a triplet of matrices L_1, L_2, L_3 (some basis in $R^{(3)}$) which satisfies relation (3) or (what amounts to the same thing), the relation for the matrices in the invariant equations. Conversely, a triplet of matrices L_1, L_2, L_3 satisfying relation (21) generates in S a three-dimensional subspace, invariant and irreducible with respect to the representation $g \to \tau_g$.

In this manner the problem of searching for the matrices L_1, L_2, L_3 has been reduced to the separation in the space of all matrices (the space S) of the irreducible subspaces with weight $l = 1$ with respect to the representation $g \to \tau_g$, and the construction of the canonical bases L_0, L_+, L_- in these subspaces.

We recall that the canonical basis L_0, L_+, L_- in a three-dimensional irreducible subspace is connected with the basis L_1, L_2, L_3 by the relations

$$L_0 = L_3,$$
$$L_+ = L_1 + iL_2,$$
$$L_- = L_1 - iL_2.$$

We shall now show that the problem posed is related to the problem of the resolution of the product of two representations into irreducible components. We recall (see the note on p. 49) that if $g \to T_g^{(p)}$ and $g \to T_g^{(q)}$ are two representations acting in the spaces $R^{(p)}$ and $R^{(q)}$ (p and q are the dimensions of the spaces), then the product of these representations $T_g^{(p)} \times T_g^{(q)}$, acting in the space $R^{(p)} \times R^{(q)}$ is equivalent to the representation acting in the space of rectangular matrices A with dimensions $p \times q$ (p is the number of rows and q the number of columns) according to the formula

$$\tau_g A = U_g A V_g^{\mathrm{T}},$$

where U_g is the matrix of the representation $g \to T_g^{(p)}$ expressed in some basis of the space R^p, V_g is the matrix of the representation $g \to T_g^{(q)}$, expressed in some basis of the space $R^{(q)}$, and V_g^{T} denotes the transpose of the matrix V_g.

Now let the representation $g \to T_g$ of the rotation group act in the space R.

We shall denote by U_g by the matrices of the representation $g \to T_g$ in some basis. In R we set up one further representation $g \to T_g$ whose matrices in the same basis are given by

$$V_g = (U_g^{\mathrm{T}})^{-1}. \tag{23}$$

(The reader can easily convince himself that the correspondence $g \to V_g$ is in fact a representation).

Making use of the above remarks, we see that the product of the two representations $T_g \times \hat{T}_g$ can be realized in the space of square matrices L according to the formula

$$\tau_g L = U_g L V_g^{\mathrm{T}}$$

or

$$\tau_g L = U_g L U_g^{-1},$$

which coincides with formula (22).

In this way we see that representation (22) in the space of matrices S is equivalent to the product of the two representations T_g and $\hat{T}_g$ acting in R.

Our problem now is to select in the product representation $T_g \times \hat{T}_g$ an irreducible three-dimensional subspace and to find its canonical basis.

As usual, we denote by ξ^{τ}_{lm} the canonical basis of the representation $g \to T_g$ in the space R. We shall moreover suppose that the matrices U_g and V_g of the representations $g \to T_g$ and $g \to \hat{T}_g$, in this basis, are connected by relation (23) $V_g = (U_g^{\mathrm{T}})^{-1}$. We note here that the basis $\{\xi^{\tau}_{lm}\}$ is not canonical for the representation $g \to T_g$. However it differs from canonical only in the numbering of the vectors: the vector ξ^{τ}_{lm} is the characteristic vector of the operator $\hat{H}_3$ for the representation $g \to T_g$ with the characteristic value $-m$,

$$\hat{H}_3 \xi^{\tau}_{lm} = -m \xi^{\tau}_{lm},$$

and therefore the basis $\{\eta^{\tau}_{lm}\} = \{\xi^{\tau}_{l,-m}\}$ is canonical for the representation $g \to \hat{T}_g$. In connexion with this we pick for the space $R \times R$ the basis $\{\xi^{\tau}_{lm}\eta^{\tau'}_{l',-m'}\} = \{\xi^{\tau}_{lm}\xi^{\tau'}_{l'm'}\}$ which is the product of the canonical bases of the representations $g \to T_g$ and $g \to \hat{T}_g$. The vector h from $R \times R$ written in terms of this basis is, clearly,

$$h = \sum c^{\tau\tau'}_{lml'm'} \xi^{\tau}_{lm} \xi^{\tau'}_{l'm'} = \sum c^{\tau\tau'}_{lml'm'} \xi^{\tau}_{lm} \eta^{\tau'}_{l',-m'}.$$

By the note on p. 49 under the representation $T_g \times \hat{T}_g$ in $R \times R$ the matrix $\left\| c_{lm\,l'm'}^{\tau\tau'} \right\|$ is transformed according to formula (22).

Finally, let R^3 be the three-dimensional irreducible subspace, i.e. the subspace in $R \times R$ in which the representation with weight $l = 1$ acts, and let $\tilde{L}_0$, $\tilde{L}_+$, $\tilde{L}_-$ be the canonical basis in it.

We write the vectors $\tilde{L}_0$, $\tilde{L}_+$, $\tilde{L}_-$ in terms of the basis $\xi_{lm}^{\tau} \eta_{l',\,-m'}^{\tau'}$:

$$\left.\begin{aligned} \tilde{L}_0 &= \sum a_{lm\,l'm'}^{\tau\tau'\,(0)} \xi_{lm}^{\tau} \eta_{l',\,-m'}^{\tau'}, \\ \tilde{L}_+ &= \sum a_{lm\,l'm'}^{\tau\tau'\,(+)} \xi_{lm}^{\tau} \eta_{l',\,-m'}^{\tau'}, \\ \tilde{L}_- &= \sum a_{lm\,l'm'}^{\tau\tau'\,(-)} \xi_{lm}^{\tau} \eta_{l',\,-m'}^{\tau'}. \end{aligned}\right\} \tag{24}$$

The numbers $a_{lml'm'}^{\tau\tau'\,(0,+,-)}$ are clearly elements of the matrices L_0, L_+, L_- of the invariant equation. We shall find the general form of these numbers. As before, we assume that the space R in which the representation $g \to T_g$ acts, resolves into a sum of invariant subspaces R_l^{τ} in each of which the representation $g \to T_g$ generates an irreducible representation of weight l. (We shall use the symbol τ to distinguish subspaces with the same l).

$$R = \sum_{(l\tau)} R_l^{\tau}. \tag{25}$$

The basis $\{\xi_{l,\,-l}^{\tau}, \xi_{l,\,-l+1}^{\tau} \ldots \xi_{l,\,l-1}^{\tau} \xi_{ll}^{\tau}\}$ is canonical in the subspace R_l^{τ}. A similar resolution of R also occurs for the representation $g \to \hat{T}_g$

$$R = \sum R_{l'}^{\tau'} \text{*}). \tag{26}$$

The basis $\{\eta_{l'm'}^{\tau'}\}$ $(m' = -l, \ldots, l)$ is canonical in the subspace $R_{l'}^{\tau'}$.

In this way, the product, $R \times R$ of the space with itself is the sum of all possible subspaces of the form $R_l^{\tau} \times R_{l'}^{\tau'}$

$$R \times R = \sum_{(l,\,\tau)(l'\tau')} R_l^{\tau} \times R_{l'}^{\tau'}.$$

In the subspace $R_l^{\tau} \times R_{l'}^{\tau'}$ which is invariant with respect to the representation $\tau_g \to T_g \times T_g$ the latter acts as the product of two irreducible representations with weights l and l'. If now in each of the subspaces $R_l^{\tau} \times R_{l'}^{\tau'}$ we extract (whenever possible) a three-dimensional irreducible subspace and its canonical basis

$$g_0(l\tau;\, l'\tau'), \qquad g_+(l\tau;\, l'\tau'), \qquad g_-(l\tau;\, l'\tau'),$$

then with the aid of linear combinations of the form

$$\left.\begin{aligned} \tilde{L}_0 &= \sum d_{ll'}^{\tau\tau'} g_0(l\tau;\, l'\tau'), \\ \tilde{L}_+ &= \sum d_{ll'}^{\tau\tau'} g_0(l\tau;\, l'\tau'), \end{aligned}\right\}$$

* The resolutions (25) and (26) may be chosen identical.

$$\tilde{L}_- = \sum d_{ll'}^{\tau\tau'} g_- (l\tau;\ l'\tau')\Big\} \tag{27}$$

we obtain a canonical basis in a three-dimensional subspace, of $R \times R$, irreducible with respect to the representation $T_g \times \hat{T}_g$.

Thus the problem has been reduced to that of finding in the product $R_l^\tau \times \times R_{l'}^\tau$ the canonical basis of a three-dimensional irreducible representation ($l = 1$).

We notice immediately that such a representation exists only when the weights l and l' differ at most by one:

$$l = l' - 1, \quad l', \quad l' + 1$$

i.e. in subspaces of the form

$$R_{l-1}^\tau \times R_l^{\tau'}, \quad R_l^\tau \times R_l^{\tau'}, \quad R_{l+1}^\tau R_l^{\tau'}.$$

We write for $R_{l-1}^\tau \times R_l^{\tau'}$:

$$g_0\,(l-1,\ \tau;\ l\tau') = \sum B_{l-1,\,m;\,l,\,-m}^{10}\,\xi_{l-1,\,m}^\tau \eta_{l,\,-m}^{\tau'},$$

$$g_+\,(l-1,\ \tau;\ l\tau') = \sum B_{l-1,\,m+1;\,l,\,-m}^{11}\,\xi_{l-1,\,m+1}^\tau \eta_{l,\,-m}^{\tau'},$$

$$g_-\,(l-1,\ \tau;\ l\tau') = \sum B_{l-1,\,m-1;\,l,\,-m}^{1,\,-1}\,\xi_{l-1,\,m-1}^\tau \eta_{l,\,-m}^{\tau'};$$

for $R_l^\tau \times R_l^{\tau'}$:

$$g_0\,(l\tau;\ l\tau') = \sum B_{l,\,m;\ ,\,-m}^{10}\,\xi_{lm}^\tau \eta_{l,\,-m}^{\tau'},$$

$$g_+\,(l\tau;\ l\tau') = \sum B_{l,\,m+1;\,l,\,-m}^{11}\,\xi_{l,\,m+1}^\tau \eta_{l,\,-m}^{\tau'},$$

$$g_-\,(l\tau;\ l\tau') = \sum B_{l,\,m-1;\,l,\,-m}^{1,\,-1}\,\xi_{l,\,m-1}^\tau \eta_{l,\,-m}^{\tau'};$$

and for $R_{l+1}^\tau \times R_l^{\tau'}$:

$$g_0\,(l+1,\ \tau;\ l\tau') = \sum B_{l+1,\,m;\,l,\,-m}^{10}\,\xi_{l+1,\,m}^\tau \eta_{l,\,-m}^{\tau'},$$

$$g_+\,(l+1,\ \tau;\ l\tau') = \sum B_{l+1,\,m+1;\,l,\,-m}^{11}\,\xi_{l+1,\,m+1}^\tau \eta_{l,\,-m}^{\tau'},$$

$$g_-\,(l+1,\ \tau;\ l\tau') = \sum B_{l+1,\,m-1;\,l,\,-m}^{1,\,-1}\,\xi_{l+1,\,m-1}^\tau \eta_{l,\,-m}^{\tau'}.$$

The coefficients $\left\{B_{l+k,\,m+s;\,l,\,-m}^{1s}\right\}$ ($s = 1,\ 0,\ -1;\ k = 1,\ 0,\ -1$) are the Clebsch-Gordan coefficients. Using formulae (22) Sect.10, paragraph 3, we have:

$$g_0\,(l-1,\ \tau;\ l\tau') = \alpha^{(-1)}(l)\sqrt{2}\sum_m (-1)^{m+1}\sqrt{(l+m)(l-m)}\,\xi_{l-1,\,m}^\tau \eta_{l,\,-m}^{\tau'},$$

$$g_+\,(l-1,\ \tau;\ l\tau') = \alpha^{(-1)}(l)\sum_m (-1)^m\sqrt{(l-m)(l-m-1)}\,\xi_{l-1,\,m+1}^\tau \eta_{l,\,-m}^{\tau'},$$

$$g_-\,(l-1,\ \tau;\ l\tau') = \alpha^{(-1)}(l)\sum_m (-1)^m\sqrt{(l+m-1)(l+m)}\,\xi_{l-1,\,m-1}^\tau \eta_{l,\,-m}^{\tau'}.$$

where

$$\alpha^{(-1)}(l) = \sqrt{3}(-1)^l \sqrt{\frac{1}{2l(2l+1)(2l-1)}}.$$

similarly for $R_l^{\tau} \times R_l^{\tau'}$

$$g_0(l\tau;\ l\tau') = \alpha^{(0)}(l)\sqrt{2}\sum(-1)^{m+1} m\xi^{\tau}_{lm}\eta^{\tau'}_{l,-m},$$

$$g_+(l\tau;\ l\tau') = \alpha^{(0)}(l)\sum(-1)^m\sqrt{(l+m+1)(l-m)}\,\xi^{\tau}_{l,\ m+1}\eta^{\tau'}_{l,-m},$$

$$g_-(l\tau;\ l\tau') = \alpha^{(0)}(l)\sum(-1)^{m+1}\sqrt{(l+m)(l-m+1)}\,\xi^{\tau}_{l,\ m-1}\eta^{\tau'}_{l,-m},$$

$$\alpha^{(0)}(l) = \sqrt{3}(-1)^l\sqrt{\frac{1}{(2l+1)\,2l\,(l+1)}}$$

and for $R^{\tau}_{l+1} \times R^{\tau'}_{l}$

$$g_0\ (l+1,\ \tau;\ l\tau') =$$
$$= \alpha^{(+)}(l)\sqrt{2}\sum(-1)^m\sqrt{(l+m+1)(l-m+1)}\,\xi^{\tau}_{l+1,\ m}\eta^{\tau'}_{l,-m},$$
$$g_+(l+1,\ \tau;\ l\tau') =$$
$$= \alpha^{(+)}(l)\sum(-1)^m\sqrt{(l+m+1)(l+m+2)}\,\xi^{\tau}_{l+1,\ m+1}\eta^{\tau'}_{l,-m},$$
$$g_-(l+1,\ \tau;\ l\tau') =$$
$$= \alpha^{(+)}(l)\sum(-1)^m\sqrt{(l-m+1)(l+m+2)}\,\xi^{\tau}_{l+1,\ m-1}\eta^{\tau'}_{l,-m},$$

$$\alpha^{(+)}(l) = (-1)^l\sqrt{3}\sqrt{\frac{1}{(2l+1)(2l+2)(2l+3)}}.$$

As we have already stated, the following combinations of these vectors of the form (27) are again a canonical basis in some three-dimensional irreducible space:

$$\left.\begin{aligned}
\tilde{L}_0 &= \sum d^{\tau\tau'}_{l-1,\,l}\,g_0(l-1,\ \tau;\ l\tau') + \\
&\qquad + d^{\tau\tau'}_{ll}\,g_0(l\tau;\ l\tau') + d^{\tau\tau'}_{l+1,\,l}\,g_0(l+1,\ \tau;\ l\tau'), \\
\tilde{L}_+ &= \sum d^{\tau\tau'}_{l-1,\,l}\,g_+(l-1,\ \tau;\ l\tau') + \\
&\qquad + d^{\tau\tau}_{ll}\,g_+(l\tau;\ l\tau') + d^{\tau\tau'}_{l+1,\,l}\,g_+(l+1,\ \tau;\ l\tau'), \\
\tilde{L}_- &= \sum d^{\tau\tau'}_{l-1,\,l}\,g_-(l-1,\ \tau;\ l\tau') + \\
&\qquad + d^{\tau\tau'}_{ll}\,g_-(l\tau;\ l\tau') + d^{\tau\tau}_{l+1,\,l}\,g_-(l+1,\ \tau;\ l\tau')
\end{aligned}\right\} \quad (27')$$

And conversely the canonical basis in any irreducible subspace in $R \times R$ with weight $l = 1$ has the form (27'). In this way the required vectors $\tilde{L}_0$, $\tilde{L}_+$, $\tilde{L}_-$ from $R \times R$ may be written as

$$\tilde{L}_0 = \sum_{m,\,\tau,\,\tau',\,l}(-1)^m\left[c^{\tau\tau}_{l-1,\,l}\sqrt{l^2-m^2}\,\xi^{\tau}_{l-1,\ m}\eta^{\tau'}_{l,\ -m} + \right.$$
$$\left. + c^{\tau\tau'}_{ll}\,m\xi^{\tau}_{lm}\eta^{\tau}_{l,\ -m} + c^{\tau\tau'}_{l+1,\,l}\sqrt{(l+m+1)(l-m+1)}\,\xi^{\tau}_{l+1,\ m}\eta^{\tau'}_{l,-m}\right].$$

$$\tilde{L}_+ = \frac{1}{\sqrt{2}} \sum_{m,\tau,\tau',l} (-1)^m \Big[-c^{\tau\tau'}_{l-1,l}\sqrt{(l-m)(l-m-1)}\,\xi^{\tau}_{l-1,\,m+1}\eta^{\tau'}_{l,-m} - $$
$$-c^{\tau\tau}_{ll}\sqrt{(l+m+1)(l-m)}\,\xi^{\tau}_{l,\,m+1}\eta^{\tau'}_{l,-m} +$$
$$+ c^{\tau\tau'}_{l+1,l}\sqrt{(l-m+1)(l-m+2)}\,\xi^{\tau}_{l+1,\,m+1}\eta^{\tau}_{l,-m}\Big],$$

$$\tilde{L}_- = \frac{1}{\sqrt{2}} \sum (-1)^m \Big[-c^{\tau\tau'}_{l-1,l}\sqrt{(l+m-1)(l+m)}\,\xi^{\tau}_{l-1,\,m-1}\eta^{\tau'}_{l,-m} +$$
$$+ c^{\tau,\tau+1}_{ll}\sqrt{(l+m)(l-m+1)}\,\xi^{\tau}_{l,\,m-1}\eta^{\tau'}_{l,-m} +$$
$$+ c^{\tau\tau'}_{l+1,l}\sqrt{(l-m+1)(l-m+2)}\,\xi^{\tau}_{l+1,\,m-1}\eta^{\tau'}_{l,-m}\Big].$$

Here the following notation is used

$$c^{\tau\tau'}_{l-1,l} = -d^{\tau\tau'}_{l-1,l}\alpha^{(-1)}(l)\sqrt{2},$$
$$c^{\tau\tau'}_{ll} = -d^{\tau\tau'}_{ll}\alpha^{(0)}(l)\sqrt{2},$$
$$c^{\tau\tau'}_{l+1,l} = d^{\tau\tau'}_{l+1,l}\alpha^{(+1)}(l)\sqrt{2}.$$

Hence for the elements of the matrices $L_0 = L_3$ in the invariant equations we obtain:

$$c^{\tau\tau'}_{l-1,\,m;\,lm} = c^{\tau\tau'}_{l-1,l}\sqrt{l^2-m^2},$$
$$c^{\tau\tau'}_{lm;\,lm} = c^{\tau\tau'}_{ll}\,m,$$
$$c^{\tau\tau'}_{l+1,\,m;\,lm} = c^{\tau\tau'}_{l+1,l}\sqrt{(l+m+1)(l-m+1)},$$

which coincide with the formulae (8).

Similarly we may write out the elements of the matrices $L_1 = \frac{L_+ + L_-}{2}$ and $L_2 = \frac{L_+ - L_-}{2i}$.

§ 7. Invariant Equations with $\varkappa = 0$

All the foregoing discussion and results concerned the equation of the form (21) with the constant $\varkappa \neq 0$. They are, obviously, also applicable to the case $\varkappa = 0$. But it happens that for $\varkappa = 0$ there arise really new possibilities, for the construction of invariant equations.

We note in passing that invariant equations with $\varkappa = 0$ are rarely encountered. We expound this case here on the grounds of preparing the reader for the similar case of the relativistic-invariant equations with $\varkappa = 0$ in the second part of the book - these latter equations being important in theoretical physics.

We shall explain first of all the nature of the conditions for the invariance of equations with $\varkappa = 0$.

Let us assume that a system of equations is given:

$$L_1 \frac{\partial \psi}{\partial x_1} + L_2 \frac{\partial \psi}{\partial x_2} + L_3 \frac{\partial \psi}{\partial x_3} = 0. \tag{28}$$

We notice from the very beginning that in contra-distinction to the case $\varkappa \neq 0$ the matrices L_1, L_2, L_3 in this system are not necessarily square; in other words the number of equations in system (28) need not be the same as the number of components of the function. Let us assume as we have done before that

$$\psi'(x) = T_g \psi(g^{-1}x).$$

Then

$$\frac{\partial \psi'(x)}{\partial x_i} = T_g \sum \frac{\partial \psi(x')}{\partial x'_k} g_{ik},$$

if

$$x' = g^{-1}x.$$

In this manner we obtain the result that the function $\psi'(x)$ satisfies the equation:

$$\sum L_k T_g^{-1} g_{ik} \frac{\partial \psi'}{\partial x_i} = 0. \tag{29}$$

We now assume that there exists some (as yet undetermined) transformation V_g such that

$$V_g \sum L_k T_g^{-1} g_{ik} \frac{\partial \psi'(x)}{\partial x_i} \equiv \sum L_i \frac{\partial \psi'}{\partial x_i} \quad (\text{for all } \psi(x))$$

i.e.

$$\sum_k V_g L_k T_g^{-1} g_{ik} = L_i.$$

If a transformation V_g exists, then equation (29) is equivalent to the equation

$$\sum L_i \frac{\partial \psi'}{\partial x_i} = 0,$$

i.e. after the change $x = gx'$, $\psi'(x) = T_g \psi(x')$ equation (28) has remained unchanged. In this connexion an equation of the form (28) is invariant with respect to the rotation group, if under the simultaneous substitutions $x = gx'$ and $\psi'(x) = T_g \psi(x')$ the transformed equation coincides, to within an arbitrary transformation V_g with the initial one.

We note that in the case of the equation with $\varkappa \neq 0$ the transformation V_g by necessity coincides with the transformation T_g. For the matrices L_1, L_2, L_3 in the invariant equation we obtain the relation

$$V_g L_k T_g^{-1} g_{ik} = L_i,$$

or

$$V_g L_k T_g^{-1} = \sum g_{ki} L_i,$$

where T_g is the matrix of the representation $g \to T_g$, V_g is the matrix of the undetermined transformation. It is easily verified that the correspondence $g \to V_g$ is a representation acting in some space $\tilde{R}$, generally speaking different from the space R in which the representation $g \to T_g$ acts.

In this manner we have obtained the result that the equation

$$\sum L_i \frac{\partial \psi}{\partial x_i} = 0$$

is invariant with respect to the rotation group, if side by side with the representation $g \to T_g$ transforming the functions ψ there exists a representation $g \to V_g$, transforming the same equation such that

$$V_g L_k T_g^{-1} = \sum g_{ki} L_i.$$

This is also the condition for the invariance of equation (28).

From this condition it is clear that the matrices L_1, L_2, L_3 are obtained from the analysis into irreducible components of the two representations

$$V_g \times \hat{T}_g,$$

where $g \to \hat{T}_g$ is the representation contragredient to $g \to T_g$ (the matrices of contragredient representations U_g and $\hat{U}_g$ are connected in some basis by the relation $\hat{U}_g = (U_g^T)^{-1}$).

This case may be developed with the aid of the case $\varkappa \neq 0$. We see that in our case $\varkappa = 0$ the representations V_g and T_g are in no way related to each other whilst for $\varkappa \neq 0$ they must coincide.

We shall now find the general form of the matrices L_1, L_2, L_3 in an invariant equation of the form (28).

We note that the vector ψ of the space R in which the representation $g \to T_g$ acts, is transformed by the matrices L_k $(k = 1, 2, 3)$ into the vector ξ of the space $\tilde{R}$ where the representation $g \to \tilde{T}_g$ acts. We write this as follows,

$$\xi = L_k \psi. \tag{30}$$

Let σ^{τ}_{lm} be the canonical basis of the representation $g \to T_g$ in space $\tilde{R}$ and ξ^{τ}_{lm} the canonical basis of the representation $g \to T_g$ in space R; with this notation $\eta^{\tau'}_{l',-m'} = \xi^{\tau'}_{l'm'}$ is the canonical basis of the representation $g \to \hat{T}_g$.

We now arrange for ξ to refer to the basis σ^{τ}_{lm} and the vector ψ to the basis ξ^{τ}_{lm}. Equation (30) may then be rewritten as

$$x^{\tau}_{lm} = \sum c^{\tau\tau'\,(k)}_{lml'm'} y^{\tau'}_{l'm'},$$

where x^{τ}_{lm} are the coordinates of ξ in the basis $\{\sigma^{\tau}_{lm}\}$ and $y^{\tau'}_{l'm'}$ are the coordinates of ψ in the basis $\{\xi^{\tau'}_{l'm'}\}$. The numbers $c^{\tau\tau'\,(k)}_{lml'm'}$ also form the elements of the matrix L_k,

$$L_k = \| c^{\tau\tau'\,(k)}_{lml'm'} \|.$$

It is possible to define how to assign to each matrix L_k a vector h_k from the product $R \times R$ of the space with itself; thus

$$L_k \sim h_k = \sum c^{\tau\tau'}_{lml'm'} \sigma^{\tau}_{lm} \xi^{\tau'}_{l'm'} = \sum c^{\tau\tau'}_{lml'm'} \sigma^{\tau}_{lm} \eta^{\tau'}_{l',-m'}.$$

Hence it is clear that the search for the matrices L_1, L_2, L_3 (or what amounts to the same thing, L_0, L_+, L_-) boils down to the writing down of the basis in a three-dimensional subspace of $R \times R$ invariant with respect to the representation $\tau_g = T_g \times \hat{T}_g$ by means of the vector $\{\sigma^{\tau}_{lm} \eta^{\tau}_{l',-m'}\}$. This problem we have already solved in a foregoing section. If we utilize the results of that section then we obtain the result that the elements of the matrices L_1, L_2, L_3 — the numbers $c^{\tau\tau'\,(k)}_{lml'm}$, $(k = 1, 2, 3)$ are given by formulæ (8), (11) and (12) of paragraph 3.

Section 10. Analysis of the Product of Two Representations Clebsch-Gordan Coefficients

§ 1. Evaluation of the Clebsch-Gordan Coefficients

In § 3 of Sect.4 we considered into what kind of representations the product of two irreducible representations of rotation groups may be analysed. Here we shall actually perform this analysis, i.e., we express the vectors of the canonical bases in each of the irreducible subspaces into which the space $R_1 \times R_2$ analyses in terms of the vector $e_{m_1} f_{m_2}$.

Let $\{g^l_m\}$ be the orthonormal canonical basis in the subspace $R_l \subset R_1 \times R_2$, in which acts an irreducible representation of weight l $(|l_1 - l_2| \leqslant$

$\leqslant l \leqslant l_1 + l_2$).

We write g^l_m in the form of a linear combination of vectors of the form $e_{m_1} f_{m_2}$:

$$g^l_m = \sum \delta_{m,\, m_1+m_2} B^{lm}_{l_1 m_1;\, l_2 m_2} e_{m_1} f_{m_2}. \tag{1}$$

Our problem consists of determining the coefficients $B^{lm}_{l_1 m_1;\, l_2 m_2}$, known as the Clebsch-Gordan coefficients.

We conduct their evaluation in several stages.

(i) We magnify the vectors of the canonical basis $\{e_{m_1}\}$ in the space R_1, i.e., we change to the vector

$$\xi^{m_1} = \gamma^{l_1}_{m_1} e_{m_1} \tag{2}$$

such that in the new basis $\{\xi^{m_1}\}$ the operators H_+ and H_- have the form

$$\left.\begin{aligned} H_+ \xi^{m_1} &= (l_1 - m_1)\, \xi^{m_1+1}, \\ H_- \xi^{m_1} &= (l_1 + m_1)\, \xi^{m_1-1}. \end{aligned}\right\} \tag{3}$$

(The operator H_3 clearly acts as before, $H_3 \xi^{m_1} = m_1 \xi^{m_1}$.)

We note that with such a choice of the operators H_+, H_-, H_3 the commutation relations (11) of Sect. 2 are obeyed. This means that the indicated substition of the basis may actually be performed.

We now find the coefficients $\gamma^{l_1}_{m_1}$. We have:

$$H_- \xi^{m_1+1} = (l_1 + m_1 + 1)\, \xi^{m_1} = (l_1 + m_1 + 1)\, \gamma^{l_1}_{m_1} e_{m_1}.$$

On the other hand

$$H_- \xi^{m_1+1} = \gamma^{l_1}_{m_1+1} H_{-1} e_{m_1+1} = \gamma^{l}_{m_1+1} \alpha_{m_1+1} e_{m_1}.$$

Hence we obtain the following system of equations

$$\gamma^{l_1}_{m_1+1} \alpha_{m_1+1} = (l_1 + m_1 + 1)\, \gamma^{l_1}_{m_1}.$$

From it we obtain

$$\gamma^{l_1}_{m_1} = \gamma^{l_1}_{l_1} \frac{\alpha_{m_1+1} \cdot \alpha_{m_1+2} \cdots \alpha_{l_1}}{(l_1 + m_1 + 1)(l_1 + m_1 + 2) \ldots 2l_1}.$$

Substituting the value of α_s from (17) Sect.2 and taking $\gamma^{l_1}_{l_1} = 1$ we finally obtain

$$\gamma^{l_1}_{m_1} = \sqrt{\frac{(l_1+m_1)!\,(l_1-m_1)!}{(2l_1)!}}. \tag{4}$$

We perform a similar modification of the basis in the space R_2, taking

$$\eta^{m_2} = \gamma^{l_2}_{m_2} f_{m_2}. \tag{5}$$

(ii) In a space R_l corresponding to an irreducible representation of weight l, we consider, together with the orthonormal canonical basis $\{g^l_m\}$, the basis $\{x^l_s\}$, which is constructed in the following fashion:

$$x^l_0 = g^l_l, \quad x^l_1 = H_- x^l_0, \quad x^l_2 = H_- x^l_1 = H^2_- x^l_0, \ \ldots, \ x^l_s = H^s_- x^l_0. \tag{6}$$

It is clear that

$$x^l_s = \sigma^l_m g_m \qquad (m = l - s). \tag{7}$$

We now find the coefficients σ^l_m:

$$H_- x^l_{s-1} = x^l_s = \sigma^l_m g^l_m.$$

On the other hand

$$H_- x^l_{s-1} = \sigma_{m+1} H_- g^l_{m+1} = \sigma^l_{s-1} \alpha_{m+1}.$$

Hence we obtain the following system of equations

$$\sigma^l_m = \sigma^l_{m+1} \alpha_{m+1}$$

or

$$\sigma^l_m = \sigma^l_l \alpha_{m+1} \alpha_{m+2} \ldots \alpha_l.$$

Substituting α_s from (17) Sect. 2 and taking $\sigma^l_l = 1$, we obtain

$$\sigma^l_m = \sqrt{\frac{(2l)!\,(l-m)!}{(l+m)!}}. \tag{8}$$

(iii) Thus in place of the orthonormal bases $\{e_{m_1}\}$ and $\{f_{m_2}\}$ in the spaces R_1 and R_2 we have selected the bases $\{\xi^{m_1}\}$ and $\{\eta^{m_2}\}$ and in place of the basis $\{g^l_m\}$ in R_l we chose the basis $\{x^l_s\}$. It so happens that the vectors x^l_s may be expressed simply in terms of the vectors ξ^{m_1}, η^{m_2}.

Let $l = l_1 + l_2 - \alpha$ $[0 \leqslant \alpha \leqslant \max (2l_1, 2l_2)]$.

We write the vector $x_0^{l_1+l_2-\alpha}$ in the following form:

$$x_0^{l_1+l_2-\alpha}=\sum_{k=0}^{k=\alpha} a_k^{\alpha}\xi^{l_1-k}\eta^{l_2-\alpha+k}.$$

We now find the coefficients a_k^{α}. It is clear that

$$H_+x_0^{l_1+l_2-\alpha}=0.$$

On the other hand, from formulae (3) we obtain

$$H_+\left(\sum a_k^{\alpha}\xi^{l_1-k}\eta^{l_2-\alpha+k}\right)=$$
$$=\sum a_k^{\alpha}\left[k\xi^{l_1-k+1}\eta^{l_1-\alpha+k}+(\alpha-k)\,\xi^{l_1-k}\eta^{l_2-\alpha+k+1}\right].$$

Hence

$$a_0^{\alpha}\alpha+a_1^{\alpha}=0,$$
$$a_1^{\alpha}(\alpha-1)+2a_2^{\alpha}=0,$$
$$\cdots\cdots\cdots\cdots\cdots\cdots$$
$$a_k^{\alpha}(\alpha-k)+(k+1)\,a_{k+1}^{\alpha}=0.$$

From these equations we find

$$a_k^{\alpha}=a_0^{\alpha}(-1)^{-k}C_{\alpha}^{k},$$

where C_{α}^{k} is the number of combinations of α elements taken k at a time. Hence

$$x_0^{l_1+l_2-\alpha}=a_0^{\alpha}\sum(-1)\;C_{\alpha}^{k}\xi^{l_1-k}\eta^{l_2-\alpha+k}. \tag{9}$$

If the indices of the vectors ξ^p and η^t are formally understood as the powers of the variables ξ and η, then the vector $x_0^{l_1+l_2-\alpha}$ may be written in the following form.

$$x_0^{l_1+l_2-\alpha}=a_0^{\alpha}(\xi-\eta)^{\alpha}\xi^{l_1-\alpha}\eta^{l_2-\alpha}. \tag{10}$$

In order to obtain the vectors $x_1^{l_1+l_2-\alpha}$, $x_2^{l_1+l_2-\alpha}$ etc., it is necessary to apply the operator H_- successively. To this end we now find the law according to which the operator H_- acts upon the vector $\xi^p\eta^t$ $(l_1\geqslant p\geqslant -l_1,\ l_2\geqslant t\geqslant \geqslant -l_2)$,

$$H_-\xi^p\eta^t=(p+l_1)\,\xi^{p-1}\eta^t+(t+l_2)\,\xi^p\eta^{t-1}.$$

If the expression $\xi^p\eta^t$ is again formally understood as a product of powers of the variables ξ, η, then the operator H_- may clearly be written as

$$H_{-}\xi^p\eta^t = \xi^{-l_1}\eta^{-l_2}\left(\frac{\partial}{\partial\xi}+\frac{\partial}{\partial\eta}\right)(\xi^{l_1}\eta^{l_2}\xi^p\eta^t).$$

and in general, if we have the expression

$$P(\xi, \eta) = \sum A_{p,t}\xi^p\eta^t \qquad (l_1 \geqslant p \geqslant -l_1,\ l_2 \geqslant t \geqslant -l_2),$$

then

$$H_{-}P(\xi, \eta) = \xi^{-l_1}\eta^{-l_2}\left(\frac{\partial}{\partial\xi}+\frac{\partial}{\partial\eta}\right)[\xi^{l_1}\eta^{l_2}P(\xi, \eta)].$$

Moreover, clearly

$$H_{-}^{s}P(\xi, \eta) = \xi^{-l_1}\eta^{-l_2}\left(\frac{\partial}{\partial\xi}+\frac{\partial}{\partial\eta}\right)^{s}[\xi^{l_1}\eta^{l_2}P(\xi, \eta)].$$

In this manner we have for $x_s^{l_1+l_2-\alpha}$:

$$x_s^{l_1+l_2-\alpha} = H_{-}^{s}x_0^{l_1+l_2-\alpha} = a_0^{\alpha}H_{-}^{s}(\xi-\eta)^{\alpha}\xi^{l_1-\alpha}\eta^{l_2-\alpha},$$

or

$$x_s^{l_1+l_2-\alpha} = a_0^{\alpha}\xi^{-l_1}\eta^{-l_2}\left(\frac{\partial}{\partial\xi}+\frac{\partial}{\partial\eta}\right)^{s}[(\xi-\eta)^{\alpha}\xi^{2l_1-\alpha}\eta^{2l_2-\alpha}].$$

We note that

$$\left(\frac{\partial}{\partial\xi}+\frac{\partial}{\partial\eta}\right)(\xi-\eta)^{\alpha} = 0.$$

Therefore

$$x_s^{l_1+l_2-\alpha} = a_0^{\alpha}\xi^{-l_1}\eta^{-l_2}(\xi-\eta)^{\alpha}\left(\frac{\partial}{\partial\xi}+\frac{\partial}{\partial\eta}\right)^{s}(\xi^{2l_1-\alpha}\eta^{2l_2-\alpha}) =$$

$$= a_0^{\alpha}\xi^{-l_1}\eta^{-l_2}(\xi-\eta)^{\alpha}\left(\sum_{p=0}^{p=s} C_s^p \frac{\partial^s\xi^{2l_1-\alpha}\eta^{2l_2-\alpha}}{(\partial\xi)^p(\partial\eta)^{s-p}}\right) =$$

$$= a_0^{\alpha}(\xi-\eta)^{\alpha}\Big[\sum C_s^p(2l_1-\alpha)\ldots(2l_1-\alpha-p+1)(2l_2-\alpha)\ldots$$
$$\ldots(2l_2-\alpha-s+p+1)\cdot\xi^{l_1-\alpha-p}\eta^{l_2-\alpha-s+p}\Big].$$

We note that

$$(2l_1-\alpha)\ldots(2l_1-\alpha-p+1) = \frac{(2l_1-\alpha)!}{(2l_1-\alpha-p)!},$$

$$(2l_2-\alpha)\ldots(2l_2-\alpha-s+p+1) = \frac{(2l_2-\alpha)!}{(2l_2-\alpha+s-p)!}.$$

It is necessary here to write $\frac{1}{k!} = 0$ for $k < 0$. We may further write

$$x_s^{l_1+l_2-\alpha} = a_0^\alpha\left(\sum(-1)^k C_\alpha^k \xi^{\alpha-k}\eta^k\right)\left(\sum C_s^p \frac{(2l_1-\alpha)!\,(2l_2-\alpha)!}{(2l-\alpha-p)!\,(2l_2-\alpha+s-p)!}\times\right.$$
$$\left.\times\,\xi^{l_1-\alpha-p}\eta^{l_2-\alpha-s+p}\right)$$

or

$$x_s^{l_1+l_2-\alpha} = a_0^\alpha\sum(-1)^k C_\alpha^k C_s^p \frac{(2l_1-\alpha)!\,(2l_2-\alpha)!}{(2l_1-\alpha-p)!\,(2l_2-\alpha+s-p)!}\,\xi^{l_1-p-k}\times$$
$$\times\,\eta^{l_2-\alpha-s+p+k}.$$

We let $k+p=n$. We then obtain

$$x_s^{l_1+l_2-\alpha}=$$
$$= a_0^\alpha\sum(-1)^{n-p}\frac{\alpha!\,s!\,(2l_1-\alpha)!\,(2l_2-\alpha)!\,\xi^{l_1-n}\eta^{l_2-\alpha-s+n}}{(n-p)!\,(\alpha+p-n)!\,p!\,(s-p)!\,(2l_1-\alpha-p)!\,(2l_2-\alpha+s-p)!} \tag{11}$$

or

$$x_s^{l_1+l_2-\alpha} = \sum T_n^{s,\alpha}\xi^{l_1-n}\eta^{l_2-\alpha-s+n}, \tag{12}$$

where

$$T_n^{s,\alpha}=$$
$$= a_0^\alpha\sum_p(-1)^{n-p}\frac{\alpha!\,s!\,(2l_1-\alpha)!\,(2l_2-\alpha)!}{(n-p)!\,(\alpha+p-n)!\,p!\,(s-p)!\,(2l_1-\alpha-p)!\,(2l_2-\alpha-s+p)!}. \tag{13}$$

The sum is taken over all those values of p for which none of the brackets in the denominator is negative.

(iv) We now revert to the original orthonormal bases

$$\{e_{m_1}\},\quad \{f_{m_2}\},\quad \{g_m^l\}.$$

We let $l_1-n=m_1$, $l_2-\alpha+n-s=m_2$, $l_1+l_2-\alpha=l$ and $s=l-m$. Formula (12) now reads:

$$x_s^l = \sum T_{l_1-m_1}^{l-m,\,l_1+l_2-l}\xi^{m_1}\eta^{m_2}\delta_{m,\,m_1+m_2}.$$

In place of x_s^l, ξ^{m_1} and η^{m_2} we substitute their expressions from (7), (2) and (5):

$$\sigma_m^l g_l^m = \sum T_{l_1-m_1}^{l-m,\,l_1+l_2-l}\gamma_{m_1}^{l_1}\gamma_{m_2}^{l_2}e_{m_1}f_{m_2}\delta_{m,\,m_1+m_2}.$$

Hence we obtain our required coefficients

$$B_{l_1m_1;\,l_2m_2}^{lm} = \frac{T_{l_1-m_1}^{l-m,\,l_1+l_2-l}\gamma_{m_1}^{l_1}\gamma_{m_2}^{l_2}}{\sigma_m}, \tag{14}$$

where $T^{l-m,\, l_1+l_2-l}_{l_1-m_1}$ is defined by formula (13).

We must transform the expression obtained so that it depends clearly upon the numbers $l,\ l_1,\ l_2,\ m,\ m_1,\ m_2$.

First of all we write out the coefficients $T^{l-m,\, l_1+l_2-l}_{l_1-m_1}$.

According to (13) we have:

$$T^{l-m,\, l_1+l_2-l}_{l_1-m_1} =$$

$$= a_0^{\alpha} \sum_p \frac{(-1)^{l_1-m_1-p}(l_1+l_2-l)!\,(l-m)!\,(l+l_1-l_2)!\,(l+l_2-l_1)!}{(l_1-m_1-p)!\,p!\,(l_2+m_1+p-l)!\,(l-m-p)!\,(l+l_1-l_2-p)!\,(l_2-l_1+m+p)!}$$

or, taking $l_1-m_1-p=z$,

$$T^{l-m,\, l_1+l_2-l}_{l_1-m_1} =$$

$$= a_0^{\alpha} \sum \frac{(-1)^z (l_1+l_2-l)!\,(l-m)!\,(l+l_1-l_2)!\,(l+l_2-l_1)!}{z!\,(l_1-m_1-z)!\,(l_2+l_1-l-z)!\,(l-l_1-m_2+z)!\,(l-l_2+m_1+z)!\,(l_2+m_2-z)!}. \tag{15}$$

Hence substituting in (14) the values $\gamma^{l_1}_{m_1}$, $\gamma^{l_2}_{m_2}$, σ^l_m and $T^{l-m,\, l_1+l_2-l}_{l_1-m_1}$ from (4), (8) and (15) we obtain

$$B^{lm}_{l_1m_1l_2m_2} =$$

$$a_0^{\alpha}\left(\sum \frac{(-1)}{z!(l_1-m_1-z)!(l_2+l_1-l-z)!(l-l_1-m_2+z)!(l-l_2+m_1+z)!(l_2+m_2-z)!}\right)\times$$

$$\times\sqrt{\frac{(l_1+m_1)!\,(l_1-m_1)!\,(l_2+m_2)!\,(l_2-m_2)!\,(l+m)!\,(l-m)!}{(2l_1)!\,(2l_2)!\,(2l)!}}\times$$

$$\times (l_1+l_2-l)!\,(l+l_2-l_1)!\,(l+l_1-l_2)!. \tag{16}$$

It remains to determine the constant a_0^{a}. It must be chosen such that the basis $\{g^l_m\}$ is orthonormal.

We return to formula (9) for $x_0^{l_1+l_2-\alpha} = g^l_l$

$$g^l_l = x_0^{l_1+l_2-\alpha} = a_0^{\alpha} \sum (-1)^k C^k_{\alpha} \xi^{l_1-k} \eta^{l_2-\alpha+k},$$

or

$$g^l_l = a_0^{\alpha} \sum (-1)^k C^k_{\alpha} \sqrt{\frac{k!\,(2l_1-k)}{(2l_1)!}} \sqrt{\frac{(\alpha-k)!\,(2l_2-\alpha+k)!}{(2l_2)!}}\, e_{l_1-k} f_{l_2-\alpha+k}.$$

Using the fact that the vectors $e_{l_1-k} f_{l_2-\alpha+k}$ are orthonormal and $\|g^l_l\| = 1$, we obtain

$$a_0^{\alpha} = \left(\sum (C^k_{\alpha})^2 \frac{k!\,(2l_1-k)!\,(\alpha-k)!\,(2l_2-\alpha+k)!}{(2l_1)!\,(2l_2)!}\right)^{-\frac{1}{2}}$$

or

$$a_0^\alpha = \sqrt{\frac{(2l_1)!\,(2l_2)!}{(\alpha!)^2 \sum \frac{(2l_1-k)!\,(2l_2-\alpha+k)!}{k!\,(\alpha-k)!}}}.$$

For the evaluation of the sums in the denominator we make use of the following identity

$$\sum_{k=0}^{k=\alpha} \frac{(m+\alpha-k)!\,(n+k)!}{k!\,(\alpha-k)!} = \frac{m!n!\,(m+n+\alpha+1)!}{\alpha!\,(m+n+1)!} *$$

Taking $m = 2l_1 - \alpha$, $n = 2l_2 - \alpha$, we obtain:

$$\sum \frac{(2l_1-k)!\,(2l_2-\alpha+k)!}{k!\,(\alpha-k)!} = \frac{(2l_1-\alpha)!\,(2l_2-\alpha)!\,(2l_1+2l_2-\alpha+1)!}{\alpha!\,(2l_1+2l_2-2\alpha+1)!}$$

Finally using the fact that $\alpha = l_1 + l_2 - l$ we obtain for a_0^α the expression

$$a_0^\alpha = \sqrt{\frac{(2l_1)!\,(2l_2)!\,(2l+1)!}{(l+l_1-l_2)!\,(l+l_2-l_1)!\,(l_1+l_2+l+1)!\,(l_1+l_2-l)!}}$$

Substituting a_0^α in the expression for the Clebsch-Gordan coefficients we have finally:

$$B^{lm}_{l_1m_1l_2m_2} = \sqrt{\frac{(2l+1)}{(l_1+l_2+l+1)}\,(l_1+l_2-l)!\,(l+l_1-l_2)!\,(l+l_2-l_1)!} \times$$

---

* This identity may be rewritten in the form

$$\sum_{k=0}^{k=\alpha} P_{m+\alpha-k} P_{n+k} C_\alpha^k = A^\alpha_{m+n+\alpha+1} P_m P_n$$

(P_s is the number of permutations of s elements, A_l^k is the number of arrangements of l elements taken k at a time, C_α^k is the number of combinations from α elements taken k at a time) and admits of a simple combinatorial proof.

We imagine three sets of elements, N, A, M, containing n, α and m elements respectively. We divide the set A into two parts containing k and α-k elements. Such a division, clearly, may be performed in C_α^k ways. After this we add k of the elements to set N and obtain aggregate I of $n+k$ elements, and from the remainder and set M we form aggregate II containing $m+\alpha-k$ elements. We consider the various sequences Γ of elements of aggregate I (the number of such sequences being P_{n+k}) and the various sequences D of the second aggregate (their number is equal to $P_{m+\alpha-k}$). We now form all possible sequences of the type $\Gamma \times D$ (after the sequence Γ the sequence D is written). The total number of such sequences is

$$\sum_{k=0}^{k=\alpha} P_{m+\alpha-k} P_{n+k} C_\alpha^k.$$

We now derive this number somewhat differently. We take a series of

$$\times \sqrt{(l_1+m_1)!\,(l_1-m_1)!\,(l_2+m_2)!\,(l_2-m_2)!\,(l+m)!\,(l-m)!}\times$$
$$\times\left(\sum_z \frac{(-1)^z}{z!\,(l_1-m_1-z)!\,(l_2+l_1-l-z)!\,(l-l_1-m_2+z)!\,(l-l_2+m_1+z)!\,(l_2+m_2-z)!}\right).$$

§ 2. The Clebsch-Gordan Coefficients for the Case when One of the Representations has Weight 1 or $\frac{1}{2}$*

We write out the values of the Clebsch-Gordan coefficients for the case when $l_2 = 1$ or $l_2 = {}^1/_2$. This case has already been considered at length in Sect. 4

I. Let $l_2 = 1$ (and $l_1 \geqslant 1$). Then l takes the values $l = l_1 - 1,\ l_1,\ l_1 + 1$. Upon fixing m, m_1 takes not more than three values: $m_1 = m - 1,\ m,\ m + 1$. We write down the matrix

$$C^m = \begin{pmatrix} B^{l_1-1,\,m}_{l_1,\,m-1;\,11} & B^{l_1-1,\,m}_{l_1m;\,10} & B^{l_1-1,\,m}_{l_1m+1;\,1,-1} \\ B^{l_1m}_{l_1,\,m-1;\,11} & B^{l_1m}_{l_1m;\,10} & B^{l_1m}_{l_1m+1;\,1,-1} \\ B^{l_1+1,\,m}_{l_1,\,m-1;\,11} & B^{l_1+1,\,m}_{l_1m;\,10} & B^{l_1+1,\,m}_{l_1,\,m+1;\,1,-1} \end{pmatrix}.$$

---

* (Footnote contd. from p.149)
$m + n + \alpha + 1$ cells arranged one after another. In an arbitrary fashion we distribute the elements of set A therein. This may be done in $A^{\alpha}_{m+n+\alpha+1}$ ways. After this we count off n vacant cells from the left-hand end and distribute in them the elements of set N (this may be done in P_n ways). From the right-hand end of the series we count off m cells and place therein the elements of set M (in P_m ways). One cell thus remains vacant. Everything situtated to the left of it we call sequence Γ and to the right sequence D.
In this way we arrive anew at a sequence of the form $\Gamma \times D$. Their number will clearly be

$$A^{\alpha}_{m+n+\alpha+1}P_mP_n.$$

Thus we have proved the identity

$$\sum_{k=0}^{k=\alpha} P_{m+\alpha-k}P_{n+k}C^k_{\alpha} = A^{\alpha}_{m+n+\alpha+1}P_mP_n,$$

or

$$\sum_{k=0}^{k=\alpha} \frac{(m+\alpha-k)!\,(n+k)!\,\alpha!}{k!\,(\alpha-k)!} = \frac{m!n!\,(m+n+\alpha+1)!}{(m+n+1)!}.$$

* We recall that these coefficients have already been evaluated in Part I, Chapter 2, Sect.4, paragraph 4. We quote them here once again for the completeness of the exposition; here the coefficients $B^{l_1-k,\,m}_{l_1,\,m-i;\,1i}$ $(k = -1,\ 0,\ 1;\ i = -1,\ 0)$ correspond to the coefficients C^m_{pr} $(p = 1, 2, 3;\ r = 1, 2, 3)$ in the notation of the paragraph referred to.

Evaluating according to formula (17) gives ($l_1 = l$);

$$C^m = \begin{Vmatrix} \sqrt{\frac{(l-m)(l-m+1)}{2l(2l+1)}} & -\sqrt{\frac{(l+m)(l-m)}{l(2l+1)}} & \sqrt{\frac{(l+m)(l+m+1)}{2l(2l+1)}} \\ -\sqrt{\frac{(l+m)(l-m+1)}{2l(l+1)}} & \sqrt{\frac{m}{l(l+1)}} & \sqrt{\frac{(l+m+1)(l-m)}{2l(l+1)}} \\ \sqrt{\frac{(l+m)(l+m+1)}{(2l+1)(2l+2)}} & \sqrt{\frac{(l+m+1)(l-m+1)}{(2l+1)(l+1)}} & \sqrt{\frac{(l-m)(l-m+1)}{(2l+1)(2l+2)}} \end{Vmatrix}. \quad (18)$$

II. Now let $l_2 = {}^1/_2$ $(l_1 \geqslant {}^1/_2)$. In this case l and m_1 take the values:

$$l = l_1 - {}^1/_2 \quad \text{or} \quad l = l_1 + {}^1/_2 \text{and} \, m_1 = m + {}^1/_2 \quad \text{or} \quad m_1 = m - {}^1/_2.$$

For the matrix

$$C^m = \begin{pmatrix} B^{l_1 - 1/2,\, m}_{l_1,\, m-1/2;\, 1/2,\, 1/2} & B^{l_1 - 1/2,\, m}_{l_1,\, m+1/2;\, 1/2,\, -1/2} \\ B^{l_1 + 1/2,\, m}_{l_1,\, m-1/2;\, 1/2,\, 1/2} & B^{l_1 + 1/2,\, m}_{l_1,\, m+1/2;\, 1/2,\, -1/2} \end{pmatrix}$$

we obtain the following expression

$$C^m = \begin{Vmatrix} -\sqrt{\frac{l - m + 1/2}{2l + 1}} & \sqrt{\frac{l + m + 1/2}{2l + 1}} \\ \sqrt{\frac{l + m + 1/2}{2l + 1}} & \sqrt{\frac{l - m + 1/2}{2l + 1}} \end{Vmatrix}. \quad (19)$$

§ 3. The Symmetry of the Clebsch-Gordan Coefficients

We note that Clebsch-Gordan coefficients possess two symmetry relations with respect to the pairs (l, m), (l_1, m_1),(l_2, m_2).

I. It is quite clear that

$$B^{lm}_{l_1 m_1;\, l_2 m_2} = B^{lm}_{l_2 m_2;\, l_1 m_1}.$$

II. Far less trivial are the relations

$$\left.\begin{aligned} B^{l_1 m_1}_{lm;\, l_2, -m_2} &= (-1)^{l - l_1 - m_2} \sqrt{\frac{2l_1 + 1}{2l + 1}}\, B^{lm}_{l_1 m_1;\, l_2 m_2}, \\ B^{l_2 m_2}_{l_1, -m_1;\, lm} &= (-1)^{l - l_2 - m_1} \sqrt{\frac{2l_2 + 1}{2l + 1}}\, B^{lm}_{l_1 m_1;\, l_2 m_2}, \end{aligned}\right\} \quad (20)$$

which are obtained by interchanging a lower pair of indices with the upper pair. These relations may be obtained directly from formula (17). We shall employ them to evaluate the coefficients:

$$\{B^{1s}_{l+k;\, m+s;\, l, -m}\} \quad (k = -1,\ 0,\ 1; \quad s = -1,\ 0,\ 1).$$

These coefficients are encountered on separating out an irreducible representation of weight $l = 1$ from the product of two representations of weights l and $l+k$ $(k = -1, 0, 1)$.

We have:

$$B^{1s}_{l+k,\, m+s;\, l,-m} = (-1)^{l+k-1-m} \sqrt{\frac{3}{2(l+k)+1}}\, B^{l+k,\, m+s}_{lm;\, 1s}.$$

The coefficients

$$\{B^{l+k,\, m+s}_{lm;\, 1s}\} \qquad (k = -1, 0, 1; \quad s = -1, 0, 1)$$

have already been calculated in a preceding section.

We can now write out the matrix

$$\left\| \begin{matrix} B^{1,-1}_{l-1,\, m-1;\, l,-m} & B^{10}_{l-1,\, m;\, l,-m} & B^{11}_{l-1,\, m+1;\, l,-m} \\ B^{1,-1}_{l,\, m-1;\, l,-m} & B^{10}_{lm;\, l,-m} & B^{11}_{l,\, m+1;\, l,-m} \\ B^{1,-1}_{l+1,\, m-1;\, l,-m} & B^{10}_{l+1,\, m;\, l,-m} & B^{11}_{l+1,\, m+1;\, l,-m} \end{matrix} \right\|. \tag{21}$$

Calculation with the aid of relation (20) gives for this matrix the following expression

$$(-1)^{l-m}\sqrt{3} \left\| \begin{matrix} \sqrt{\frac{(l+m-1)(l+m)}{2l(2l+1)(2l-1)}} & -\sqrt{\frac{(l+m)(l-m)}{(2l-1)(2l+1)l}} & \sqrt{\frac{(l-m)(l-m-1)}{(2l-1)(2l+1)2l}} \\ -\sqrt{\frac{(l+m)(l-m+1)}{(2l+1)2l(l+1)}} & -\frac{m}{\sqrt{(2l+1)l(l+1)}} & \sqrt{\frac{(l+m+1)(l-m)}{2l(l+1)(2l+1)}} \\ \sqrt{\frac{(l-m)(l-m+2)}{(2l+1)(2l+2)(2l+3)}} & \sqrt{\frac{(l+m+1)(l-m+1)}{(2l+1)(l+1)(2l+3)}} & \sqrt{\frac{(l+m+1)(l+m+2)}{(2l+1)(2l+2)(2l+3)}} \end{matrix} \right\|. \tag{22}$$

§ 4. The Transformation from a Canonical Basis in $R_1 \times R_2$ to the Basis $\{e_i f_k\}$

In conclusion we shall quote the formulae expressing the basis $\{e_{m_1} f_{m_2}\} = \{\xi_{l_1 m_1} \xi_{l_2 m_2}\}$ in terms of the basis $\{g_{lm}\}$.

We have

$$\xi_{l_1 m_1} \xi_{l_2,\, m-m_1} = \sum A^{lm}_{l_1 m_1;\, l_2 m_2} g_{lm}.$$

It is possible to express the coefficients $A^{lm}_{l_1 m_1;\, l_2 m_2}$ in terms of Clebsch-Gordan coefficients.

Let there be, in the product of the spaces $R_{l_1} \times R_l$ a subspace of eigenvectors of the operator H_3, with the eigenvalue m.

In this subspace we select two orthonormal bases

$$\{g_{l_1+l_2-k,\, m}\} = g_k \qquad (k = 0, \ldots, l_1 + l_2 - |m|)$$

and

$$\{\xi_{l_1, l_1-s}\xi_{l_2, m-l_1+s}\} = \eta_s \qquad (s = 0,\ 1,\ \ldots,\ l_1 + l_2 - |m|).$$

We have

$$g_{l_1+l_2-k,\, m} = \sum B^{l_1+l_2-k,\, m}_{l_1,\, l_1-s;\, l_2,\, m-l_1+s}\,\xi_{l_1,\, l_1-s}\,\xi_{l_2,\, m-l_1+s},$$

or

$$g_k = \sum c_{ks}\eta_s,$$

where c_{ks} is defined by

$$c_{ks} = B^{l_1+l_2-k,\, m}_{l_1,\, l_1-s;\, l_2,\, m-l_1+s}.$$

Because of the orthonormality of the matrix $\|c_{ks}\|$

$$\eta_s = \sum c_{sk} g_k,$$

or

$$\xi_{l_1,\, l_1-s}\,\xi_{l_2,\, m-l_1+s} = \sum B^{l_1+l_2-s,\, m}_{l_1,\, l_1-k;\, l_2,\, m-l_1+k}\, g_{l_1+l_2-k,\, m}.$$

We now return to the previous notation

$$l_1 + l_2 - k = l, \qquad k = l_1 + l_2 - l,$$
$$l_1 - s = m_1, \qquad s = l_1 - m_1;$$

then

$$l_1 + l_2 - s = l_2 + m_1, \qquad l_1 - k = l - l_2.$$

In this manner

$$\xi_{l_1 m_1}\,\xi_{l_2,\, m-m_1} = \sum B^{l_2+m_1,\, m}_{l_1,\, l-l_2;\, l_2,\, m+l_2-l}\, g_{lm},$$

i.e.

$$A^{lm}_{l_1 m_1;\, l_2 m_2} = B^{l_2+m_1,\, m}_{l_1,\, l-l_2;\, l_2,\, m+l_2-l}.$$

§ 5. Racah Coefficients

We consider the product of three irreducible representations which act in the subspaces R_{l_1}, R_{l_2}, R_{l_3} respectively:

$$\tau_g = T_g^{l_1} \times T_g^{l_2} \times T_g^{l_3},$$

τ_g acts in the product of the spaces

$$R = R_{l_1} \times R_{l_2} \times R_{l_3}.$$

The product of the spaces may be written in two ways:

$$R = [R_{l_1} \times R_{l_2}] \times R_{l_3}$$

and

$$R = R_{l_1} \times [R_{l_2} \times R_{l_3}].$$

We note that it is possible to analyse the space R into irreducible subspaces with respect to the representation τ_g using either of the above two factorizations for R. Actually we use the first and analyse the product of the spaces $R_{l_1} \times R_{l_2}$ into irreducible components with respect to the representation $T_g^{l_1} \times T_g^{l_2}$ of the subspace. We denote the weight of the irreducible representations obtained in this manner by l_{12}, the irreducible subspaces by $R_{l_{12}}$ and the irreducible representations themselves by $T_g^{l_{12}}$ (the indices 1 and 2 indicate the fact that these representations have arisen from the product of the representations $T_g^{l_1}$, $T_g^{l_2}$). We multiply each of the spaces $R_{l_{12}}$ obtained and analyse the product obtained into subspaces R_l which are irreducible with respect to the product of the representation $T_g^{l_{12}} \times T_g^{l_3}$. The spaces R_l clearly belong to the whole space $R = R_{l_1} \times R_{l_2} \times R_{l_3}$ and are irreducible with respect to the representation τ_g. We denote a canonical basis in the space R_l by $\{g_m^l(l_1, l_2, l_{12}, l_3)\}$ (the numbers in brackets indicate the sequence in which the space R_l and its canonical basis were obtained). The set of canonical bases in all the spaces R_l clearly forms a basis in the entire space R.

Analogously to the above we may construct an analysis of the space R into irreducible subspaces, using the expression

$$R = R_{l_1} \times [R_{l_2} \times R_{l_3}].$$

We denote the canonical basis of the irreducible subspaces $R_{l'}$ obtained in this manner by $\{g_m^{l'}(l_2, l_3, l_{23}, l_1)\}$. The set of all such bases forms a basis of the entire space R.

We note that our two analyses of the space R into irreducible subspaces are, generally speaking, different. In fact, the space R may contain isotypic invariant subspaces * and these may be analysed into irreducible subspaces in

---

* For example, the product of the representations $T_g^1 \times T_g^1 \times T_g^1$ contains three irreducible representations of weight 1 and two representations of weight 2.

different ways. Correspondingly the bases

$$\{g_m^l(l_1, l_2, l_{12}, l_3)\} \text{ and } \{g_m^{l'}(l_2, l_3, l_{23}, l_1)\}$$

are also different.

In this manner, using the two different methods for forming the product of the spaces R_{l_1}, R_{l_2} and R_{l_3}: $[R_{l_1} \times R_{l_2}] \times R_{l_3}$, $R_{l_1} \times [R_{l_2} \times R_{l_3}]$, we construct two different analyses of the space R into irreducible subspaces, and obtain two different bases in R:

$$\{g_m^l(l_1, l_2, l_{12}, l_3)\} \text{ and } \{g_m^{l'}(l_2, l_3, l_{23}, l_1)\}.$$

Our problem is to find how these bases may be expressed in terms of each other. We note, first of all, that two spaces R_l and $R_{l'}$ for different l and l' belong to different invariant subspaces of the space R and are consequently orthogonal.

In this manner, the vectors $g_m^l(l_1, l_2, l_{12}, l_3)$ are expressed only in terms of the vector $g_m^{l'}(l_2, l_3, l_{13}, l_1)$ with $l' = l$.

We write

$$g_m^l = \sum K^{l,\ l_1,\ l_2,\ l_{12},\ l_3,\ m}_{\quad l_2,\ l_3,\ l_{23},\ l_1,\ m'} g^l_{m'}. \tag{23}$$

It may be shown that the coefficients $K^{l,\ l_1,\ l_2,\ l_{12},\ l_3,\ m}_{\quad l_2,\ l_3,\ l_{23},\ l_1,\ m'}$ depend upon m and m' in the following fashion

$$K^{l,\ l_1,\ l_2,\ l_{12},\ l_3,\ m}_{\quad l_2,\ l_3,\ l_{23},\ l_1,\ m'} = K^{l,\ l_1,\ l_2,\ l_{12},\ l_3}_{\quad l_2,\ l_3,\ l_{23},\ l_1} \delta_{mm'} \text{*}$$

and formula (23) assumes the form

---

* This circumstance is general: if the space R^l in which an isotypic representation of weight l acts, is analysed in the two ways into the irreducible spaces $R_1, \ldots, R_k$ and $\tilde{R}_1, \ldots, \tilde{R}_k$ with canonical bases $\{e_{ms}\}$ and $\{\tilde{e}_{ms}\}$ respectively then the matrix of the transformation from the basis $\{e_{ms}\}$ to the basis $\{\tilde{e}_{ms}\}$

$$\tilde{e}_{ms} = \sum_{m',\ s'} a_{msm's'} e_{m's'}$$

has the form

$$a_{msm's'} = a_{ss'} \delta_{mm'}.$$

The proof of this assertion is contained, in essence, in the second part of the book in Sect. 2, paragraph 9.

$$g^{l}_{m}(l_1, l_2, l_{12}, l_3) = \sum K^{l,\ l_1,\ l_2,\ l_{12},\ l_3}_{\ \ l_2,\ l_3,\ l_{23},\ l_3}\ g^{l}_{m}(l_2, l_3, l_{23}, l_1).$$

The summation is extended over all permissible values of the weight l_{23}. We consider the numbers

$$W^{l,\ l_{12},\ l_{23}}_{l_1,\ l_2,\ l_3} = \frac{R^{l,\ l_1,\ l_2,\ l_{12},\ l_3}_{\ \ l_2,\ l_3,\ l_{23},\ l_1}}{\sqrt{(2l_{12}+1)(2l_{23}+1)}}.$$

The numbers $W^{l,\ l_{12},\ l_{23}}_{l_1,\ l_2,\ l_3}$ are known as the Racah Coefficients. The Racah Coefficients may be expressed in terms of the Clebsch-Gordan coefficients. We omit the proof and merely quote the final result:

$$W^{l,\ l_{12},\ l_{23}}_{l_1,\ l_2,\ l_3} = \frac{1}{(2l+1)\sqrt{(2l_{12}+1)(2l_{23}+1)}} \sum_{m_1+m_2+m_3+m=0} B^{l_{12}\ m_{12}}_{l_1 m_1\ l_2 m_2} \times$$
$$\times B^{l;-m}_{l_{12}m_{12}l_3m_3}\ B^{l_{23}\ m_{23}}_{l_2m_2\ l_3m_3}\ B^{l;-m}_{l_1m_1l_{23}m_{23}}. \tag{24}$$

Racah found an explicit expression for the coefficients $W^{l,\ l_{12},\ l_{23}}_{l_1,\ l_2,\ l_3}$. The corresponding formula and also various useful relations between the Racah coefficients are contained, for example, in the book by G.Ya.Lyubarskii "Group Theory and its Application to Physics".

PART II

REPRESENTATIONS OF THE LORENTZ GROUP

CHAPTER I

THE LORENTZ GROUP AND ITS REPRESENTATIONS

1. The Lorentz Group

The representations of the Lorentz group are next in importance for theoretical physics only to the representations of the three-dimensional rotation group.

1. Definition of the Lorentz Group

We consider the quadratic form

$$S^2(x) = x_1^2 + x_2^2 + x_3^2 - x_0^2, \tag{1}$$

depending on the vectors $x = (x_1 x_2 x_3 x_0)$ of a four-dimensional space $R^{(4)}$. A linear transformation $x' = gx$ which does not change this quadratic form, i.e. such that $S^2(x') = S^2(x)$ is known as a general Lorentz transformation.

We denote by I the matrix of the quadratic form $S^2(x)$:

$$I = \begin{Vmatrix} 1 & 0 & 0 & 0 \\ 0 & 1 & 0 & 0 \\ 0 & 0 & 1 & 0 \\ 0 & 0 & 0 & -1 \end{Vmatrix}.$$

Under any linear transformation with matrix g the matrix I of the quadratic form transforms to $g^* I g$, where g^* is the transpose matrix of matrix g. Consequently for a general Lorentz transformation we have the equation

$$g^* I g = I. \tag{2}$$

This clearly implies that $\det g = \pm 1$ so that the transformation g has an inverse g^{-1}. It is clear that g^{-1} is also a general Lorentz transformation. The product of two general Lorentz transformations is clearly again a general Lorentz transformation. In this manner the set of general Lorentz transformations forms a group - the general Lorentz group.

The equation $S^2(x) = x_1^2 + x_2^2 + x_3^2 - x_0^2 = 0$ defines in $R^{(4)}$ a cone (called the light cone) whose axis is the x_0-axis (the time axis)*.

The light cone divides the whole of the space $R^{(4)}$ into three regions: an external region, where $S^2(x) > 0$, and two internal regions $S^2(x) < 0, x_0 > 0$ and $S^2(x) < 0, x_0 < 0$. Any general Lorentz transformation transforms the light cone and its internal region (i.e. the region where $S^2(x) < 0$) into themselves. A general Lorentz transformation under which each region of the light cone also remains in place we shall call simply a Lorentz transformation. It is clear that Lorentz transformations do not alter the positive direction of the time axis. The Lorentz transformations also form a group, known as the complete Lorentz group. We shall call Lorentz transformations with a determinant equal to 1 proper Lorentz transformations. They also form a group - the proper Lorentz group. We note that the complete Lorentz group is obtained from the proper group by the addition of a special transformation - a spatial reflection s with the matrix

$$s = \begin{Vmatrix} -1 & 0 & 0 & 0 \\ 0 & -1 & 0 & 0 \\ 0 & 0 & -1 & 0 \\ 0 & 0 & 0 & 1, \end{Vmatrix},$$

and also all possible transformations of the form sg, where is an element of the proper Lorentz group.

Similarly the general Lorentz group is obtained from the complete Lorentz group by the addition of the so-called "temporal reflection" i.e., of the transformation t with the matrix

$$t = \begin{Vmatrix} 1 & 0 & 0 & 0 \\ 0 & 1 & 0 & 0 \\ 0 & 0 & 1 & 0 \\ 0 & 0 & 0 & -1 \end{Vmatrix}$$

and of all possible transformations of the form tg, where g is an element of the complete Lorentz group.

Let $g = |g_{ik}|$ be the matrix of a rotation of three-dimensional space. We consider the following transformation in $R^{(4)}$:

$$\begin{aligned} x'_1 &= g_{11}x_1 + g_{12}x_2 + g_{13}x_3, \\ x'_2 &= g_{21}x_1 + g_{22}x_2 + g_{23}x_3, \\ x'_3 &= g_{31}x_1 + g_{32}x_2 + g_{33}x_3, \\ x'_0 &= x_0. \end{aligned} \tag{3}$$

It is clear that this is a proper Lorentz transformation.

If we identify each three-dimensional rotation with the corresponding proper Lorentz transformation indicated above, then we may say that the three-dimensional rotations form a subgroup of the proper Lorentz group.

---

* The terminology stems from the physical interpretation of the four-dimensional space $R^{(4)}$, the quantity $S^2(x)$ and the Lorentz group.

Finally we make an observation concerning spatial and temporal reflections.

We associate with each proper Lorentz transformations g another Lorentz transformation according to the formula

$$\tilde{g} = sgs^{-1} \text{ (s is the spatial reflection)} \tag{4}$$

It is clear that $\tilde{g}$ is again a proper Lorentz transformation.

The correspondence $\tilde{g} \sim g$, clearly satisfies the following

1) $e \sim e$ (e is the unit transformation)

2) If $\tilde{g}_1 \sim g_1$ and $\tilde{g}_2 \sim g_2$ then $\tilde{g}_1\tilde{g}_2 \sim g_1 g_2$. Any one-one correspondence $\tilde{g} \sim g$ between elements of one and the same group which satisfies these conditions is known as an automorphism of the group. In this manner a spatial reflection produces according to formula (4) an automorphism of the proper Lorentz group. The temporal reflection t also generates an automorphism

$$\tilde{g} = tgt^{-1}. \tag{4'}$$

This automorphism coincides with the previous one, since it is easily seen that

$$tgt^{-1} = sgs^{-1}.$$

We note that the matrix of the transformation t coincides with the matrix I of the quadratic form $S^2(x)$. From equation (2) it follows therefore that

$$g^{*-1} = IgI^{-1} = tgt^{-1}.$$

In this manner, the matrix $\tilde{g}$ of the transformation $sgs^{-1} = tgt^{-1}$ is equal to

$$\tilde{g} = g^{*-1}.$$

(Henceforth g^* denotes the transpose of the matrix g.)

Let g be an arbitrary element of some group, and g_0 be a fixed element of this same group. It is clear that the correspondence $g \sim g_0 g g_0^{-1}$ is an automorphism of the group. Such an automorphism is called inner. Any other automorphism is called outer. Automorphism (4) of the proper Lorentz group

$$\tilde{g} = sgs^{-1} = g^{*-1},$$

generated by the spatial reflection s, cannot be represented in the form

$$\tilde{g} = g_0 g g_0^{-1},$$

where g_0 is an element of the proper group. This simple circumstance can easily be verified by the reader. In this way we see that the automorphism

$$g = sgs^{-1}$$

is an outer automorphism of the proper group (for the complete and the general group this automorphism is, obviously, inner).

It can be proved that any outer automorphism of the proper Lorentz group is given in the form

$$\tilde{g} = g_0 sgs^{-1} g_0^{-1},$$

where g_0 is a proper Lorentz transformation. This means that the automorphism $g = sg_0s^{-1}$ is in some sense the only outer automorphism of the proper Lorentz group.

As we shall see below, the automorphism (4) plays an important role in the study of the representations of the complete and the general Lorentz group.

§ 2. Orthogonal Coordinate Systems

Upon transferring from the coordinate system $(x_0 x_1 x_2 x_3)$ to the coordinates $(x_0' x_1' x_2' x_3')$ with the aid of the linear transformation g the matrix I of the quadratic form

$$S^2(x) = x_1^2 + x_2^2 + x_3^2 - x_0^2$$

transforms, as we know, to

$$I' = g^* I g.$$

Here the matrix I' of the quadratic form $S^2(x)$ in the coordinate system $(x_0' x_1' x_2' x_3')$ coincides with the matrix I if, and only if, g is a general Lorentz transformation. The coordinate systems $(x_0' x_1' x_2' x_3')$ in which the quadratic form $S^2(x)$ is represented by the matrix I are known as orthogonal coordinate systems in the four-dimensional space $R^{(4)}$.

It is clear that a linear transformation effecting a transformation from one orthogonal coordinate system to another, is a general Lorentz transformation. Conversely, any general Lorentz transformation carries an orthogonal coordinate system into another (orthogonal coordinate system). In the following we shall use only orthogonal coordinate systems, without explicitly stating this each time.

§ 3. Surfaces in Four-Dimensional Space which are Transitive with respect to Lorentz Groups. The Connected Components of a Lorentz Group

It is well known that any rotation of three-dimensional space carries every sphere centre the origin into itself; and that any point on such a sphere may be carried to any other point by a suitable rotation.

To describe this we say that spheres (with centre at the origin) are transitive surfaces with respect to the group of rotations.

In general, if a group G of transformations acts in some space R then a surface is known as a surface of transitivity for the group G, provided that any transformation from G carries this surface into itself and that any of its points can be carried by some transformation from G to any other.

We determine those surfaces in four-dimensional space which are surfaces of transitivity for the Lorentz group.

Inasmuch as the form

$$s^2(x) = x_1^2 + x_2^2 + x_3^2 - x_0^2$$

is not altered under a Lorentz transformation, the surfaces*

$$x_0^2 - x_1^2 - x_2^2 - x_3^2 = \text{const} \tag{5}$$

transform into themselves under Lorentz transformations.

The surfaces (5) are of the following types:

I. $s^2(x) = c < 0$, $x_0 > 0$ is the upper branch of a hyperboloid of two sheets.

II. $s^2(x) = c < 0$, $x_0 < 0$ is the lower branch of this hyperboloid.

III. $s^2(x) = 0$, $x_0 > 0$ is the upper branch of the light cone.

IV. $s^2(x) = 0$, $x_0 < 0$ is the lower branch of the light cone.

V. $s^2(x) = c > 0$ is a hyperboloid of one sheet.

VI. The origin of coordinates $x_0 = x_1 = x_2 = x_3 = 0$.

We shall now show that each of these surfaces is a transitive surface with respect to the proper Lorentz group.

We note, first of all, that it is possible to carry any point A $(x_0 x_1 x_2 x_3)$ by a rotation (i.e. by a proper Lorentz transformation which does not alter the fourth coordinate x_0), to the right-hand side of the plane $(x_0 x_3)$, $x_3 > 0$

---

* Strictly speaking these are three-dimensional hyper-surfaces. However, we shall call them surfaces for the sake of simplicity.

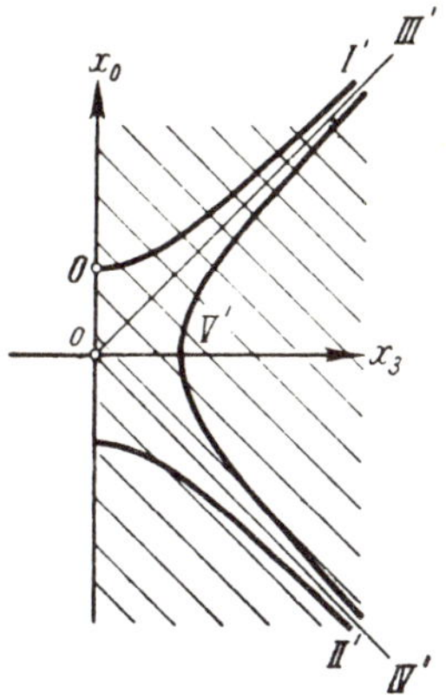

Fig. 8

(see Fig.8, in which the right-hand side of the plane (x_0, x_3) is shaded). To do this it is clearly sufficient to rotate the ray passing through the point (x_1, x_2, x_3) in three-dimensional space $(x_0 = 0)$ so that it coincides with the positive direction of the x_3 axis.

We now consider the intersection of the surfaces (I-VI) enumerated above with the right-hand half plane (x_0, x_3), $x_3 > 0$. We obtain, clearly, six curves (see Fig.8):

I'. The upper branch of a hyperbola: $x_0^2 - x_3^2 = c > 0$, $x_0 > 0$.

II'. Its lower branch: $x_0^2 - x_3^2 = c > 0$, $x_0 < 0$.

III'. The upper asymptote: $x_0 = x_3$, $x_0 > 0$.

IV'. The lower asymptote: $x_0 = -x_3$, $x_0 < 0$.

V'. The right-hand branch of the hyperbola: $x_0^2 - x_3^2 = c < 0$, $x_3 > 0$.

VI'. The origin of coordinates $x_0 = x_3 = 0$.

Any proper Lorentz transformation which acts only in the plane (x_0, x_3) (i.e. which does not alter the coordinates x_1, x_2) leaves each of the curves (I'-VI') in place. Now any two points on one and the same curve can be interchanged by some such transformation; in other words, curves (I'-VI') are transitive curves with respect to proper Lorentz transformations which act in the plane (x_0, x_3). These transformations are known as hyperbolic screws in the plane (x_0, x_3). In the following we shall occasionally denote them by g_{03}. We note that not only the plane (x_0, x_3), but also any general plane S which passes through the axis x_0, intersects the surfaces (I-VI) along curves of the same types as (I'-VI'). In this case, hyperbolic screws in the plane S (i.e. proper Lorentz transformations which leave this plane in place) act transitively on such curves.

Now let A_1 and A_2 be two points in four-dimensional space which lie on one and the same surface (I-VI).

We rotate each of these so that they coincide with the points B_1 and B_2 of the right-hand half plane $(x_0 x_3)$: $B_1 = u_1A_1$, $B_2 = u_2A_2$ (u_1 and u_2 are rotations). Under the rotations each of the surfaces (I-VI) is carried into itself. It follows that the points B_1 and B_2 lie on one and the same curve (I'-VI') and this means that each can be carried by a proper transformation g_{03} in the plane (x_0, x_3) to the other:

$$B_2 = g_{03}B_1.$$

It is clear that the transformation $g = u_2^{-1}g_{03}u_1$ carries A_1 to A_2. In this manner, we see that the surfaces I-VI are surfaces of transitivity for the proper Lorentz group.

It is clear that the spatial reflection s transforms each of the surfaces I-VI into itself. This means that the surfaces of transitivity for the complete Lorentz group are the same as for the proper group. Temporal reflection interchanges the two branches of the hyperboloid of two sheets and branches of the light cone. Therefore the surfaces of transitivity for the general Lorentz group are of only four types:

I. Hyperboloid of two sheets: $x_0^2 - x_1^2 - x_2^2 - x_3^2 = c > 0$.

II. The light cone: $x_0^2 - x_1^2 - x_2^2 - x_3^2 = 0$.

III. Hyperboloid of one sheet: $x_0^2 - x_1^2 - x_2^2 - x_3^2 = c < 0$.

IV. The origin of coordinates: $x_0 = x_1 = x_2 = x_3 = 0$.

Now we make a few observations that are important for the sequel. As we have shown, any point A of the upper branch of the hyperboloid

$$x_0^2 - x_1^2 - x_2^2 - x_3^2 = 1, \quad x_0 > 0, \tag{6}$$

can be carried by proper Lorentz transformations to any other point on this branch, in particular to its apex O (1, 0, 0, 0).

The simplest of such transformations is the hyperbolic screw g_{OA} in the plane (x_0, A), which passes through the point A and the x_0 axis. But there is no unique proper Lorentz transformation which transfers the point A to the point O.

It is clear that any two such transformations differ from one another by a transformation u which leaves the point O in place: $uO = O$. Any transformation u which leaves O (and together with it the x_0 axis as well) in place is, clearly, a rotation.

In this manner we see that any proper Lorentz transformation g which transfers point A to the point O has the form

$$g = ug_{OA}$$

where u is a rotation and g_{OA} is a hyperbolic screw in the plane (x_0, A).

Hence we see that in order to specify a proper Lorentz transformation it is necessary to indicate a point A on the upper branch of the hyperboloid (6) which is transferred by this transformation to the apex O of the hyperboloid, and then with the aid of a hyperbolic screw in the plane (A, x_0) to transfer the point A to the point O, and finally to effect a rotation u. In other words, each proper Lorentz transformation is defined by a pair $g \sim (u, A)$ where u is a rotation and A is a point on the hyperboloid (6). One can easily verify that different transformations are defined by different pairs.

From this observation it follows immediately that

1) Each element of the proper Lorentz group is given by six independent parameters (that is, the proper Lorentz group is a six-parameter group). In fact, the point A on the hyperboloid provides three independent parameters (e.g. its coordinates x_1, x_2, x_3) and the rotation u a further three parameters (e.g. the Eulerian angles);

2) The proper Lorentz group is connected, i.e., any two of its elements g_1 and g_2 may be connected by a continuous path. In fact let $g_1 \sim (u_1, A_1)$ and $g_2 \sim (u_2, A_2)$. If now the rotations u_1 and u_2 are joined by a continuous path (this is always possible since as we have seen in the first part of the book, the rotation group is connected) and so are A_1 and A_2 (the upper branch of the hyperboloid is also connected), then g_1 itself will be joined by a continuous path to g_2.

With regard to the latter observation we now determine the number of connected components of the complete and general Lorentz groups*.

The proper group is a connected component of the general Lorentz group. In fact any Lorentz transformation g not included in the proper group either alters the positive direction of the time axis x_0, or satisfies $\det g = -1$ and consequently it cannot be joined by a continuous path to any proper Lorentz transformation. In this manner we see that the proper group is connected but that any extension of it is not, i.e., the proper Lorentz group forms a connected component of the general group

It is clear that all transformations of the form sg where (s is a spatial reflection and g is a proper transformation, also form a connected component. This means that the complete Lorentz group consists of two components.

---

* By a connected component of a continuous group, we understand a connected subset which is not contained in any larger connected subset.

Temporal reflection t produces another two components: a component consisting of the elements of the form tg, and a component consisting of the elements of the form $tsg = jg$ (j is a complete reflection in $R^{(4)}$).

In this manner the general group consists of four connected components:

1) The proper group, which we denote by G_0.

2) The component sG_0 consisting of elements of the form sg (g a proper transformation).

These two components form the complete Lorentz group.

3) The component tG_0 (of elements tg).

4) The component tsG_0 (in which all elements stg occur).

§ 4. The Relation of the Lorentz Group to the Group of Complex Matrices of the Second Order with Determinant Equal to Unity

In the study of the representations of the group of three-dimensional rotations an important role was played by the fact that each rotation may correspond to a unique bilinear transformation of the complex plane:

$$z \to \frac{\alpha z + \beta}{\gamma z + \delta}$$

having a unitary matrix $a = \begin{Vmatrix} \alpha & \beta \\ \gamma & \delta \end{Vmatrix}$ with $\det a = 1$. To each rotation g was assigned a unitary matrix of the second order of undefined sign $\pm a = \pm \begin{Vmatrix} \alpha & \beta \\ \gamma & \delta \end{Vmatrix}$ with determinant equal to unity. On the other hand, to each unitary matrix a with determinant equal to unity there corresponded a completely defined rotation g_a, $a \to g_a$ such that

1) to the product $a_1 a_2$ of two matrices corresponds to the product $g_{a_1} g_{a_2}$ of the rotations $g_{a_1} g_{a_2} = g_{a_1 a_2}$;

2) the unit matrix $\begin{pmatrix} 1 & 0 \\ 0 & 1 \end{pmatrix}$ defines the unit rotation e;

3) two distinct matrices a_1 and a_2 define one and the same rotation g if, and only if, these matrices differ only in sign, $a_1 = -a_2$.

This correspondence between the group U of unitary matrices of the second order with determinant equal to 1 and the group of rotations allowed us to consider any representation $g \to T_g$ of the group of rotations as a representation of the group U, $a \to Tg_a = T_a$; and conversely, to consider the representation $a \to T_a$ of the group U, generally speaking, as a two-valued representation of the group of rotations.

It turns out that there is an analogous correspondence between proper Lorentz transformations and complex matrices of the second order. We shall now establish this. Incidentally, we shall obtain the correspondence between rotations and unitary matrices once again and in a simpler fashion. We consider the set of hermitian matrices of the second order

$$c = \begin{Vmatrix} x_0 - x_3 & x_2 - ix_1 \\ x_2 + ix_1 & x_0 + x_3 \end{Vmatrix}. \tag{7}$$

With each such matrix c we associate a vector x from $R^{(4)}$ with the coordinates x_0, x_1, x_2, x_3:

$$c \longleftrightarrow x. \tag{7'}$$

We note that

$$\det c = x_0^2 - x_1^2 - x_2^2 - x_3^2 = -S^2(x).$$

The correspondence between the matrices c and the vectors x is one-one, bi-unique and linear. Therefore, any linear transformation in the space of the matrices c may be considered as a linear transformation in R^4.

We specify a linear transformation in the space of the matrices c with the aid of the formula

$$c' = aca^*, \tag{8}$$

where a is a matrix of the second order with determinant equal to 1 (the asterisk denotes the conjugate transpose). It is clear that $(c')^* = ac^*a^* = aca^* = c'$, so that c' is a hermitian matrix.

We shall denote by g_a the corresponding linear transformation in $R^{(4)}$ obtained with the aid of formula (7').

Since $\det c' = \det c$ ($\det a = \det a^* = 1$), $S^2(x') = S^2(x)$, i.e. the transformation g_a is a general Lorentz transformation.

The correspondence $a \sim g_a$ clearly satisfies $g_{a_1} g_{a_2} = g_{a_1 a_2}$, i.e. to the product of the matrices $a_1 a_2$ there corresponds the product $g_{a_1} g_{a_2}$ of the Lorentz transformations specified by them. We shall find which matrices a corresponds to the identity transformation.

It is clear that any such matrix must satisfy the equation

$$c = aca^* \tag{9}$$

for any c.

If we take $c = \begin{Vmatrix} 1 & 0 \\ 0 & 1 \end{Vmatrix} = E$, then we obtain

$$aa^* = E$$

or

$$a^* = a^{-1}.$$

We may now rewrite equation (9) in the form

$$c = aca^{-1}.$$

Hence it is apparent that

$$ac = ca,$$

i.e.the matrix a is permutable with every hermitian matrix. Such a matrix is necessarily a multiple of the unit matrix

$$a = \lambda\begin{pmatrix} 1 & 0 \\ 0 & 1 \end{pmatrix}.$$

Since $\det a = 1$, $\lambda = \pm 1$.

In this manner, the identity Lorentz transformation corresponds to the two matrices $a = \pm\begin{pmatrix} 1 & 0 \\ 0 & 1 \end{pmatrix}$, which differ only in sign.

We shall now prove that to two matrices a_1 and a_2 there corresponds one and the same Lorentz transformation if, and only if, $a_1 = \pm a_2$. In fact, let $g_{a_1} = g_{a_2}$. This means that for all c

$$a_1 c a_1^* = a_2 c a_2^*$$

or

$$a_2^{-1} a_1 c \left(a_2^{-1} a_1\right)^* = c.$$

It follows that the matrix $a_2^{-1} a_1$ corresponds the identity transformation.

Hence

$$a_2^{-1} a_1 = \pm E$$

or

$$a_2 = \pm a_1.$$

Thus, to each complex matrix a of the second order with determinant equal to 1 we have associated a Lorentz transformation g_a; the correspondence $a \longleftrightarrow g_a$ possesses the following properties: -

1) $\begin{pmatrix} 1 & 0 \\ 0 & 1 \end{pmatrix} \sim e,$

2) $g_{a_1} g_{a_2} = g_{a_1 a_2},$

3) Two distinct matrices a_1 and a_2 correspond to one and the same transformation $g_{a_1} = g_{a_2}$ if, and only if, these matrices differ only in sign $a_1 = -a_2$.

From the first two properties it follows that the set of transformations constitute a subgroup of the general Lorentz group.

We denote it by G_a. We now show that this subgroup coincides with the proper Lorentz group.

We note that the group of all complex matrices of the second order with determinant equal to 1 is connected *.

In this situation the subgroup G_a is also connected. Consequently this subgroup is contained in that connected component of the general Lorentz group which contains the identify transformation e. As we saw in the previous paragraph, this component is the proper Lorentz group. Thus the subgroup Ga of transformations g_a is contained in the proper Lorentz group. We now prove that they coincide. To this end we derive the number of independent parameters by which the elements of the group $\mathfrak{A}$ are defined (the dimension of the group $\mathfrak{A}$).

Each complex matrix is specified by eight real numbers. Since the requirement that $\det a = 1$ imposes two conditions upon these numbers: $\mathrm{Re}\det a = 1, \mathrm{Im}\det a = 0$, then six of them remain independent.

In this way each element of the group $\mathfrak{A}$ and consequently, of the subgroup G_a as well, is specified by six independent parameters. An element of the proper Lorentz group as we have seen also depends upon six independent parameters. Hence it follows that the subgroup of transformations G_a and the proper group have the same dimension, and since the first group is contained within the second, they coincide. **

* We prove here the connectedness of the group $\mathfrak{A}$. We consider the eight-dimensional real space $R^{(8)}$ of all complex matrices of the second order. The equation $\det a = 0$ selects a six-dimensional surface in this space ($\det a = 0$ signifies the two conditions $\mathrm{Re}\det a = \mathrm{Im}\det a = 0$). Since the dimension of the surface is two units less than the dimensionality of the space, it does not separate the space $R^{(8)}$. In this manner, any two matrices a_1 and a_2 with determinant different from zero may be connected with each other by a continuous curve $a(t)$ which does not pass through the surface $\det a = 0$: $a(0) = a_1$, $a(1) = a_2$ and $\det a(t) \neq 0$.
Now let a_1 and a_2 belong to the group $\mathfrak{A}$, i.e. $\det a_1 = \det a_2 = 1$; we shall deform our curve $a(t)$ as follows

$$a'(t) = \frac{1}{\det a(t)} a(t).$$

It is clear that the curve $a'(t)$ is continuous, connects a_1 and a_2 and belongs entirely to the group $\mathfrak{A}$. In this way we see that any two matrices a_1 and a_2 of the group $\mathfrak{A}$, can be connected with each other by a continuous curve which also belongs to the group $\mathfrak{A}$. In other words, the group $\mathfrak{A}$ is connected.

** In fact, inasmuch as the identity transformation e (unity of the Lorentz group) belongs to the subgroup G_a, then in virtue of the coincidence of dimension, G_a also contains an entire neighbourhood of e. It is easily proved that the connected component of any neighbourhood of the identity of a group is determined uniquely (see, e.g. L.S.Pontryagin, Continuous Groups - Gostekhizdat 1954 Ch.III, p.138). From this it follows that G_a coincides with the entire proper Lorentz group.

We sum up our results as follows:

We have constructed a correspondence $a \sim g_a$ between the proper Lorentz group and the group $\mathfrak{A}$ of complex matrices a of the second order ($\det a = 1$) such that to each matrix a there corresponds one proper Lorentz transformation g_a and to each such transformation g are related two matrices differing only in sign, $+a$ and $-a$. The correspondence constructed is such that the unit matrix corresponds to the Lorentz identity transformation, and to the product of the matrices $a_1 \, a_2$ there corresponds the product of the Lorentz transformations $a_1 a_2 \sim g_{a_1} g_{a_2}$.

We now make two important observations:

I) A spatial reflection s does not belong to the proper Lorentz group and consequently no matrix a corresponds to it. However, we may associate with the reflection s some transformation (automorphism) of the 2-rowed complex matrices. In fact, we have seen above that with the aid of a reflection s it it is possible to construct an automorphism of the proper Lorentz group

$$sgs^{-1} = (g^*)^{-1}.$$

This automorphism of the proper group carries over naturally into the group of complex matrices a with unit determinant, to wit, if to a proper Lorentz transformation $g_{\pm a}$ there correspond matrices of the second order $\pm a$, then to the proper transformation $sg_{\pm a}s^{-1}$ there correspond the matrices $\pm (a^*)^{-1}$.

In other words

$$sg_a s^{-1} = g_{(a^*)^{-1}}$$

that is to say,

$$(g_a^*)^{-1} = g_{(a^*)^{-1}}.$$

In fact, as we have just seen, a proper Lorentz transformation may be considered as a transformation in the space of second-order hermitian matrices given by the formula

$$c' = aca^*, \tag{10}$$

where a is a complex second-order matrix, and $\det a = 1$.

We find now how the matrices c transform upon spatial reflection

$$x_0 \to x_0, \quad x_1 \to - x_1, \quad x_2 \to - x_2, \quad x_3 \to - x_3.$$

It is clear that the reflection carries the matrix c into the matrix c' as follows:

$$c = \left\| \begin{matrix} x_0 - x_3 & x_2 - ix_1 \\ x_2 + ix_1 & x_0 + x_3 \end{matrix} \right\| \to \left\| \begin{matrix} x_0 + x_3 & -x_2 + ix_1 \\ -x_2 - ix_1 & x_0 - x_3 \end{matrix} \right\| = c'.$$

It is easily verified that c' may be written in the form

$$c' = \tau \bar{c} \tau^{-1}, \tag{11}$$

where $\tau = \begin{Vmatrix} 0 & 1 \\ -1 & 0 \end{Vmatrix}$ and the bar denotes the complex conjugate. In this manner, a spatial reflection generates the transformation (11) in the space of hermitian matrices.

Let there correspond to a proper transformation g_a the matrices $\pm a$. We determine now the matrices which correspond to the transformation $g^{*-1} = s g_a s^{-1}$. To this end, we use formula (11) and transform the matrix c by s^{-1}, g_a and s in turn. We obtain

$$c' = \tau \overline{\left[a\left(\tau^{-1}\bar{c}\tau\right) a^*\right]} \tau^{-1}$$

or

$$c' = \tau \bar{a} \tau^{-1} c \tau \bar{a}^* \tau^{-1}.$$

Hence we see that the transformation sgs^{-1} corresponds to the matrix $\tau\bar{a}\tau^{-1}$, i.e.

$$s g_a s^{-1} = g_{\tau \bar{a} \tau^{-1}}.$$

It is easily verified that if $\det a = 1$, then

$$\tau \bar{a} \tau^{-1} = (a^*)^{-1}. \tag{11'}$$

In this manner we obtain:

$$s g_a s^{-1} = g_{(a^*)^{-1}}$$

that is

$$g_a^{*-1} = g_{(a^*)^{-1}}.$$

II) The rotations $\tilde{g}$ in the three-dimensional space $x_0 = 0$ form, as we know, a subgroup of the proper Lorentz group. Hence, it follows that those complex matrices a, which in our correspondence $g_a \sim a$ correspond to the rotations $\tilde{g}$, also form a subgroup in the group of all complex, second-order matrices with unit determinant.

We shall now prove that this subgroup coincides with the group of all unitary matrices of the second order with determinant equal to 1. In other words in our constructed correspondence $g_a \sim a$ between complex matrices of the second order with determinant equal to 1, and the proper Lorentz transformations, the unitary matrices a correspond to the rotations $\tilde{g}_a$ in the three-dimensional space $x_0 = 0$, and conversely to each rotation $\tilde{g}$ there correspond two unitary matrices $\pm a$ differing only in sign, with determinant equal to 1.

In fact, let the complex matrix a be unitary, i.e. $a^{*-1} = a$. Then the transformation (10) in the space of the hermitian matrices may be written in the form

$$c = aca^{-1} \tag{12}$$

But under all possible transformations of the form (12) the trace (the sum of the diagonal elements) is preserved (remains constant) i.e.

$$(x_0' + x_3') + (x_0' - x_3') = (x_0 + x_3) + (x_0 - x_3),$$

hence

$$x_0' = x_0.$$

Consequently, the corresponding Lorentz transformations do not alter the fourth coordinate x_0 and are rotations in the space $x_0 = 0$. Thus we have proved that unitary matrices correspond to rotations in the three-dimensional space $x_0 = 0$.

We shall prove now that, conversely, to any rotation $\tilde{g}$ there correspond two unitary matrices of the second order $\pm a$ with unit determinant. To this end we consider those rotations which correspond to unitary matrices. It is clear that all such rotations $\tilde{g}_a$ form a subgroup $\tilde{G}_a$ of the group of rotations. The dimension (the number of independent parameters) of this subgroup is clearly equal to three, since it coincides with the dimension of the group of unitary matrices of the second order with determinant equal to one. The dimension of the group of rotations in three-dimensional space, as was shown in Part I, is also equal to three.

In this manner, the subgroup $\tilde{G}_a$ has the same dimension as the whole group of rotations, and consequently coincides with it (in virtue of the fact that the group of rotations is connected). Thus, to each rotation there correspond two matrices, differing only in sign, which are of the second order, with determinant equal to 1.

§ 5. The Relation Between the Proper Lorentz Group and the Group of Complex Matrices of the Second Order with Unit Determinant (Other Considerations)*

In the previous section we established a relation between the proper Lorentz group and the group of complex second-order matrices a, $\det a = 1$, such that with each such matrix was associated a proper Lorentz transformation. We now proceed to the converse: we set each proper Lorentz transformation in correspondence with two matrices (differing only in sign) of the second

---

* This section may be omitted at the first reading

order with determinant equal to 1. We shall construct this correspondence in a purely geometrical manner.

We recall first of all how the correspondence between rotations in three-dimensional space and bilinear transformations of the complex plane was set up. To this end we used the stereographic projection of a sphere on to the complex plane. Then it was shown that any rotation of the sphere generated in the complex plane a bilinear transformation with a unitary matrix. We shall examine an analogous construction for the Lorentz group.

We cut the light cone $x_0^2 - x_1^2 - x_2^2 - x_3^2 = 0$ with the hyperplane $x_0 = \frac{1}{2}$. The intersection is a three-dimensional sphere I of diameter 1: $x_1^2 + x_2^2 + x_3^2 = \frac{1}{4}$, $x_0 = \frac{1}{2}$. We specify a projective transformation in the hyperplane in the following manner. Any ray leaving the origin of coordinates and piercing this hyperplane is carried by any proper Lorentz transformation to a ray which again pierces this hyperplane (because the sense of the time axis is not altered). Thus each ray corresponds to a point of the hyperplane $x_0 = \frac{1}{2}$ namely the point of its intersection with it. In this way each transformation of the ray space, in particular each proper Lorentz transformation, specifies a transformation Γ of the points of the hyperplane $x_0 = \frac{1}{2}$. Since the transformation Γ carries a straight line into a straight line, and a plane into a plane, the transformation Γ is projective. Since a Lorentz transformation carries the light cone into itself, this means that the transformation in the hyperplane leaves the sphere I, defined by the equations $x_1^2 + x_2^2 + x_3^2 = \frac{1}{4}$, $x_0 = \frac{1}{2}$ in place. Thus with each transformation g from the proper Lorentz group is associated a transformation $\tilde{g}$ of the sphere I into itself. Since the transformation $\tilde{g}$ is induced by a projective transformation of three-dimensional space, it clearly carries a circle on the sphere into a circle and does not alter the orientation on it.

As before, we consider the stereographic projection of the sphere onto the plane T, which touches it at the point $\left(0, 0, -\frac{1}{2}\right)$ situated on the x_3 axis; (in the three-dimensional hyperplane $x_0 = \frac{1}{2}$ the coordinates are introduced in the usual manner). Under the stereographic projection of the sphere onto the plane T a circle is mapped onto a circle or a straight line and conversely; any circle or straight line on the plane T is the image of some circle on the sphere.

Any transformation $\tilde{g}$ of the sphere which carries a circle into a circle and preserves orientation (in particular a transformation produced by a proper Lorentz transformation) determines, by means of the stereographic projection, a transformation a of the plane T which carries a circle and a straight line into a circle or a straight line (and preserves orientation). If we consider the plane T as the plane of the complex variable z, then according to the theory of functions of a complex variable, such a transformation is a bilinear transformation of the plane z:

$$z' = \frac{\alpha z + \beta}{\gamma z + \delta}.$$

In this manner each transformation $\tilde{g}$ of the sphere I, and consequently each proper Lorentz transformation g, corresponds to a bilinear transformation of the plane z with matrix $a = \begin{Vmatrix} \alpha & \beta \\ \gamma & \delta \end{Vmatrix}$ and, ipso facto, the matrix a is uniquely determined except for an arbitrary multiplier. We select the multiplier such that the determinant of the matrix equals unity. In this way the matrix a is determined except for sign. It is impossible to exclude this latter indeterminacy. Thus the correspondence $g \sim \pm a$ has been constructed between Lorentz transformations and the complex second-order matrices $(\det a = 1)$ defined except for sign.

We note that in our geometric construction the rotations of the three-dimensional space $(x_1 x_2 x_3)$ correspond simply to rotations of the sphere I. And a rotation of the sphere I induces in the plane, under stereographic projection, a bilinear transformation with a unitary matrix as was proved in Sect.2 of Part I. In this manner the rotations g correspond to the unitary matrices a.

§ 6. The Lorentz Group as the Group of Motions in Lobachevskian Space

In the previous section we proved that the proper Lorentz group is isomorphic to the group of projective transformations of three-dimensional space, which transform some sphere into itself. Using this result, we shall now show that the Lorentz group may be considered as a group of motions in Lobachevskian space.

To this end we consider the model of Lobachevskian geometry adopted by Beltrami and Klein. In this model a point in Lobachevskian space is represented by an internal point of some sphere I of three-dimensional Euclidian space, a straight line by a chord of this sphere, and a plane by a region of a plane inside these sphere. The distance between the points A and B is determined thus: let P and Q be the points of intersection of the chord AB with the sphere I. Then the distance $\rho(A, B)$ between A and B is defined as the logarithm of the cross ratio of the points:

$$\rho(A, B) = \ln \left(\frac{AP}{PB} : \frac{AQ}{QB} \right).$$

The motions in Lobachevskian space, i.e., those transformations which do not alter distances in our model, correspond to projective transformations of three-dimensional space which carry the sphere I into itself.

In this manner, the group of motions of Lobachevskian space is isomorphic with the group of projective transformations of three-dimensional space which transform some sphere into itself. As was proved in the previous section the latter is isomorphic with the proper Lorentz group. Thus we obtain the result that the group of motions of Lobachevskian space is isomorphic with the

Lorentz group.

This last fact naturally suggests the construction of a model of Lobachevskian space on a surface of transitivity of the proper Lorentz group It appears that such a model may be constructed on the upper branch of the hyperboloid:

$$x_0^2 - x_1^2 - x_2^2 - x_3^2 = 1. \tag{13}$$

We shall briefly describe this model.

By a "straight line" in this model we understand any hyperbolic section of the hyperboloid (13) by a plane through the origin; a "plane" in this Lobachevskian space is the intersection of a three-dimensional hyperboloid passing through the origin with our hyperboloid.

If two planes, h_1 and h_2 in R^4, which pass through the origin intersect in a straight line l inside the light cone and then l has a common point with the hyperboloid (13) and so the "straight lines" H_1 and H_2 corresponding to the planes h_1 and h_2 intersect on the hyperboloid. If the line l lies on the cone then the "straight lines" H_1 and H_2 on the hyperboloid are parallel; if finally l lies outside the cone then the "straight lines" H_1 and H_2 are diverging. Parallelism and divergence of "planes" in Lobachevskian space are defined analogously in our model.

It is clear that it is possible to introduce a coordinate system (ξ, η, ζ) on the hyperboloid (13).

$$\xi = \frac{x_1}{x_0}, \quad \eta = \frac{x_2}{x_0}, \quad \zeta = \frac{x_3}{x_0}.$$

These coordinates (ξ, η, ζ) are called the Beltrami coordinates in our model of the geometry of Lobachevskian space.

The distance between two points $A(\xi_1, \eta_1, \zeta_1)$ and $B(\xi_2, \eta_2, \zeta_2)$ in Lobachevskian space in Beltrami coordinates ξ, η, ζ is given by the formula

$$\rho(A, B) = \frac{k}{2} \ln \frac{1 + \sqrt{1 - \tau^2}}{1 - \sqrt{1 - \tau^2}},$$

where

$$\tau = \frac{\sqrt{1 - \xi_1^2 - \eta_1^2 - \zeta_1^2} \cdot \sqrt{1 - \xi_2^2 - \eta_2^2 - \zeta_2^2}}{1 - \xi_1\xi_2 - \eta_1\eta_2 - \zeta_1\zeta_2},$$

and k is a fixed parameter.

It is easily verified that this metric differs only by a multiplier (factor) from the metric which is induced on the hyperboloid by the quadratic form $S^2(x) = x_0^2 - x_1^2 - x_2^2 - x_3^2$.

It would be possible to pursue other conceptions of Lobachevskian geometry but we shall restrict ourselves to the above.

§ 7. Definition of the Representations of the Lorentz Group and Fundamental Concepts of the Theory of Representations

In paragraph 5, Sect.1 of Part I we introduced the fundamental concepts of the theory of representations with particular reference to finite-dimensional representations of groups. Since all the irreducible representations of the rotation group are finite-dimensional, and any other of its representations can be resolved into a linear sum of irreducible constituents, consideration of finite-dimensional representations was sufficient.

Matters are different for the Lorentz group. As we shall see below, some of its irreducible representations are infinite dimensional. In this connexion we define afresh the basic concepts of the theory of representations of groups, in order that they may be applicable to the case of infinite dimensional representations.

Definition. Let R be a normed space, and suppose that to each element g of the group G we assign a bounded linear operator T_g in R in such a way that the following conditions are fulfilled:

1) $T_e = E$ (e is the identity of the group G, E is the unit operator in R);

2) $T_{g_1 g_2} = T_{g_1} \cdot T_{g_2}$;

3) continuity: if $F(f)$ is a bounded linear functional on R, then for any fixed f, $F(T_g f)$ depends continuously upon g.

Then the correspondence $g \to T_g$ is called a linear representation of the group G in the space R. The representation is called finite if the space R is finite (dimensional).

Unitary Representations. The representation $g \to T_g$ is called unitary if the space R is a Hilbert space and the scalar product (ξ, η) in R is invariant with respect to the operator T_g, i.e. if

$$(T_g\xi, T_g\eta) = (\xi, \eta).$$

In other words, a representation is unitary if it acts in a Hilbert space and if all the operators of the representation are unitary.

Irreducible Representations: We recall (see Part I, Sect.1) that a finite representation $g \to T_g$ is called irreducible if the space R in which it acts has no invariant subspaces apart from R itself, and the zero space. Such a definition becomes inconvenient in the infinite case. For this reason we shall define irreducibility by means of a rather more exacting condition than the simple absence of invariant subspaces. A representation $g \to T_g$, acting in the space R is irreducible if, firstly R has no closed subspaces invariant under all the operators T_g, and if secondly, any bounded operator A, permutable with all the operators T_g is a multiple of the unit operator: $A = \lambda E$.

For finite representations we may confine ourselves to either one of these two requirements, since in this case the two requirements are equivalent.* In the infinite case this is not always so.

If the representation $g \to T_g$ acting in space R is reducible then, as a rule, the space R may be decomposed into a linear sum of invariant subspaces $R_k : R = \sum^{\cdot} R_k$, in each of which the representation $g \to T_g$ induces an irreducible representation of G.

We denote the representation generated in the space R_k by $T_g^{(k)}$. The representations $g \to T_g^{(k)}$ are called the irreducible components of the representation $g \to T_g$.

Equivalent Representations. The finite representations $g \to T_g^{(1)}$ and $g \to T_g^{(2)}$ acting in the spaces $R^{(1)}$ and $R^{(2)}$ respectively are called equivalent if there exists an operator B which maps $R^{(1)}$ onto $R^{(2)}$ in a one-one biunique manner, such that for any element g of the group

$$BT_g^{(1)} = T_g^{(2)}B. \tag{14}$$

More graphically, this means that the representations are equivalent, if it is possible to establish a one-one linear correspondence $h^{(1)} \longleftrightarrow h^{(2)}$ between the elements $h^{(1)}$ of space $R^{(1)}$ and the elements $h^{(2)}$ of space $R^{(2)}$ such that if $h^{(1)} \longleftrightarrow h^{(2)}$ then $T_g^{(1)}h^{(1)} \longleftrightarrow T_g^{(2)}h^{(2)}$. The general definition of equivalence of representations, applicable both to the finite and the infinite case, differs little from that presented above. The representations $g \to T_g^{(1)}$ and $g \to T_g^{(2)}$ acting in the spaces $R^{(1)}$ and $R^{(2)}$ are said to be equivalent, if $R^{(1)}$ and $R^{(2)}$ contain everywhere dense linear manifolds $\tilde{R}^{(1)}$ and $\tilde{R}^{(2)}$ which are invariant under the operators $T_g^{(1)}$ and $T_g^{(2)}$ respectively, and there is a closed operator B which maps $\tilde{R}^{(1)}$ onto $\tilde{R}^{(2)}$ bi-uniquely and satisfies the equation:

$$T_g^{(1)}B = BT_g^{(2)}. \tag{14'}$$

This definition of equivalence amounts to the condition that it is possible to choose bases in the spaces $R^{(1)}$ and $R^{(2)}$ where the equivalent representations $g \to T_g^{(1)}$ and $g \to T_g^{(2)}$ act so that the operators $T_g^{(1)}$ and $T_g^{(2)}$ are written in terms of them by one and the same matrix.

It is clear that mutually equivalent representations are not substantially different. In the theory of representations one usually considers representations to within equivalence.

In conclusion we formulate an important proposition related to the

* This fact is the substance of the so-called Schur Lemma: every linear operator in a finite-dimensional space which commutes with a set of operators is a multiplie of the unit operator if and only if this set is irreducible. For the proof of this assertion see the footnote on p. 179.

definition of equivalent representations. Let there act in the finite-dimensional spaces $R^{(1)}$ and $R^{(2)}$ the irreducible representations $g \to T_g^{(1)}$ and $g \to T_g^{(2)}$ of some group. If there exists a linear operator B which maps the space $R^{(1)}$ into $R^{(2)}$ and satisfies the relation

$$T_g^{(2)} B = B T_g^{(1)}, \tag{15}$$

then either B maps $R^{(1)}$ onto $R^{(2)}$ bi-uniquely, and consequently the representations are equivalent, or $B = 0$.

This proposition in representation theory is usually called the general Schur Lemma. It is easy to deduce from it the assertion which we earlier designated as the Schur Lemma * (see the footnote on p. 177).

Representations equivalent to unitary representations. Such representations clearly possess the following property.

The representation $g \to T_g$ in the normed space R is equivalent to a unitary representation if there exists in the space R a positive definite hermitian bilinear form which is invariant under the operators T_g (this form may be defined either on the entire space R or on some everywhere dense linear submanifold R', also invariant under the operators T_g.)

In fact if by means of the invariant positive-definite hermitian form (ξ, η) we specify a scalar product in the space R and complete R with respect to this scalar product, then we obtain a Hilbert space $\tilde{R}$. It is possible to extend the operators T_g in R to unitary operators $\tilde{T}_g$ in $\tilde{R}$. It is clear that the representation $g \to T_g$ is equivalent to the unitary representation $g \to \tilde{T}_g$.

§ 8. The Relation between Representations of the Proper Lorentz Group and Representations of the Group of Complex Second-Order Matrices (Two-valued Representations of the Proper Lorentz Group)

Above we have studied in detail the correspondence $g_a \to \pm a$ between the proper Lorentz transformations and the group $\mathfrak{A}$ of complex second-order matrices a ($\det a = 1$). This correspondence $g_a \to \pm a$ obviously permits any representation $g \to T_g$ of the proper group to be considered as a representation of the group $\mathfrak{A}$: $a \to T_a \equiv T_{g_a}$. Here the equation $T_a = T_{-a}$ is satisfied. Conversely it is clear that any representations of the group $\mathfrak{A}$ $a \to T_a$

* For the statement of this lemma see the footnote on p. 177. We shall now give its proof.
Let the operator A commute with the operators of the irreducible representation T_g: $T_g A = A T_g$. Let λ be any eigenvalue of the operator A. It is clear that the operator $A - \lambda E$ commutes with T_g. $T_g(A - \lambda E) = (A - \lambda E) T_g$. Since the operator $(A - \lambda E)$ maps the whole of R into a proper subspace of it, then, in virtue of the general Schur lemma, it is equal to zero: $A - \lambda E = 0$ or finally $A = \lambda E$. Q.E.D.

such that $T_a = T_{-a}$ may be considered as a representation of the proper Lorentz group: $g_a \to T_{g_a} \equiv T_a$. If a representation of the group $\mathfrak{A}$ does not possess the property that $T_a = T_{-a}$ then it is not possible, strictly speaking, to consider it as a representation of the Lorentz group, since in this case each element $g = g_a$ is placed in correspondence with two distinct operators T_a and T_{-a}. We shall, however, consider these representations of the group $\mathfrak{A}$ on a par with those representations which satisfy the condition $T_a = T_{-a}$. For the sake of consistency in terminology we shall designate the representations of the group $\mathfrak{A}$ for which $T_a \neq T_{-a}$ as two-valued representations of the Lorentz group, and those for which $T_a = T_{-a}$ as unique representations of this group.

It can be proved that it is impossible to make a two-valued representation of the proper group unique by choosing one operator from each pair, T_a and T_{-a}, in such a way that the correspondence $g \to T_g$ so obtained is continuous.

We now prove that if a two-valued representation of the Lorentz group is irreducible then to each element of the group there correspond precisely two operators which differ only in sign; the situation is thus similar to the simpler case of the two-dimensional representation $g_a \to \pm a$. In fact, $T_{-a} = T_{-e}T_a$ where $-e = \begin{pmatrix} -1 & 0 \\ 0 & -1 \end{pmatrix}$. Since the matrix $-e$ commutes with all matrices a, the operator T_{-e} also commutes with all the operators T_a. In virtue of the irreducibility of the representation it follows that $T_{-e} = \lambda E$ where E is the unit operator.

Since, on the other hand $(T_{-e})^2 = T_{(-e)^2} = T_e = E, \lambda^2 = 1$; hence it follows that $\lambda = +1$ or $\lambda = -1$. In the first case $T_{-e} = +E$ $(\lambda = 1)$, $T_a = T_{-a}$ and we have a unique representation ; in the second case $T_{-e} = -E$ and $T_{-a} = -T_a$, i.e., the representation is two-valued and the operators T_{-a}, and T_a differ in sign only.

We note finally that the above reasoning clearly implies that the representation of a Lorentz group is two-valued or unique together with the representation of the group of rotations induced by it. In other words, each irreducible component of the representation of the group of rotations induced by an irreducible representation of the proper Lorentz group $g \to T_g$, is unique or two-valued simultaneously with this representation. This observation will be used in the following section to determine the unique and two-valued representations of the proper Lorentz group.

It may be verified easily that a two-valued representation of the proper Lorentz group, considered as a representation of the group $\mathfrak{A}$ of complex unimodular matrices with determinant equal to 1, is a faithful representation of this group, i.e. it is such that $T_{a_1} \neq T_{a_2}$ if $a_1 \neq a_2$.

Unique representations of the proper Lorentz group are not faithful representations of the group $\mathfrak{A}$ since $T_a = T_{-a}$. It is easy to prove, however, that in this case $T_{a_1} = T_{a_2}$ only when $a_1 = \pm a_2$.

§ 9. Two-valued Representations of the General Lorentz Group

The general Lorentz group is obtained from the proper Lorentz group G_0 by the addition of three reflections s, t, j (s is the spatial, t the temporal and j the complete reflection) and all possible elements of the form sg', tg', jg', where g' is an element of the proper group.

We note that the transformations e, s, t, j (e is the identity transformation) form a commutative group with the following multiplication table:

	e	s	t	j
	e	s	t	j
s	s	e	j	t
t	t	j	e	s
j	j	t	s	e

(16)

We shall call this group the group of reflections.

Now suppose that a representation $g \to T_g$ of the general group is specified. This representation generates a representation $g' \to T_{g'}$ of the proper group, and a representation $\tau \to T_\tau$ ($\tau = e, s, t, j$) of the group of reflections.

We consider, first of all, the case when the representation $g' \to T_{g'}$ of the proper group (generated by the representation of the general group) is two-valued $g' \to \pm T_{g'}$. Naturally the representation of the group of reflections is also two-valued:

$$e \to \pm E, \qquad s \to \pm S, \qquad t \to \pm T, \qquad j \to \pm J.$$

The operators S, T, J clearly combine in the following manner:

$$ST = \pm J, \qquad SJ = \pm T, \qquad TJ = \pm J,$$
$$S^2 = \pm E, \qquad T^2 = \pm E, \qquad J^2 = \pm E.$$

From these equations it follows easily that the operators T, S, J either commute with each other:

$$TS = ST, \qquad JS = SJ, \qquad TJ = JT,$$

or they anticommute:

$$TS = -ST, \qquad JS = -SJ, \qquad TJ = -JT.$$

Corresponding to this we consider two cases.

First Case. The operators, S, T, J commute. In this case, we can choose the sign for these operators so that they multiply according to table (16).

$$TS = ST = J, \quad JS = S, \quad J = T, \quad JT = T, \quad J = S,$$
$$S^2 = T^2 = J^2 = E.$$

It is clear that in this case the operators E, S, T, J specify a unique representation of the group of reflections $e \to E$, $s \to S$, $t \to T$, $j \to J$. A representation of the general group which leads to this unique representation of the group of reflections will be called a unique representation of the general group (the two-valued nature of this representation is related only with the two-valued nature of the representation of the proper group).

Second Case. The operators, S, T, J anticommute.

It is easily verified that by a choice of sign for the operators it is possible to secure that they combine according to the table:

	E	S	T	J
E	E	S	T	J
S	S	E	J	T
T	T	$-J$	E	$-S$
J	J	$-T$	S	$-E$

(17)

It is easily seen that it is not possible to select four operators out of the eight operators $\pm E$, $\pm S$, $\pm T$, $\pm J$ which form a unique representation of the group of reflections; in other words the representation $e \to \pm E$, $s \to \pm S$, $t \to \pm T$, $j \to \pm J$ of this group is essentially two-valued.

A representation of the general group, leading to such a two-valued representation of the group of reflections is said to be a two-valued * representation of the general group (its two-valued nature is connected not only with the two-valued nature of the representation of the proper group, but also with the two-valued nature of the representation of the group of reflections).

We note that a representation of the general group can be two-valued even if the representation of the proper group generated by it is unique; it suffices that the representation of the group of reflections is two-valued. We shall describe the necessary construction below (see Sect. 3).

In conclusion we note that, just as a two-valued representation of the proper group can be considered as a faithful, unique representation of the group $\mathfrak{A}$ of complex second-order matrices with determinant equal to 1, a two-valued representation of the group of reflections may be considered as a faithful, unique representation of a group composed of eight elements, e, e', s, s', t, t',

---

* Such two-valued representations, as we shall see below, are encountered in the physical applications of the theory of representations of the Lorentz group

j, j' with the following multiplication table:

$$e'^2 = s^2 = s'^2 = t^2 = t' = e, j^2 = j'^2 = e',$$
$$se' = e's = s', \quad te' = e't = t', \quad je' = e'j = j',$$
$$st = t's = j, \quad sj = js' = t, \quad ts = s't = j'.$$

The remaining relations are determined by those already written out (see (17)). Unique representations of the group of reflections are not faithful representations of this group of eight elements but are such that $T_{e'} = T_e$, $T_{s'} = T_s$ etc. This connexion between representations of the group of reflections and the group of eight elements that we have constructed is completely analogous to that which exists between the representations of the proper Lorentz group and the group of complex unimodular matrices of the second order.

§ 10. The Basic Differences between the Representations of the Group of Rotations of Three-dimensional Space and the Lorentz Group

In Part I of this book we saw that any representation of the group of rotations can be made unitary by a suitable choice of the scalar product, and moreover, any irreducible representation of this group is finite-dimensional. Neither of these facts hold for the Lorentz group, for it has no unitary representations (the defining representation itself of this group in four-dimensional space is not unitary), and, as we shall see below, it has infinite, irreducible, representations.

These essential differences between the representations of the group of rotations and the Lorentz group are tied up with the fact that the first of them is compact, i.e., from any sequence of rotations it is possible to select a converging subsequence. The second group is non-compact: it is possible to give a sequence of Lorentz transformations, of which no subsequence converges. We may convince ourselves of the compactness of the group of rotations as follows: every rotation is described by an orthogonal matrix of the third order; hence the group of rotations forms a closed manifold G in the nine-dimensional space of all matrices of the third order. Since the sum of the squares of all the elements of an orthogonal matrix $\|g_{ik}\|$ is equal to

$$\sum_{i,k} g_{ik}^2 = 3,$$

the closed manifold G in the nine-dimensional space is bounded and consequently compact.

In order to convince ourselves that the Lorentz group is non-compact we proceed thus: we choose an infinite sequence of points $A_1, A_2, A_3, \ldots, A_n, \ldots$, diverging to infinity, on the hyperboloid $S^2(x) = 1$. If we now take a sequence of Lorentz transformations g_n such that $g_n O = A_n$ (O is the vertex of the hyperboloid) then it is easily seen that the sequence $\{g_n\}$ has no converging subsequence.

From the theory of topological groups we know that it is possible to introduce an invariant integral on any compact group; any bounded function $f(g)$ on the compact group G has a finite integral $\int f(g)\,dg$, which is not altered by right and left translations of the group, i.e. such that

$$\int f(gg_0)\,dg = \int f(g_0g)\,dg = \int f(g)\,dg \text{ for all } g_0\text{*}$$

We recall how the existence of integral ensures the finiteness and unitary nature of all representations of a compact group. In the space R, in which the representation $g \to T_g$ of the compact group acts, we select a positive-definite bilinear hermitian form (ψ_1, ψ_2). We construct a new form

$$(\psi_1, \psi_2)_1 = \int (T_g\psi_1, T_g\psi_2)_2\,dg$$

(the function $T_g\psi_1, T_g\psi_2$ is bounded like any continuous function on a compact manifold). It is clear that the hermitian form $(\psi_1, \psi_2)_1$ is positive-definite and invariant under the operators T_g

$$(T_g\psi_1, T_g\psi_2)_1 = (\psi_1, \psi_2)_1.$$

If we use the form $(\psi_1, \psi_2)_1$ to specify a scalar product in the space R then the representation $g \to T_g$ will be unitary. The argument just cited fails for non-compact groups, and in particular for the Lorentz group, for two reasons. Firstly, although it is possible to introduce an invariant double-sided integral on the Lorentz group, this is however not defined over all bounded functions on this group. Secondly, the function $(T_g\psi_1, T_g\psi_2)$ by means of which we constructed the invariant scalar product, can even turn out to be unbounded, so that it is all the more impossible to integrate it.

Both these circumstances are consequences of the non-compactness of the Lorentz group.

We prove, finally, that any irreducible representation of a compact group is finite-dimensional.

Let there act in space R the irreducible representation $g \to T_g$ of a compact group. By the preceding remarks we may assume the representation is unitary. This means that if h is a vector from R and $\|h\| = 1$ then $\|T_gh\| = 1$ as well, i.e. the transforms of the unit vector h in R lie on a sphere of radius 1. It is clear that the compactness of the group implies the compactness of the set $\{T_gh\}$ of all transforms of the vector h. But any compact subset of the unit sphere in Hilbert space is contained in a finite subspace. On the other hand, in virtue of the irreducibility of the representation the linear envelope of the set $\{T_gh\}$ coincides with R. Thus R is finite. It is obvious that this argument which rests on the compactness of the group is quite useless in the case of the Lorentz group.

* In Part I, (Sect.1) for example, the invariant measure for the group of rotations was calculated.

Section 2. Infinitesimal Operators and Representations of the Proper Lorentz Group

§ 1. The Basic One-parameter Subgroups of the Lorentz Group

In Sect.2 of Part I we introduced for each representation of the group of rotations the (operators) matrices A_1, A_2, A_3, acting in the space R describing infinitesimal rotations about the axes x_1, x_2 and x_3. It was there proved that these three matrices A_k uniquely determine the representation of the rotation group. We shall construct the analogous operators for the proper Lorentz group.

As was proved in Part I, any rotation of three-dimensional space can be brought about by the successive performance of three rotational displacements: in the plane (x_1, x_2) (about the x_3 axis); in the plane (x_1, x_3) (about the x_2 axis) and finally, again in the plane (x_1, x_2) (about the x_3 axis).

Similarly each transformation of the Lorentz group is equivalent to the successive performance of six transformations of a special type: a transformation in the plane (x_1, x_2) which does not alter the coordinates x_3 and x_4, and the analogous transformations in the planes (x_1, x_3), (x_2, x_3), (x_1, x_0), (x_2, x_0), (x_3, x_0). We shall consider these transformations in greater detail.

We write down the transformation in the plane (x_1, x_2):

$$\begin{aligned} x_1^1 &= g_{11}x_1 + g_{12}x_2, \\ x_2^1 &= g_{21}x_1 + g_{22}x_2, \\ x_3^1 &= \qquad\qquad\quad x_3, \\ x_0^1 &= \qquad\qquad\quad x_0. \end{aligned}$$

It clearly does not alter the quadratic form $x_1^2 + x_2^2$. Consequently, it is a rotation in the plane (x_1, x_2) through an angle φ. The matrix of such a transformation has the form:

$$g_{12}(\varphi) = \begin{Vmatrix} \cos\varphi & \sin\varphi & 0 & 0 \\ -\sin\varphi & \cos\varphi & 0 & 0 \\ 0 & 0 & 1 & 0 \\ 0 & 0 & 0 & 1 \end{Vmatrix}$$

Similarly the transformations in the planes (x_1, x_3) and (x_2, x_3) are described by the matrices:

$$g_{13} = \begin{Vmatrix} \cos\varphi & 0 & \sin\varphi & 0 \\ 0 & 1 & 0 & 0 \\ -\sin\varphi & 0 & \cos\varphi & 0 \\ 0 & 0 & 0 & 1 \end{Vmatrix},$$

$$g_{23} = \begin{Vmatrix} 1 & 0 & 0 & 0 \\ 0 & \cos\varphi & \sin\varphi & 0 \\ 0 & -\sin\varphi & \cos\varphi & 0 \\ 0 & 0 & 0 & 1 \end{Vmatrix},$$

where φ is the angle of the corresponding rotation.

A transformation in the plane (x_3, x_0) does not change the first two coordinates or the quadratic form $x_3^2 - x_0^2$. The matrix of such a transformation * may be written analogously to the foregoing only in terms of hyperbolic functions

$$g_{03} = \begin{Vmatrix} 1 & 0 & 0 & 0 \\ 0 & 1 & 0 & 0 \\ 0 & 0 & \text{ch}\,\varphi & \text{sh}\,\varphi \\ 0 & 0 & \text{sh}\,\varphi & \text{ch}\,\varphi \end{Vmatrix}.$$

The matrices g_{01} and g_{02} are similar in appearance.

We note that for given i, k, the matrices $g_{ik}(\varphi)$ form a subgroup in the Lorentz group which depends upon one parameter (a so-called one-parameter subgroup). In fact, using the addition theorem for circular and hyperbolic functions, we easily see that

$$g_{ik}(\varphi_1)\, g_{ik}(\varphi_2) = g_{ik}(\varphi_1 + \varphi_2) \qquad (i,\ k = 0,\ 1,\ 2,\ 3).$$

§ 2. The Representation of the Elements of the Proper Lorentz Group in the Form of Products of Elements from the Basic One-parameter Subgroups

In Sect.1 of Part I it was proved that each rotation in three-dimensional space may be represented in the form of a product of three rotations

$$g = g_{\theta_1} g_{\varphi} g_{\theta_2},$$

where $g_{\theta_1} g_{\theta_2}$ are rotations about the z axis through the angles θ_1 and θ_2 respectively and g_φ is a rotation about the x axis through an angle φ. Similarly each element of the Lorentz group may be presented in the form of a product:

$$g = u_1 g_{03} u_2,$$

where u_1 and u_2 are rotations and g_{03} is a hyperbolic screw in the plane $x_0 x_3$.

In fact in paragraph 2, Sect.1, we saw that any proper Lorentz transformation g has the form

$$g = u g_{0A},$$

where u is a rotation, g_{0A} is a hyperbolic screw in the plane $(x_0 A)$, passing through the x_0 axis and the point image A of the apex O of the hyperboloid $S^2(x) = -1$ under the transformation g. It is clear that the hyperbolic screw g_{0A} is expressible in the form

$$g_{0A} = u_{0A}^{-1} g_{03}(t)\, u_{0A},$$

* A transformation in the plane (x, y) which does not alter the quadratic form $x^2 - y^2$ and preserves the sense of the axes has already been called a hyperbolic screw.

where u_{0A} is the rotation which carries the plane (x_0, A) into the plane (x_0, x_3).

Thus

$$g = uu_{0A}^{-1} g_{03}(t)\, u_{0A} = u_1 g_{03}(t)\, u_2,$$

where

$$u_1 = uu_{0A}^{-1}, \quad u_2 = u_{0A}.$$

Writing the rotations u_1 and u_2 as products of three plane rotations

$$u_1 = g_{12}(\theta_1')\, g_{13}(\varphi')\, g_{12}(\theta_2')$$

and

$$u_2 = g_{12}(\theta_1'')\, g_{13}(\varphi'')\, g_{12}(\theta_2''),$$

we finally obtain for a proper Lorentz transformation g:

$$g = g(\theta_1')\, g(\varphi')\, g(\theta_2')\, g_{03}(t)\, g_{12}(\theta_1'')\, g_{13}(\varphi'')\, g_{12}(\theta_2''). \tag{1}$$

In this manner any proper Lorentz transformation is factorized into a product of screws from the basic one-parameter subgroups g_{12}, g_{13}, g_{03}.

§ 3. Definition of the Infinitesimal Operators

Let some representation $g \to T_g$ of the proper Lorentz group be specified. Then the operators $Tg_{ik}(\varphi) = T_{ik}(\varphi)$ are functions of the parameter φ. We take any vector f from R. The screw $g_{ik}(\varphi)$ through the angle φ carries f over into the vector $T_{ik}(\varphi) f$: Its increment, therefore, is equal to

$$T_{ik}(\varphi) f - f = [T_{ik}(\varphi) - E] f.$$

For the vector f suppose that the limits $\lim\limits_{\varphi \to 0} \dfrac{(T_{ik}(\varphi) - E) f}{\varphi} = h_{ik}$ exist for all the pairs (i, k) $(i, k = 0, 1, 2, 3;\ i < k)$*

In this way the operators A_{ik}, B_i, acting on such vectors f are defined, according to the formulae

$$A_{ik} f = \lim_{\varphi \to 0} \frac{(T_{ik}(\varphi) - E) f}{\varphi} = h_{ik} \qquad (i, k = 1, 2, 3),$$

$$B_i f = \lim_{\varphi \to 0} \frac{(T_{0i}(\varphi) - E) f}{\varphi} = h_{0i} \qquad (i = 1, 2, 3).$$

---

* The totality of vectors f for which these limits exist forms a linear, everywhere dense subspace R' of R. The proof of the fact that R' is everywhere dense in R is essentially contained in the supplement to Sect.2 of Part I (in the finite case, R' coincides with R).

The operators A_{ik} and B_i and their linear combinations are known as infinitesimal operators of the representation $g \to T_g$. The operators A_{ik} and B_i describe infinitely small screws respectively in the planes (x_i, x_k) (ordinary screw) and (x_i, x_0) (hyperbolic screw). Three of these operators A_{12}, A_{13}, A_{23} corresponding only to three-dimensional rotations were considered in Sect. 2 of Part I*. There, commutation relations were established between them

$$[A_{12}, A_{13}] = -A_{23}, \quad [A_{12}, A_{23}] = A_{13}, \quad [A_{13}, A_{23}] = -A_{12} \qquad \text{(I-III)}$$

The remaining commutation relations are as follows:**

$$\left.\begin{aligned}
&[A_{12}, B_1] = B_2, && [A_{13}, B_1] = -B_3, && [A_{23}, B_1] = 0,\\
&[A_{12}, B_2] = -B_1, && [A_{13}, B_2] = 0, && [A_{23}, B_2] = B_3,\\
&[A_{12}, B_3] = 0, && [A_{13}, B_3] = B_1, && [A_{23}, B_3] = -B_2,
\end{aligned}\right\} \text{(IV-XII)}$$

$$[B_1, B_2] = -A_{12}, \quad [B_1, B_3] = A_{13}, \quad [B_2, B_3] = -A_{23}. \quad \text{(XIII-XV)}$$

For convenience we introduce instead of the operators A_{ik} and B_i the following combinations:

$$\left.\begin{aligned}
H_+ &= iA_{23} - A_{13}, & H_- &= iA_{23} + A_{13}, & H_3 &= iA_{12};\\
F_+ &= iB_1 - B_2, & F_- &= iB_1 + B_2, & F_3 &= iB_3.
\end{aligned}\right\} \qquad (2)$$

It is easy to obtain the commutation relations between these operators:

$$[H_+, H_3] = -H_+, \quad [H_-, H_3] = H_-, \quad [H_+, H_-] = 2H_3; \quad \text{(I'-III')}$$

$$\left.\begin{aligned}
&[F_+, H_+] = [H_-, F_-] = [H_3, F_3] = 0,\\
&[H_+, F_3] = -F_3, \qquad [H_-, F_3] = F_-,\\
&[H_+, F_-] = -[H_-, F_+] = 2F_3,\\
&[F_+, H_3] = -F_+, \qquad [F_-, H_3] = F_-;
\end{aligned}\right\} \quad \text{(IV'-XII')}$$

$$[F_+, F_3] = H_+, \quad [F_-, F_3] = -H_-, \quad [F_+, F_-] = -2H_3. \quad \text{(XIII'-XV')}$$

In conclusion we note that any subspace R' of R which is invariant under the representation $g \to T_g$ is also invariant under all the infinitesimal operators H_+, H_-, H_3, F_+, F_-, F_3; and conversely, a subspace R' which is invariant under the infinitesimal operators is also invariant under the representa-

* We denoted these operators in Part I by A_1, A_2, A_3,

$$A_1 = A_{23}, \quad A_2 = A_{13}, \quad A_3 = A_{12}.$$

** They are obtained in precisely the same way as the foregoing (see Part I, Sect. 2).

tion $g \to T_g$ itself. It follows in particular, that the representation $g \to T_g$ is irreducible if, and only if, the space R in which it acts is irreducible with respect to its infinitesimal operators (i.e. if in R there is no subspace invariant under all the operators H_+, H_-, H_3, F_+, F_-, F_3). We shall constantly use this observation in the sequel.

§ 4. The Form of the Infinitesimal Operators for Irreducible Representations of the Proper Lorentz Group

In this paragraph we shall find the general form of the operators H_+, H_-, H_3, F_+, F_-, F_3 for an irreducible representation of the proper Lorentz group.

We note that any representation of the proper Lorentz group $g \to T_g$ which acts in the space R, generates ipso facto a representation of the subgroup of all rotations. This representation is obtained if we restrict ourselves only to those operators $T_{g'}$, which correspond to three-dimensional rotations g'. The representation $g' \to T_{g'}$ in space R is generally speaking reducible. But, as was proved in Sect.2 of Part I, R may be analysed into a linear sum of invariant subspaces R_l, in each of which an irreducible representation of weight l of the group of rotations is induced by the representation $g' \to T_{g'}$. We shall assume that whenever the representation of $g \to T_g$ of the proper Lorentz group is irreducible no two subspaces R_l with the same weight are encountered* in this analysis.

For this reason we may number these subspaces simply by the index l. In each subspace R_l we select a canonical basis ξ_{lm} (i.e. a basis consisting of eigenvectors of the operator H_3). The vectors $\{\xi_{lm}\}$ clearly form a basis for the entire space R. We shall call this basis the canonical basis in space R.

The operators H_+, H_-, H_3 (the infinitesimal operators of the representation of the group of rotations) are written in this basis in the following manner (see Part I, Sect.2 (18)):

$$\left.\begin{aligned} H_3\xi_{lm} &= m\xi_{lm}, \\ H_-\xi_{lm} &= \sqrt{(l+m)(l-m+1)}\,\xi_{l,\,m-1}, \\ H_+\xi_{lm} &= \sqrt{(l+m+1)(l-m)}\,\xi_{l,\,m+1}, \\ m &= -l,\ -l+1,\ \ldots,\ l-1,\ l. \end{aligned}\right\} \tag{3}$$

We now write out the operators F_3, F_+, F_- in the basis $\{\xi_{lm}\}$:

$$\begin{aligned} F_3\xi_{lm} &= C_l\sqrt{l^2-m^2}\,\xi_{l-1,\,m} - A_l m\xi_{lm} - C_{l+1}\sqrt{(l+1)^2-m^2}\,\xi_{l+1,\,m}, \\ F_+\xi_{lm} &= C_l\sqrt{(l-m)(l-m-1)}\,\xi_{l-1,\,m+1} - \\ &\quad - A_l\sqrt{(l-m)(l+m+1)}\,\xi_{l,\,m+1} + \end{aligned}$$

* This assumption is not arbitrary. It may be proved that it is in fact always satisfied for irreducible representations, both finite and infinite.

$$
\left.\begin{aligned}
&\qquad\qquad + C_{l+1}\sqrt{(l+m+1)(l+m+2)}\,\xi_{l+1,\,m+1},\\
F_{-}\xi_{lm} &= -C_l\sqrt{(l+m)(l+m-1)}\,\xi_{l-1,\,m-1} -\\
&\quad - A_l\sqrt{(l+m)(l-m+1)}\,\xi_{l,\,m-1} -\\
&\qquad\qquad - C_{l+1}\sqrt{(l-m+1)(l-m+2)}\,\xi_{l+1,\,m-1},
\end{aligned}\right\} \quad (3')
$$

$$
A_l = \frac{il_0 l_1}{l(l+1)}, \qquad C_l = \frac{i}{l}\sqrt{\frac{(l^2-l_0^2)(l^2-l_1^2)}{4l^2-1}},
$$

$$
m = -l, \quad -l+1, \ldots, l-1, \; l; \qquad l = l_0, \; l_0+1, \ldots
$$

Here l_1 is some complex number and formulae (3) and (3') give the form of the infinitesimal operators of an irreducible representation of the proper Lorentz group. It is apparent from these formulae that each such representation is uniquely defined by the pair of numbers l_0 and l_1; the first number, l_0 is the smallest weight participating * in the irreducible representation and consequently may be taken only as integral or half integral; the number l_1 is arbitrary.

The successive derivation of formulae (3') will be given below, but first we shall acquaint ourselves with them in a little more detail. We note that if the weight l participates in the irreducible representation then it is apparent from formula (3') that the vectors from the space R_l are carried by the operators F_+, F_-, F_3 into linear combinations of vectors from the spaces R_{l-1}, R_l and R_{l+1}; in this case the vectors R_{l-1} enter into the linear combination if, and only if, $C_l \neq 0$; similarly the vectors from R_{l+1} participate in this linear combination if, and only if, $C_{l+1} \neq 0$. In particular the vectors from the space R_{l_0} are carried by the operators F_3, F_+, F_- only into a linear combination of vectors from R_{l_0} and R_{l_0+1} $(C_{l_0} = 0)$. This entirely accords with the fact that l_0 is the smallest of the weights participating in the representation.

Thus, the operators F_3, F_+, F_- carry the vectors from the space R_{l_0} into vectors belonging to the sum of the spaces R_{l_0} and R_{l_0+1}; these, in their turn, under the action of the operators F_3, F_+, F_- transform to vectors belonging to the sum of the spaces R_{l_0}, R_{l_0+1}, R_{l_0+2}. Continuing this process we construct a finite or infinite chain of spaces

$$
R_{l_0} R_{l_0+1} R_{l_0+2} \ldots \qquad (3'')
$$

It is clear that the sum of all the spaces R_l entering into this chain

---

* We shall say that the weight l participates in a representation of the proper group $g \to T_g$ if the representation of the group of rotations, generated by the representation $g \to T_g$ contains an irreducible component of weight l.

is invariant under the operators H_3, H_+, H_-, F_3, F_+, F_- and, consequently, in virtue of the irreducibility of our representation, coincides with the entire space R. From the very construction of the chain (3") it is clear that the weights l participating in our representation run through a subsequence of the the values

$$l_0, \; l_0+1, \; l_0+2, \; \ldots$$

This is also indicated in formula (3').

We note that in the case when the chain (3") is infinite the representation is infinite. In the case when the chain terminates in some maximum weight $\tilde{l}$ the representation is finite. In the latter case it is easy to see how the maximum weight $\tilde{l}$ is related to the number l_1. In fact the chain terminates at the weight $\tilde{l}$, as we have said, only if $C_{\tilde{l}+1} = 0$. But since $\tilde{l} \geqslant l_0$ this is only possible for $(\tilde{l}+1)^2 - l_1^2 = 0$. Hence $\tilde{l} = |l_1| - 1$. The latter equality is possible only if l_1 is integral or half-integral simultaneously with l_0 and $|l_1| > l_0$. This is also the condition for the irreducible representation to be finite. Below we shall return again to this case. Thus, we see finally that any irreducible representation of the proper Lorentz group is defined by a pair of numbers, (l_0, l_1) where l_0 is an integer or half-integer and l_1 is an arbitrary complex number.

The infinitesimal operators H_+, H_-, H_3, F_3, F_+, F_- are specified in this case by the formulae (3) and (3'). The weights l participating in this representation take in turn all the values l_0+1, l_0+2 ... etc. In this case the representation is either infinite and the weights l tend to infinity, or the representation is finite and contains a maximum weight $\tilde{l}$. The latter case arises if, and only if, l_1 is integral or half-integral simultaneously with l_0 and $|l_1| > l_0$; in this case the maximum weight $\tilde{l}$ is equal to $|l_1| - 1$.

After this preliminary discussion we proceed to the derivation of formulae (3').

We turn, first of all, to the commutation relations between the A_{ik} and the B_i We note that after changes of notation $B_i \longleftrightarrow L_i$, $A_{sk} = A_j$ $(s \neq k \neq j)$, the formulae (IV-XII) coincide with formulae (5) Sect. 9 of Part I, which give the commutation relations between the matrices A_i and the matrices L_i of invariant equations. In Sect. 9 the general form of all matrices L_i in the basis $\{\xi_{lm}\}$ was found (we shall omit the index τ in virtue of our assumption that no weight is encountered more than once).

We write out the expressions obtained for the operators $F_+ = iB_1 - B_2$, $F_- = iB_1 + B_2$, $F_3 = iB_3$:

$$F_3 \xi_{lm} = d_{l-1,\,l} \sqrt{l^2 - m^2}\, \xi_{l-1,\,m} - \\ - d_{ll} m \xi_{lm} - d_{l+1,\,l} \sqrt{(l+1)^2 - m^2}\, \xi_{l+1,\,m}, \tag{4}$$

$$F_+\xi_{lm} = d_{l-1,\,l}\sqrt{(l-m)(l-m-1)}\,\xi_{l-1,\,m+1} - \\ - d_{ll}\sqrt{(l-m)(l+m+1)}\,\xi_{l,\,m+1} + \\ + d_{l+1,\,l}\sqrt{(l+m+1)(l+m+2)}\,\xi_{l+1,\,m+1}, \quad (5)$$

$$F_-\xi_{lm} = -d_{l-1,\,l}\sqrt{(l+m)(l+m-1)}\,\xi_{l-1,\,m-1} - \\ - d_{ll}\sqrt{(l+m)(l-m+1)}\,\xi_{l,\,m-1} - \\ - d_{l+1,\,l}\sqrt{(l-m+1)(l-m+2)}\,\xi_{l+1,\,m-1} \quad (6)$$

(we have somewhat altered the notation of Sect.9 of Part I, taking

$$d_{l,\,l} = -ic_{l,\,l}, \quad d_{l+1,\,l} = -ic_{l+1,\,l}).$$

We note that there is some arbitrariness in the selection of the numbers $d_{l-1,\,l}$ $d_{l,\,l}$ $d_{l+1,\,l}$ depending upon the arbitrariness in the normalization of the basis $\{\xi_{lm}\}$.

In fact if we dilate all the vectors of the basis in each of the subspaces R_l in the same manner, i.e., if we take

$$\xi'_{lm} = h(l)\,\xi_{lm},$$

where $h(l)$ is some number, then the form of the operators H_+, H_-, H_3 in the new basis is the same as before, for they act independently in each R_l. Such a change of basis carries the numbers $d_{l-1,\,l}$, $d_{l,\,l}$, $d_{l+1,\,l}$ into

$$d'_{l-1,\,l} = \frac{h(l)}{h(l-1)}\,d_{l-1,\,l}, \quad d'_{ll} = d_{ll}, \quad d'_{l+1,\,l} = \frac{h(l)}{h(l+1)}\,d_{l+1,\,l}.$$

Consequently

$$d'_{l-1,\,l}d'_{l,\,l-1} = \frac{h(l)}{h(l-1)}\,\frac{h(l-1)}{h(l)}\,d_{l-1,\,l}\,d_{l,\,l-1} = d_{l-1,\,l}\,d_{l,\,l-1},$$

i.e. the product $d_{l-1,\,l}\,d_{l,\,l-1}$ is preserved. By a suitable choice of multipliers $h(l)$ the numbers $d'_{l-1,\,l}$ and $d'_{l,\,l-1}$ can be made equal *.

We shall assume that this has already been arranged and write

$$d'_{l,\,l-1} = d'_{l-1,\,l} = C_l; \qquad d'_{ll} = A_l.$$

---

* To this end it suffices to take

$$h(l) = C\sqrt{\prod_{r=l_0}^{r=l}\frac{d_{r+1,\,r}}{d_{r-1,\,r}}},$$

where l_0 is the smallest weight participating in the representation.

Then formulae (4), (5) and (6) may be rewritten as follows:

$$F_3\xi_{lm} = C_l\sqrt{l^2 - m^2}\,\xi_{l-1,\,m} - A_l m\xi_{lm} - \\ - C_{l+1}\sqrt{(l+1)^2 - m^2}\,\xi_{l+1,\,m}, \quad (7)$$

$$F_+\xi_{lm} = C_l\sqrt{(l-m)(l-m-1)}\,\xi_{l-1,\,m+1} - \\ - A_l\sqrt{(l-m)(l+m+1)}\,\xi_{l,\,m+1} + \\ + C_{l+1}\sqrt{(l+m+1)(l+m+2)}\,\xi_{l+1,\,m+1}, \quad (8)$$

$$F_-\xi_{lm} = -C_l\sqrt{(l+m)(l+m-1)}\,\xi_{l-1,\,m-1} - \\ - A_l\sqrt{(l+m)(l-m+1)}\,\xi_{l,\,m-1} - \\ - C_{l+1}\sqrt{(l-m+1)(l-m+2)}\,\xi_{l+1,\,m-1}. \quad (9)$$

In order to determine the numbers A_l and C_l we now use the commutation relations between F_3 and F_+: $[F_+, F_3] = H_3$. Substituting F_+, F_3, H_3 therein and equating the coefficients of identical vectors, we obtain the following equations

$$\left.\begin{aligned} [A_l(l+1) - (l-1)A_{l-1}]\,C_l &= 0, \\ [A_{l+1}(l+2) - lA_l]\,C_{l+1} &= 0, \\ (2l-1)C_l^2 - (2l+3)C_{l+1}^2 - A_l^2 &= 1. \end{aligned}\right\} \quad (10)$$

The remaining relations $[F_3, F_-] = -2H_3$ and $[F_-, F_3] = -H_-$ lead to the same equations.

The evaluation of A_l. When $C_l \neq 0$ the first two of equations (10) imply that

$$A_l(l+1) = (l-1)A_{l-1},$$
$$A_{l-1}l = (l-2)A_{l-2},$$
$$\cdots\cdots\cdots\cdots\cdots\cdots$$
$$A_{l_0+1}(l_0+2) = l_0 A_{l_0},$$

where l_0 is the smallest weight participating in the representation. Hence

$$A_l = \frac{A_l\, l_0(l_0+1)}{l(l+1)}.$$

A_{l_0} may clearly be chosen arbitrarily. For the sake of symmetry in the subsequent formulae it is convenient to take $A^{l_0}(l_0+1) = il_1$. Then we obtain:

$$A_l = \frac{il_0 l_1}{l(l+1)} \quad (11)$$

Evaluation of C_l. We multiply both sides of the last of equations (10) by $2l+1$ and substitute the value of A_l just derived:

$$(4l^2-1)C_l^2-[4(l+1)^2-1]\,C_{l+1}^2=2l+1-l_0^2l_1^2\left[\frac{1}{l^2}-\frac{1}{(l+1)^2}\right].$$

If we write out all these equations from $l+1$ to l_0 and add, we obtain

$$-[4(l+1)-1]\,C_{l+1}^2=\sum_{p=l_0}^{p=l}(2p+1)-l_0^2l_1^2\left[\frac{1}{l_0^2}-\frac{1}{(l+1)^2}\right].$$

After a little manipulation we find:

$$C_{l+1}^2=-\frac{\left[(l+1)^2-l_0^2\right]\left[(l+1)^2-l_1^2\right]}{(l+1)^2\,[4(l+1)^2-1]}.$$

Finally we obtain the following expression for C_l

$$C_l=\frac{i}{l}\sqrt{\frac{(l^2-l_0^2)(l^2-l_1^2)}{4l^2-1}}\;^{*)}. \tag{12}$$

We now make the following observation. When deriving formulae (11) and (12) from equations (10) we assumed that $C_{l_0+1}, C_{l_0+2}, \ldots, C_l$ do not vanish. We shall prove that this requirement is always satisfied by an irreducible representation. Let C_l be the first of these numbers to vanish: $C_l=0$, $l=l_0+n+1$. In this case, it is clear from the formulae (3), (4) and (5), that the space $\overline{R}$ spanned by the vectors $\{\xi_{l\,m}\}$ $\{\xi_{l+1,\,m}\}$ $\ldots$ $\{\xi_{l_0+n,\,m}\}$ is invariant under the operators H_+, H_-, H_3, F_+, F_-, F_3 and, consequently, invariant under all the operators of the representation T_g. Since the representation $g\to T_g$ is irreducible, $\overline{R}$ coincides with R and the other weights $l>l_0+n$ do not participate in the representation. Thus, in fact, for any weight $l>l_0$ which participates in the representation $C_l\neq 0$. Thus we obtain finally the expression for the operators F_+, F_-, F_3:

$$F_3\xi_{lm}=C_l\sqrt{(l^2-m^2)}\,\xi_{l-1,\,m}-A_lm\xi_{l,\,m}-$$
$$-C_{l+1}\sqrt{(l+1)^2-m^2}\,\xi_{l+1,\,m}, \tag{13}$$

---

* Formula (12) defines the numbers C_l to within sign. However, by a transformation of the canonical basis ξ_{lm} of the form

$$\xi'_{lm}=(-1)^{\theta_l}\xi_{lm}\qquad(\theta_l=0,1)$$

we may always ensure that

$$|\arg C_l|\leqslant\frac{\pi}{2}.$$

In particular, if C_l is real then we may consider it to be positive.

$$F_+\xi_{lm} = C_l\sqrt{(l-m)(l-m-1)}\,\xi_{l-1,\,m+1} - $$
$$- A_l\sqrt{(l-m)(l+m+1)}\,\xi_{l,\,m+1} +$$
$$+ C_{l+1}\sqrt{(l+m+1)(l+m+2)}\,\xi_{l+1,\,m+1}, \qquad (14)$$

$$F_-\xi_{lm} = -C_l\sqrt{(l+m)(l+m-1)}\,\xi_{l-1,\,m-1} -$$
$$- A_l\sqrt{(l+m)(l-m+1)}\,\xi_{l,\,m-1} -$$
$$- C_{l+1}\sqrt{(l-m+1)(l-m+2)}\,\xi_{l+1,\,m-1}, \qquad (15)$$

$$A_l = \frac{il_0l_1}{l(l+1)}, \qquad C_l = \frac{i}{l}\sqrt{\frac{(l^2-l_0^2)(l^2-l_1^2)}{4l^2-1}}, \qquad (16)$$
$$m = -l, \quad -l+1, \ldots, \quad l-1, \ l,$$
$$l = l_0, \quad l_0+1, \ldots$$

We also write down the formulae for H_+, H_-, H_3:

$$H_3\xi_{lm} = m\xi_{lm}, \qquad (17)$$
$$H_+\xi_{lm} = \sqrt{(l+m+1)(l-m)}\,\xi_{l,\,m+1}, \qquad (18)$$
$$H_-\xi_{lm} = \sqrt{(l+m)(l-m+1)}\,\xi_{l,\,m-1}, \qquad (19)$$
$$m = -l, \quad -l+1, \ldots, \quad l-1, \quad l.$$

Thus any irreducible representation of the proper Lorentz group is defined by a pair of numbers (l_0, l_1) (l_0 is a positive integer or half-integer, l_1 is an arbitrary complex number), and its infinitesimal operators in the canonical basis $\{\xi_{lm}\}$ have the form (13)-(19).

We now note that under a simultaneous change of sign of the numbers in the pair (l_0, l_1): $(l_0, l_1) \to (-l_0, -l_1)$ the form of the formulae for the infinitesimal operators is not altered. We shall therefore specify an irreducible representation both by the pair (l_0, l_1) and by the pair $(-l_0, -l_1)$. It is clear that two equivalent irreducible representations are defined by one and the same pair $(\pm l_0, \pm l_1)$ and their infinitesimal operators in the canonical bases are written down in the same way.

To conclude this paragraph we make the following observation. The formulae (13)-(19) for the infinitesimal operators H_3, H_+, H_-, F_3, F_+, F_- of an irreducible representation are the general solution of the commuation relations (I-XV) (for the case when the space in which these operators act is irreducible with respect to them). Since the infinitesimal operators of any representation satisfy the relations (I-XV) we may be assured that the infinitesimal operators for irreducible representations of the proper group have the form (13)-(19).

The converse is not at all obvious: do any six operators specified by formulae (13)-(19) serve as the infinitesimal operators of some irreducible represent tion? In other words, for any pair (l_0, l_1) (l_0 is integral or half-integral, l_1, is an arbitrary complex number) does there, in fact exist a representation which is defined by it?

An affirmative answer will be given to this question in the following: we shall in fact construct an irreducible representation of the proper Lorentz group corresponding to each admissible pair (l_0, l_1).

§ 5. The Unique and Two-valued Representations of the Proper Lorentz Group

We now determine those pairs (l_0, l_1) which correspond to unique representations of the proper group, and the pairs corresponding to two-valued representations.

In the previous paragraph we saw that if an irreducible representation of the proper Lorentz group is two-valued or unique then simultaneously each irreducible component of the representation of the group of rotations generated by this representation of the Lorentz group is two-valued or unique. We apply this remark to the representation of the group of rotations with the lowest weight l_0.

In Part I of this book is was proved that an irreducible representation of the group of rotations specified by an integral weight is unique, and by a half-integral weight is two-valued. Thus an irreducible representation of the proper Lorentz group is unique for integral l_0 and two-valued for half-integral l_0.

§ 6. Conjugate Representations

We note that with each representation $g \to T_g$ there is closely connected another representation $g \to T_{(g^*)^{-1}}$, acting in the same space. We shall call the representation $g \to T_{(g^*)^{-1}}$ the conjugate of the representation $g \to T_g$.

Since the initial representation $g \to T_g$ is conjugate to the representation $g \to T_{(g^*)^{-1}}$, the two representations $g \to T_g$ and $g \to T_{(g^*)^{-1}}$ are mutually conjugate and we shall simply call them conjugate representations.

We now find out how the infinitesimal operators H_+, H_-, H_3, F_+, F_-, F_3 of the representation $g \to T_g$ and the infinitesimal operators $\tilde{H}_+$, $\tilde{H}_-$, $\tilde{H}_3$, $\tilde{F}_+$, $\tilde{F}_-$, $\tilde{F}_3$ of the representation $g \to T_{(g^*)^{-1}}$ are related.

To this end we recall that the transformations $g_{0i}(\varphi)_i = 1, 2, 3$, satisfy the following relations

$$g^*_{0i} = g_{0i} \tag{20}$$

and consequently, $(g^*_{0i})^{-1} = g^{-1}_{0i}$. For the transformations g_{kl} ($k, l = 1, 2, 3$) we have $g^*_{kl} = g^{-1}_{kl}$ and $(g^*_{kl})^{-1} = g_{kl}$. Hence it appears that the infinitesimal operators $\tilde{A}_{ik}$ and $\tilde{B}_i$ of the representation $g \to T_{(g^*)^{-1}}$ are related to the infinitesimal operators A_{ik} and B_i of the representation $g \to T_g$ in the following manner:

$$\left.\begin{aligned} \tilde{A}_{ik} &= A_{ik}, \\ \tilde{B}_i &= -B_i, \end{aligned}\right\} \quad i, k = 1, 2, 3, \tag{21}$$

and consequently

$$\tilde{F}_+ = -F_+, \quad \tilde{F}_- = -F_-, \quad \tilde{F}_3 = -F_3. \tag{22}$$

Actually the first of the equations follows immediately from the fact that $T_{\left(\overset{*}{g}_{ik}\right)^{-1}} = T_{g_{ik}}$ $(i, k = 1, 2, 3)$ the second is obtained from the following obvious equalities:

$$\tilde{B}_i = \frac{d}{d\varphi} T_{\left(\overset{*}{g}_{0i}\right)^{-1}(\varphi)}\bigg|_{\varphi=0} = \frac{d}{d\varphi} T_{g_{0i}^{-1}(\varphi)}\bigg|_{\varphi=0} = \frac{d}{d\varphi} T_{g_{0i}(-\varphi)}\bigg|_{\varphi=0} = -B_i.$$

It follows from equations (21) that two conjugate representations $g \to T_g$ and $g \to T_{(g^*)^{-1}}$ have exactly the same subspaces. In fact, since the operators A_{ik}, B_i differ by no more than sign alone from the operators $\tilde{A}_{ik}$, $\tilde{B}_i$, the subspaces which are invariant under the first six operators (and ipso facto also invariant with respect to the representation $g \to T_g$) are also invariant under the second sextet $\tilde{A}_{ik}$, $\tilde{B}_i$ (and, consequently, with respect to the representation $g \to T_{(g^*)^{-1}}$).

From this it follows, in particular, that the conjugate representations $g \to T_g$ and $g \to T_{(g^*)^{-1}}$ are reducible or irreducible together.

We now find how the pairs (l_0, l_1) and $(\tilde{l}_0, \tilde{l}_1)$ defining the two irreducible conjugate representations $g \to T_g$ and $g \to T_{(g^*)^{-1}}$ are related.

We note that the equations $T_{g_{ik}} = T_{\left(\overset{*}{g}_{ik}\right)^{-1}}$ $(i, k = 1, 2, 3)$ mean that the representations of the group of rotations generated in the space R by the conjugate representations $g \to T_g$ and $g \to T_{(g^*)^{-1}}$ coincide. In particular, this means that conjugate irreducible representations have their lowest weights l_0 and $\tilde{l}_0$ equal (we recall that l_0 is the lowest of the weights l of the irreducible representations of the group of rotations participating in the irreducible representation $g \to T_g$ of the proper Lorentz group).

Further, from relations (22) and from formulae (13)-(16) for the infinitesimal operators F_3, F_+, F_- of the irreducible representation $g \to T_g$ we find:

$$A_l = -A_l$$

or (see formula (16))

$$l_0 l_1 = -\tilde{l}_0 \tilde{l}_1.$$

Consequently, as $l_0 = \tilde{l}_0$,

$$l_1 = -\tilde{l}_1.$$

Thus, if the representation $g \to T_g$ is irreducible and defined by the pair (l_0, l_1), then its conjugate representation $g \to T_{(g^*)^{-1}}$ is also irreducible and is defined by the pair $(l_0, -l_1)$ or, what is the same thing, $(-l_0, l_1)$. Hence it follows that an irreducible representation $g \to T_g$ is equivalent to its conjugate $g \to T_{(g^*)^{-1}}$ if, and only if, either $l_0 = 0$ or $l_1 = 0$, i.e., if, and only if, one of the numbers of the pair (l_0, l_1) vanishes.

In what follows $\dot{\tau}$ denotes the irreducible representation conjugate to the representation τ. We will use conjugate irreducible representations of the proper Lorentz group in the next section to construct the irreducible representations of the complete (general) Lorentz group.

Notation. In the following we shall designate as the conjugate to the representation $g \to T_g$ not only the representation $g \to T_{(g^*)^{-1}}$, acting in the same space, but also any representation equivalent to the representation $g \to T_{(g^*)^{-1}}$.

§ 7. Finite Representations of the Proper Lorentz Group

Here we shall once again indicate how the pairs (l_0, l_1) define finite representations. It was observed above that if C_{l_0+n+1} is zero for the first time, then by formulae (7), (8) and (9) the irreducible representation contains all the weights l_0, $l_0+1, \ldots, l_0+n$ and only these. Conversely, for the finite representation of maximum weight $l = l_0 + n$, formulae (7), (8) and (9) show that C_{l_0+n+1} reduces to zero: $C_{l_0+n+1} = 0$. But $C_{l_0+n+1} = 0$, only if $l_1^2 = (l_0 + n + 1)^2$ (see (16)). Hence $|l_1| - 1 = l_0 + n$, i.e. l_1 is integral or half-integral (simultaneously with l_0), and $|l_1| - 1$ is the maximum weight participating in the finite representation. Thus, a representation is finite when l_0 and l_1 are simultaneously integral or half-integral and $|l_1| \geq |l_0|$ in which case all the weights from $|l_0|$ to $|l_1| - 1$ inclusive participate in the representation. In the case of other pairs (l_0, l_1) the representation is infinite. We note also that with each finite representation defined by the pair (l_0, l_1) there is closely related another, infinite representation with the pair (l_1, l_0). This representation is known as "the tail" of the finite representation (l_0, l_1) The formulae for the infinitesimal operators of "the tail" are precisely the same as for the finite representation itself, since l_0 and l_1 enter symmetrically into these formulae. However in the first case $l_0 \leqslant l \leqslant |l_1|$, and in the second $|l_1| \leqslant l < \infty$.

We shall now consider three important examples of finite representations of the proper group.

First Example. In Sect.1 we constructed a correspondence between the elements $g_a \sim a_g$ of the proper Lorentz group and the second-order complex matrices determinant equal to 1, determined except for sign. It is easily verified that this correspondence specifies an irreducible two-valued representation of the proper Lorentz group, which acts in the two-dimensional complex space $(z_0 z_1)$ according to the formula

$$\left.\begin{aligned} z'_0 &= a_{00}z_0 + a_{01}z_1, \\ z'_1 &= a_{10}z_0 + a_{11}z_1, \end{aligned}\right\} \tag{23}$$

where the matrix $\|a_{ij}\| = a_g$.

We note that if g is a rotation then a_g is a unitary matrix and formula (23) specifies an ordinary spinor representation of the group of rotations of weight $l = \frac{1}{2}$. Thus in the irreducible representation (23) there participates one weight $l = \frac{1}{2}$. Consequently the numbers of the pair (l_0, l_1) defining this representation have the form $|l_0| = \frac{1}{2}$; $|l_1| = \frac{3}{2}$. As we know, the numbers are determined to within a simultaneous change of sign. We shall therefore assume that $l_1 = \frac{3}{2}$. In Sect.4 we shall evaluate the infinitesimal operators of this representation and convince ourselves that $l_0 = \frac{1}{2}$ i.e. that the pair (l_0, l_1) is given by $(l_0, l_1) = \left(\frac{1}{2}, \frac{3}{2}\right)$.

Second Example. Besides the representation $g_a \to a_g$, there acts in the two-dimensional complex space the representation conjugate to it $g_a \to a_{(g^*)^{-1}} = (a_g^*)^{-1}$. According to the conclusion of paragraph 6 this representation is defined by the pair $\left(-\frac{1}{2}, \frac{3}{2}\right)$.

As we saw in Sect.1 we have $(a_g^*)^{-1} = \tau \bar{a} \tau^{-1}$ where $\tau = \begin{pmatrix} 0 & 1 \\ -1 & 0 \end{pmatrix}$ and $\bar{a}$ is the complex conjugate of the matrix a. This equation means that the representation $g_a \to (a_g^*)^{-1}$ is equivalent to the representation $g_a \to \bar{a}_g$, which is therefore also defined by the pair $\left(-\frac{1}{2}, \frac{3}{2}\right)$. From formulae (13)-(19) for the infinitesimal operators H_+, H_-, H_3, F_+, F_-, F_3 of an irreducible representation it follows that apart from the representations $g_a \to a$ and $g_a \to \bar{a}$, no other irreducible representation acts in two-dimensional space which is not equivalent to one of them.

Third Example. This is the identity representation $g \to g$ of the proper group, acting in the four-dimensional space $R^{(4)}$ $(x_0 x_1 x_2 x_3)$. This representation is easily seen to be irreducible.

We shall find the pair (l_0, l_1) defining it. The space $R^{(4)}$ contains two subspaces invariant under rotations: the time axis x_0 $(x_1 = x_2 = x_3 = 0)$ and the three-dimensional space $(x_1 x_2 x_3)$ $(x_0 = 0)$. Hence it follows that the numbers (l_0, l_1) defining the representation $g \to g$ are given by $l_0 = 0$; $l_1 = 2$. The canonical basis of this representation has the following form:

$$\xi_{00} = e_{x_0}, \qquad \xi_{1,-1} = \frac{e_{x_1} - ie_{x_2}}{2}, \qquad \xi_{10} = e_{x_3}, \qquad \xi_{11} = -\frac{e_{x_1} + ie_{x_2}}{2},$$

where e_{x_0}, e_{x_1}, e_{x_2}, e_{x_3} are intercepts of the coordinate directions x_0, x_1, x_2, x_3. In Sect.4 we shall again consider the examples just described.

§ 8. Unitary Irreducible Representations of the Proper Lorentz Group

The unitary nature of a representation means that in the space R, where the representation $g \to T_g$ acts, there exists a positive definite bilinear Hermitian form, invariant under all the operators T_g.*

We shall prove that in the case of a unitary representation the operators H_3 and F_3 are Hermitian, i.e.

$$(H_3 f, h) = (f, H_3 h),$$
$$(F_3 f, h) = (f, F_3 h).$$

In fact, let $g(\varphi)$ be a screw about the x_3 axis through a small angle φ and $T_{g(\varphi)}$ be the corresponding operator. From the definition of the operator H_3 we may write:

$$T_{g(\varphi)} = E + iH_3\varphi + o(\varphi).$$

Since $T_{g(\varphi)}$ is a unitary operator, then

$$T^*_{g(\varphi)} = T^{-1}_{g(\varphi)} = T_{g(-\varphi)}.$$

Hence $\quad T^*_g = E - iH^*_3\varphi + o(\varphi) = E - iH_3\varphi + o(\varphi)$

or

$$H^*_3 = H_3, \tag{24}$$

as required.

Similarly the Hermitian nature of F_3 may be proved. Moreover the operator H_- is the operator conjugate to H_+ and F_- that conjugate to F-:

$$H^*_+ = H_-, \qquad F^*_+ = F_-.$$

---

* We recall that a bilinear Hermitian form (f, h) in R is called invariant, if for any two vectors f and h and for any operator T_g of the representation

$$(T_g f, T_g h) = (f, h).$$

It is easily verified that the basis ξ_{lm} introduced in the space of the representation is orthogonal * in the case of a unitary representation.

We shall find what pairs (l_0, l_1) define unitary representations. From formula (13) it is clear that

$$(F_3\xi_{lm}, \xi_{lm}) = -mA_l(\xi_{lm}, \xi_{lm})$$

and

$$(\xi_{lm}, F_3\xi_{lm}) = -m\bar{A}_l(\xi_{lm}, \xi_{lm}),$$

whence $\bar{A}_l = A_l$ i.e. A_l is a real number. But $A_l = \frac{il_0l_1}{l(l+1)}$. Consequently, there are two possibilities:

1) l_1 is a pure imaginary (l_0 arbitrary)

2) $l_0 = 0$.

We now find the restrictions imposed in case 2)

We write:

$$\left.\begin{aligned} &(F_3\xi_{lm}, \xi_{l-1,\,m}) = C_l\sqrt{(l-m)(l+m)} \\ \text{and} \\ &(\xi_{lm}, F_3\xi_{l-1,\,m}) = -\bar{C}_l\sqrt{l^2-m^2}. \end{aligned}\right\} \quad (25)$$

Thus $C_l = -\bar{C}_l$ i.e. C_l is an imaginary number. For this to be the case we must have $\frac{(l^2-l_0^2)(l^2-l_1^2)}{4l^2-1} \geqslant 0$. But $l > l_0$ and $l > \frac{1}{2}$, consequently, $l^2 - l_1^2 \geqslant 0$. If 1) occurs, i.e. if l_1 is an imaginary number then the latter inequality is satisfied. If $l_0 = 0$ then the next weight $l = 1$ and consequently $1 - l_1^2 \geqslant 0$, i.e. either l_1 is a real number and $|l_1| \leqslant 1$ or l_1 is an imaginary number and we arrive at case 1).

Thus, an irreducible representation of the proper Lorentz group is unitary only in the following cases:

1) l_1 is an imaginary number, l_0 - arbitrary (integral or half-integral),

2) $l_0 = 0$, l_1 is real and $|l_1| \leqslant 1$.

The set of irreducible unitary representations corresponding to case 1) is known as the main series of representations. The remaining irreducible unitary representations form the supplementary series.

* This follows from the fact that $\{\xi_{lm}\}$ consists of eigenvectors of the Hermitian operator H_3 which for distinct m have different eigenvalues, and also from formulae (17), (18), (19), for the two conjugate operators H_+, H_-

We note that an irreducible unitary representation is finite only when $l_0 = 0,\ l_1 = 1$. This representation is one-dimensional; there are no other finite, irreducible, unitary representations.

§ 9. The Invariant Hermitian Bilinear Form *

In the last paragraph we have explained the circumstances under which an irreducible representation $g \to T_g$ of the proper Lorentz group is unitary, i.e. when such a representation permits a positive-definite invariant Hermitian form. We saw that this is rarely encountered. It seems that even if we drop the requirement of a positive-definite form and seek all those irreducible representations which simply admit a non-degenerate** invariant Hermitian form (generally speaking, indefinite) then the number of such representations, though larger than the number of unitary representations, is nevertheless small. Below we find all such representations.

But another problem is encountered much more frequently: under what conditions does a representation of the proper group $g \to T_g$ consisting of two irreducible components, admit an invariant, non-degenerate Hermitian form, or, what is the same thing, when is it possible to construct a non-degenerate, invariant form from two quantities which are transformed according to irreducible representations of the proper Lorentz group?

In this paragraph we shall state this problem and solve it (since the calculation is far from complicated) in a more general form; namely we examine conditions that a representation of the proper group (consisting of any number of irreducible components) should admit an invariant non-degenerate Hermitian form, and we shall also find its general form. Let the space R in which the representation $g \to T_g$ of the proper group acts, be resolved into a linear sum of irreducible subspaces R^τ in which the irreducible components τ with the pairs $\tau \sim (l_0^\tau,\ l_1^\tau)$ act. We choose in each of the R^τ a canonical basis $\{\xi_{lm}^\tau\}$. The set of all such bases gives a basis of the entire space R.

We write down the Hermitian bilinear form $(\psi_1,\ \psi_2)$ in terms of the basis $\{\xi_{lm}^\tau\}$:

$$(\psi_1,\ \psi_2) = \sum a_{lml'm'}^{\tau\tau'}\, x_{lm}^{\tau}\, \overline{y_{l'm'}^{\tau'}}, \tag{26}$$

where $x_{lm}^\tau,\ y_{l'm'}^{\tau'}$ are the coordinates of the vectors $\psi_1,\ \psi_2$ in the basis $\{\xi_{lm}^\tau\}$ and the matrix $\|a_{lml'm'}^{\tau\tau'}\| = A$ is the matrix of the bilinear form, so that $a_{lml'm'}^{\tau\tau'} = \overline{a_{l'm'lm}^{\tau'\tau}}$. In the case of a non-degenerate form $(\psi_1,\ \psi_2)$ the matrix A is also non-degenerate, i.e. no vector $\xi \neq 0$ annilihates this matrix.

---

* The results of this paragraph will only be needed in Chapter II. Therefore, at the first reading, this section may be omitted.

** The form $(\psi_1,\ \psi_2)$ is called non-degenerate if there is no vector ψ_0 in the space R such that $(\psi,\ \psi_0) = 0$ for all ψ.

We shall assume that the form (ψ_1, ψ_2) is invariant under the representation $g \to T_g$.

We shall elucidate the conditions which this requirement imposes upon the matrix A. Let an operator of the representation $g \to T_g$ be given in terms of the basis $\{\xi^{\tau}_{lm}\}$ by the matrix U_g. It is clear that the matrix A of the invariant form (ψ_1, ψ_2) must satisfy the following relation

$$U_g A U_g^* = A$$

or

$$A U_g^{*-1} = U_g A. \tag{27}$$

As we know a rotation $\tilde{g}$ is given in terms of the basis $\{\xi^{\tau}_{lm}\}$ by a unitary matrix. Hence for rotations we have

$$U_g A = A U_g, \tag{28}$$

i.e. the matrices U_g, corresponding to the rotations g are permutable with the matrix A.

We now consider the hyperbolic screw g_{03} in the plane (x_0, x_3). For small φ we expand the matrices $U_{g_{03}(\varphi)}$ and $\left(U^*_{g_{03}(\varphi)}\right)^{-1}$:

$$\left.\begin{aligned} U_{g_{03}(\varphi)} &= E + i\varphi F_3 + o(\varphi), \\ U^{*-1}_{g_{03}(\varphi)} &= U^*_{g_{03}}(-\varphi) = E + i\varphi F^*_3 + o(\varphi) \end{aligned}\right\} \tag{29}$$

(where F_3 is the matrix of the operator F_3).

Substituting relation (29) in equation (27) we obtain:

$$AF_3 = F^*_3 A. \tag{30}$$

We have thus proved that the matrix A, specifying the invariant, bilinear, Hermitian form in the canonical basis ξ^{τ}_{lm} satisfies the relations:

$$\left.\begin{aligned} &1)\quad U_g A = A U_g, \text{ if } g \text{ is an element of the group of rotations} \\ &2)\quad AF_3 - \bar{F}^{\tau}_3 A = 0 \quad \text{and} \\ &3)\quad A = A^*. \end{aligned}\right\} \tag{31}$$

It is easy to prove that, conversely, any operator A satisfying these relations specifies, in terms of the canonical basis, a form which is invariant under the given representation of the proper Lorentz group.

Now we proceed to find the general solution of relations (31) and also to determine the possible irreducible components τ which are contained in a representation permitting an invariant form.

We shall begin with the first condition and find the general form of the operator A permutable with all U_g, where g is a rotation.

We note that the matrix U_g of the representation of the group of rotations in terms of the canonical basis $\{\xi^{\tau}_{lm}\}$ may also be written in the following "block" form:

$$U_g = \begin{Vmatrix} U^{\tau_1}_{gl_1} & 0 & \dots & 0 \dots \\ 0 & U^{\tau_2}_{gl_2} \dots & & 0 \dots \\ \vdots & \vdots & \ddots & \vdots \\ 0 & 0 & \dots & U^{\tau_k}_{gl_s} \dots \\ \vdots & \vdots & & \ddots \end{Vmatrix},$$

where U^{τ}_{gl} is the matrix of the irreducible representation of the group of rotations of weight l participating in the irreducible component $g \to T_g$, of the representation of the entire proper group. We write down the matrix of the bilinear form

$$A = \| a^{\tau\tau'}_{lml'm'} \| \cdot$$

similarly in blocks

$$A = \begin{Vmatrix} A^{\tau_1\tau_1}_{l_1l_1} & A^{\tau_1\tau_2}_{l_1l_2} & \dots & A^{\tau_1\tau_k}_{l_1l_s} & \dots \\ \dots & \dots & \dots & \dots & \dots \\ \dots & \dots & \dots & \dots & \dots \\ A^{\tau_k\tau_1}_{l_sl_1} & A^{\tau_k\tau_2}_{l_sl_2} & \dots & A^{\tau_k\tau_k}_{l_sl_s} & \dots \end{Vmatrix},$$

where

$$A^{\tau\tau'}_{ll'} = \| a^{\tau\tau'}_{lml'm'} \| \qquad (m = -l, \dots, l; \quad m' = -l', \dots, l')$$

is a $(2l+1)$ by $(2l'+1)$ rectangular matrix.

From the relation

$$U_g A = A U_g$$

it follows that

$$U^{\tau}_{gl} A^{\tau\tau'}_{ll'} = A^{\tau\tau'}_{ll'} U^{\tau'}_{gl'}. \tag{31'}$$

From this equation, in virtue of the general Schur lemma (see Sect.1, paragraph it follows that the matrix $A_{ll'}^{\tau\tau'}$ is either zero or is a square, non-degenerate matrix. In the latter case the representations U_{gl}^{τ} and $U_{gl'}^{\tau'}$ are equivalent, so that $l = l'$ and the matrix $A_{ll'}^{\tau\tau'}$ has the form

$$A_{ll'}^{\tau\tau'} = A_l^{\tau\tau'} \delta_{ll'},$$

where $\delta_{ll'} = 1$ for $l = l'$ and $\delta_{ll'} = 0$ for $l \neq l'$.

Further, since the representations U_{gl}^{τ} and $U_{gl}^{\tau'}$ are equivalent, their matrices in terms of the canonical basis coincide, i.e.

$$U_{gl}^{\tau} = U_{gl}^{\tau'}.$$

Equation (31) then assumes the form

$$U_{gl}^{\tau} A_l^{\tau\tau'} = A_l^{\tau\tau'} U_{gl}^{\tau},$$

i.e. the matrix $A_l^{\tau\tau'}$ is permutable with the matrices U_{gl}^{τ} of an irreducible representation of the group of rotations. Such a matrix, as we know, must be a multiple of the unit matrix.

$$A_l^{\tau\tau'} = a_l^{\tau\tau'} E.$$

Hence for the matrix elements $a_{lml'm'}^{\tau\tau'}$ we obtain the final expression

$$a_{lml'm'}^{\tau\tau'} = a_l^{\tau\tau'} \delta_{mm'} \delta_{ll'}. \tag{32}$$

We note that in the derivation of formula (32) we used only the fact that matrix A commutes with the matrices of the representation of the group of rotations.

Thus, the matrix of any operator which commutes with the operators of a representation of the group of rotations, or, what is the same thing, with the infinitesimal operators H_+, H_-, H_3 of this representation has the form (32) in terms of the canonical basis $\{\xi_{lm}^{\tau}\}$.

We shall use the formula (32) frequently in what follows.

It now remains for us to find the form of the numbers $a_l^{\tau\tau'}$. We use relation 2) of (31), which may be rewritten thus:

$$(F_3 \xi_{lm}^{\tau}, \xi_{l'm'}^{\tau'}) = (\xi_{lm}^{\tau} F \xi_{l'm'}^{\tau'}) \tag{33}$$

for any pair of basis vectors ξ_{lm}^{τ} and $\xi_{l'm'}^{\tau'}$. Substituting in (33) the expression for F_3 (see (13)) and expanding (33) with the aid of (26) and (32), we arrive at the following equations:

$$A_l^{\tau} a_l^{\tau\tau'} = \bar{A}_l^{\tau'} a_l^{\tau\tau'}, \tag{34}$$

$$C_l^{\tau} a_l^{\tau\tau'} = -\bar{C}_l^{\tau'} a_{l-1}^{\tau\tau'}, \tag{35}$$

$$\bar{C}_l^{\tau'} a_l^{\tau\tau'} = -C_l^{\tau} a_{l-1}^{\tau\tau'}. \tag{36}$$

From these equations and from formulae (16) we see that $a_l^{\tau\tau'} \neq 0$ and $a_{l-1}^{\tau\tau'} \neq 0$ only if

$$l_0 l_1' + l_0' \bar{l}_1' = 0,$$
$$l_0^2 + l_1^2 = l_0'^2 + l_1'^2.$$

Hence it follows that

$$(l_0', l_1') = (l_0, -\bar{l}_1)\ [\text{or } (l_0', l_1') = (-l_0, \bar{l}_1)].$$

Since we are assuming that the form(ψ_1, ψ_2) is non-degenerate, for each component τ of the representation $g \to T_g$ there is a component τ^* such that $a_l^{\tau\tau^*} \neq 0$ for all l. Thus an invariant, non-degenerate, bilinear form can exist only if in the representation $g \to T_g$ together with each irreducible component τ defined by the numbers (l_0, l_1) there is contained also an irreducible component defined by the numbers $(l_0, -\bar{l}_1)$.

In particular if the representation $g \to T_g$ consists of one component, i.e. if this representation is irreducible, then an invariant bilinear form exists only under the condition $(l_0, l_1) = \pm(l_0, -\bar{l}_1)$ i.e. either

1) l_1 is purely imaginary and l_0 is any integer or half-integer or

2) l_1 is real and $l_0 = 0$ *

We also note that if the representation $g \to T_g$ (for which an invariant non-degenerate form exists) contains some mutually equivalent components $\tau_1, \ldots, \tau_n$, then it must contain as many mutually equivalent components $\tau_1^*, \ldots, \tau_n^*$.

We now determine the numbers $a_l^{\tau\tau^*}$. From (35) we have:

$$a_l^{\tau\tau^*} = -\frac{\bar{C}_l^{\tau^*}}{C_l^{\tau}} a_{l-1}^{\tau\tau^*}. \tag{37}$$

Since $-\dfrac{\bar{C}_l^{\tau^*}}{C_l^{\tau}} = \pm 1$ (see formula (16)), then

$$a_l^{\tau\tau^*} = \pm a^{\tau\tau^*}. \tag{38}$$

---

* It is interesting to compare the result just obtained with the result of the previous paragraph, where we sought conditions for an irreducible representation to be unitary. The unitary nature means essentially the existence of an invariant Hermitian positive-definite form. The supplementary restriction obtained there in the second case ($l_0 = 0$, l_1 is real and $|l_1| \leqslant 1$) is connected with the positive definiteness of the form specifiying the scalar product.

We note that in the case of a finite representation (l_1 real and $l < |l_1|$) the ratio $\frac{C_l^{\tau^*}}{C_l^{\tau}}$ may be considered positive (see footnote p.194), i.e. for a finite representation $a_l^{\tau\tau^*} = - a_{l-1}^{\tau\tau^*}$.

The numbers $a^{\tau\tau^*} = a^{-\tau^*\tau}$ may be anything. To different sets of such numbers there correspond different bilinear invariant forms.

We now summarize the above discussion:

A representation $g \to T_g$ of the proper Lorentz group admits an invariant non-degenerate Hermitian bilinear form if, and only if, the number of irreducible components τ, defined by the pair (l_0, l_1), coincides with the number of irreducible components τ^*, defined by the pair $(l_0, -\bar{l}_1)$ (or if both numbers are infinite). In this case, the invariant non-degenerate Hermitian form is given, in terms of the canonical basis $\{\xi_{lm}^{\tau}\}$ of the representation $g \to T_g$ by

$$(\psi_1, \psi_2) = \sum a^{\tau\tau^*} s_l^{\tau\tau^*} x_{lm}^{\tau} \bar{y}_{lm}^{\tau^*}, \tag{39}$$

where $\{x_{lm}^{\tau}\}$ and $\{y_{lm}^{\tau^*}\}$ are the coordinates of the vectors ψ_1 and ψ_2 in the canonical basis. Here $s_l^{\tau\tau^*} = \pm 1$ and for finite representations it may be taken that $s_l^{\tau\tau^*} = (-1)^{[l]}$; the numbers $a^{\tau\tau^*} = a^{-\tau^*\tau}$ are arbitrary with the condition however that the matrix $\|a^{\tau_i \tau_j^*}\|$, where $\tau_1, \tau_2, \ldots, \tau_n$ is a complete set of mutually equivalent components (equally the set $\tau_1^*, \ldots, \tau_n^*$), be non-degenerate.

In the particular but important case when the representation $g \to T_g$ of the proper group consists of two irreducible components $\tau \sim (l_0, l_1)$ and $\tau' \sim (l_0', l_1')$, an invariant, non-degenerate, Hermitian form exists if, and only if, the pairs (l_0, l_1) and (l_0', l_1') are connected by the relation $(l_0', l_1') = \pm (l_0, -\bar{l}_1)$; in other words, from two quantities which are transformed according to the irreducible representations τ and τ' of the proper Lorentz group, it is possible to construct a non-degenerate, invariant, Hermitian form if, and only if, their representations are specified by the pairs $\tau \sim (l_0, l_1)$, $\tau' = \tau^* \sim (l_0, -\bar{l}_1)$.

We note that by passing in R to a new system of coordinates we may reduce our form to a simpler form.

In fact, let $\tau_1, \ldots, \tau_n$ be mutually equivalent components defined by the pair (l_0, l_1) and $\tau_1^*, \ldots, \tau_n^*$ be the equivalent components corresponding to the pair $(l_0, -\bar{l}_1)$. An invariant, non-degenerate form can exist only if there are the same number of both components (the number may also be infinite).

Selecting the vectors

$$\xi'^{\tau_i}_{lm} = \sum_j a(\tau_i \tau_j)\, \xi^{\tau_j}_{lm},$$

$$\xi'^{\overset{*}{\tau}_i}_{lm} = \sum_j a(\overset{*}{\tau}_i \overset{*}{\tau}_j)\, \xi^{\overset{*}{\tau}_j}_{lm}.$$

as the new coordinate vectors in the space, by stretching the vectors $\left\{\xi^{\tau_j}_{lm}\right\}$ and $\left\{\xi^{\overset{*}{\tau}_k}_{lm}\right\}$, we can always ensure that

$$a^{\tau_i \overset{*}{\tau}_j} = \delta_{ij}. \tag{40}$$

More precisely, if $\tau_1, \ldots, \tau_n$ is a set of equivalent components for which τ_k is equivalent to $\overset{*}{\tau}_k$, then, introducing the new vectors

$$\xi'^{\tau_i}_{lm} = \sum a(\tau_i \tau_j)\, \xi^{\tau_j}_{lm},$$

we can ensure that

$$a^{\tau_i \tau_j} = \pm\, \delta_{ij}. \tag{41}$$

The reduction of the bilinear form to the form just described is completely analogous to the reduction of a quadratic form to a sum of squares.

Section 3. Representations of the Complete and General Lorentz Groups*

§ 1. Preliminary Remarks

We recall that the complete Lorentz group is obtained from the proper Lorentz group by the addition of the spatial reflection, i.e., the transformation s with the matrix

$$S = \begin{Vmatrix} -1 & 0 & 0 & 0 \\ 0 & -1 & 0 & 0 \\ 0 & 0 & -1 & 0 \\ 0 & 0 & 0 & 1 \end{Vmatrix}.$$

and all possible products of the form sg' where g' is a proper Lorentz transformation. Transformations of the form sg' will be called improper Lorentz transformations. Suppose a representation $g \to T_g$ of the complete Lorentz group is specified. Ipso facto there also arises a representation $g' \to T_{g'}$ of the proper Lorentz group. Let S denote the operator corresponding to the reflection s, $s \to S$ $(S^2 = E)$. Then to each improper Lorentz transformation sg' there corresponds the operator $ST_{g'}$. As above, let H_+, H_-, H_3, F_+,

---

* This section is inserted here for the sake of a logical exposition; we recommend the reader to omit it at the first reading and pass immediately to the fourth section.

F_-, F_3 be the infinitesimal operators of the representation $g' \to T_{g'}$ of the proper group. These operators determine the representation $g' \to T_{g'}$ uniquely. In order to obtain the representation $g \to T_g$ of the complete group we must also know how the operator S acts. Thus a representation of the complete group is specified by the infinitesimal operators H_+, H_-, H_3, F_+, F_-, F_3 and the operator S corresponding to the reflection.

We now write down the commutation relations between the operators H_+, H_-, H_3, F_+, F_-, F_3 and S. Since the reflection is permutable with each rotation the operator S commutes with the operators corresponding to rotations. Consequently the following relations hold.

$$SH_+S^{-1} = H_+, \quad SH_-S^{-1} = H_-, \quad SH_3S^{-1} = H_3. \qquad (1)$$

If we examine the transformations in the planes (x_0, x_1), (x_0, x_2), (x_0, x_3) then we may easily satisfy ourselves that these transformations satisfy the equations $s^{-1}g_{0k}s = g_{0k}^{-1}$ $(k = 1, 2, 3)$ (g_{0k} is a proper Lorentz transformation in the plane (x_0, x_k)).

Similar equations hold for the operators $T_{g_{0k}}$ and S

$$S^{-1}T_{g_{0k}}S = T_{g_{0k}}^{-1}.$$

Hence for the infinitesimal operators we obtain:

$$SF_+S^{-1} = -F_+, \quad SF_-S^{-1} = -F_-, \quad SF_3S^{-1} = -F_3. \quad (2)$$

In the next section we shall find the general form of the operators H_+, H_-, H_3, F_+, F_-, F_3 and S for an irreducible representation of the complete Lorentz group.

We make the following preliminary observation. Let there act in the space R a (reducible or irreducible) representation $g \to T_g$ of the complete Lorentz group and let $g' \to T_{g'}$ be the representation of the proper group in R generated by it; let S be the operator corresponding to reflection. We recall that the elements g' of the proper group satisfy the equations $sg's^{-1} = (g')^{*-1}$ (s is spatial reflection).

A similar equation also occurs for the operators of the representation $g' \to T_{g'}$ and S

$$ST_{g'}S^{-1} = T_{(g')^{*-1}}. \qquad (3)$$

We recall that the representation $g' \to T_{(g')^{*-1}}$ of the proper Lorentz group is said to be conjugate to the representation $g' \to T_{g'}$.

Equation (3) means that the representation $g' \to T_{g'}$ of the proper Lorentz group is equivalent to the representation $g' \to T_{(g')^{*-1}}$.

Thus, a representation $g' \to T_{g'}$ of the proper Lorentz group generated by a representation $g \to T_g$ of the complete Lorentz group is equivalent to its conjugate representation $g' \to T_{(g')^{*-1}}$.

After this observation we pass on to a description of the irreducible representations of the complete Lorentz group.

§ 2. The Irreducible Components of a Representation of the Proper Lorentz Group Generated by an Irreducible Representation of the Complete Group

In a space R, where an irreducible representation $g \to T_g$ of the complete group acts, there is specified ipso facto a representation $g' \to T_{g'}$ of the proper group (generally speaking, reducible).

We shall prove that the representation $g' \to T_{g'}$ of the proper Lorentz group in space R is either irreducible or decomposes into a sum of two irreducible representations (i.e. the space R breaks up into two subspaces, irreducible under the representation $g' \to T_{g'}$). Let R^{τ} be a subspace of R where the representation $g' \to T_{g'}$ of the proper group is irreducible and defined by the pair $\tau \sim (l_0, l_1)$. We denote by $\dot{R}^{\tau}$ the image of the subspace R^{τ} under the operator S (i.e. the set of all vectors of the form $S\xi$, where ξ is an element of R^{τ}). It is clear that $\dot{R}^{\tau}$ is a subspace in R. It so happens that $\dot{R}^{\tau}$ is invariant under the operators H_+, H_-, H_3, F_+, F_-, F_3. In fact since R is invariant under the operators H_+, H_-, H_3, F_+, F_-, F_3, we have

$$H_{\alpha}R^{\tau} = R^{\tau}, \qquad F_{\alpha}R^{\tau} = R^{\tau} \qquad (\alpha = -, +, 3).$$

But then the commutation relations (1) yield

$$H_{\alpha}\dot{R}^{\tau} = H_{\alpha}SR^{\tau} = SH_{\alpha}R^{\tau} = SR^{\tau} = \dot{R}^{\tau}$$

and

$$F_{\alpha}\dot{R}^{\tau} = F_{\alpha}SR^{\tau} = -SFR^{\tau} = -SR^{\tau} = -\dot{R}^{\tau} = \dot{R}^{\tau},$$

i.e. $\dot{R}^{\tau}$ is invariant under the operators H_{α} and F_{α} and consequently, invariant under the operators $T_{g'}$ corresponding to proper Lorentz transformations. It is easily seen that a representation $g' \to T_{g'}$ of the proper group acting in $\dot{R}^{\tau}$ is irreducible. In fact, if there be in $\dot{R}^{\tau}$ a subspace $\dot{R}'^{\tau}$ which is invariant under the representation $g' \to T_{g'}$ of the proper group then $S\dot{R}'^{\tau}$ will again be invariant under this representation $g' \to T_{g'}$. But since $S\dot{R}^{\tau} = R^{\tau}$ (we recall that $S^2 = E$), then $S\dot{R}'^{\tau}$ will constitute a subspace of the space R^{τ} and the representation $g' \to T_{g'}$ in R^{τ} will be reducible, contradicting our assumption about the space R^{τ}.

Thus, in the space R where an irreducible representation of the complete group acts, together with each subspace R^{τ} in which the representation $g' \to T_{g'}$ of the proper group is irreducible, there is a subspace $R^{\dot{\tau}}$ in which the representation $g' \to T_{g'}$ of the proper group is also irreducible (here $SR^{\tau} = R^{\dot{\tau}}$). We note that the spaces R^{τ} and $R^{\dot{\tau}}$ either coincide or have no non-zero common elements. In fact the subspace $\tilde{R}$ in which the spaces R^{τ} and $R^{\dot{\tau}}$ intersect is invariant under the representation $g' \to T_{g'}$ of the proper group. In virtue of the irreducibility of the spaces R^{τ} and $R^{\dot{\tau}}$ the subspace $\tilde{R}$ either coincides with each of them (and consequently these spaces themselves coincide) or is equal to zero.

Thus two cases are possible.

1) The spaces R^{τ} and $R^{\dot{\tau}}$ coincide with each other and consequently with the entire space R where the irreducible representation of the complete Lorentz group acts. In other words, the representation $g' \to T_{g'}$ of the proper group generated by an irreducible representation of the complete group is also irreducible.

2) The spaces R^{τ} and $R^{\dot{\tau}}$ do not have non-zero elements in common. In this case it is clear that their linear sum coincides with the entire space R where the irreducible representation of the complete group acts.

We shall prove that in case 2) the representations of the proper group which act in R^{τ} and $R^{\dot{\tau}}$ are conjugate to each other and are not mutually equivalent.

We write down identity (3) in the form

$$ST_{g'} = T_{(g'^{*})^{-1}}S. \tag{3'}$$

We apply the left- and right-hand sides of (3) to a vector ξ from R^{τ}. Since $T_{g'}$ induces in R the representation $T^{\tau}_{g'}$, and in $R^{\dot{\tau}}$ the representation $T^{\dot{\tau}}_{g'}$, and since S carries R^{τ} into $R^{\dot{\tau}}$ we have:

$$ST^{\tau}_{g'}\xi = T^{\dot{\tau}}_{(g')^{*-1}}S\xi. \tag{3"}$$

In virtue of the definition of equivalent representations (Sect.1, paragraph 6), equation (3") means that the representations $T^{\tau}_{g'}$ and $T^{\dot{\tau}}_{(g')^{*-1}}$ are equivalent, i.e. the representation $T^{\tau}_{g'}$ is conjugate to the representation $T^{\dot{\tau}}_{g'}$. It remains to prove that the representations T^{τ}_{g} and $T^{\dot{\tau}}_{g}$ are not equivalent.

To do this we prove that if they were equivalent, then the representation $g \to T_g$ of the complete group in the space R is reducible.

Suppose the representations $g' \to T^{\tau}_{g'}$ and $g' \to T^{\dot{\tau}}_{g'}$ are equivalent. Using this fact we construct an operator L, in the space R, not a multiple of the unit operator, which is permutable with all the operators T_g of the representation

$g \to T_g$ of the complete group. This would mean that the representation $g \to T_g$ is reducible.

The operator L is constructed thus: we choose bases of R^{τ} and $R^{\dot{\tau}}$ such that the operators $T^{\tau}_{g'}$ and $T^{\dot{\tau}}_{g'}$ are described in terms of them by the same matrix $A_{g'}$. (This is possible since the representations $T^{\dot{\tau}}_{g'}$ and $T^{\tau}_{g'}$ are equivalent). Combining these bases we obtain a basis of the entire space R in which the operator $T_{g'}$ is described by the matrix

$$T_{g'} = \begin{Vmatrix} A_{g'} & 0 \\ 0 & A_{g'} \end{Vmatrix},$$

and the operator S which interchanges the spaces R^{τ} and $R^{\dot{\tau}}$, by the matrix

$$S = \begin{Vmatrix} 0 & \sigma \\ \tilde{\sigma} & 0 \end{Vmatrix}.$$

Since $S^2 = E$ then $\sigma\tilde{\sigma} = E$ and finally

$$S = \begin{Vmatrix} 0 & \sigma \\ \sigma^{-1} & 0 \end{Vmatrix}.$$

The relation $ST_{g'}S^{-1} = T_{(g'^*)^{-1}}$ clearly goes over into

$$\sigma A_{g'}\sigma^{-1} = A_{(g')^{*-1}}.$$

We now consider the operator

$$S_1 = \begin{Vmatrix} \sigma & 0 \\ 0 & \sigma \end{Vmatrix}.$$

In virtue of what has just been said, we have

$$S_1 T_{g'} S_1^{-1} = T_{(g')^{*-1}}. \qquad (3''')$$

Finally, we construct the operator

$$L = S_1 S = \begin{pmatrix} 0 & \sigma^2 \\ E & 0 \end{pmatrix}.$$

From (3') and (3‴) it clearly follows that $LT_{g'} = T_{g'}L$. Moreover, we find by direct computation that $LS = SL$. Thus, we have constructed an operator L, not a multiple of the unit operator, which commutes with all the operators of the representation of the complete group. Thus, it has been proved that for an irreducible representation of the complete group, the components T^{τ}_g and $T^{\dot{\tau}}_g$ of the representation of the proper group are not equivalent.

We now pass on to the calculation of the pairs (l_0, l_1) defining the irreducible components of any representation of the proper group which is induced by an irreducible representation of the complete Lorentz group.

First Case: The representation $g' \to T_{g'}$ of the proper Lorentz group (induced by the irreducible representation of the complete Lorentz group) is irreducible.

As we know, this representation is equivalent to its conjugate. We recall that, if an irreducible representation $g' \to T_{g'}$ of the proper Lorentz group is defined by the pair (l_0, l_1) then its conjugate representation $g' \to T_{(g')^{*-1}}$ is also irreducible and is defined by the pair $\pm(l_0, -l_1)$. Hence it follows that an irreducible representation, which is equivalent to its conjugate must be defined by a pair of the form $(0, l_1)$ or $(l_0, 0)$ (i.e. one of the numbers l_0 and l_1 must be zero).

Thus, if an irreducible representation $g \to T_g$ of the complete Lorentz group induces an irreducible representation $g' \to T'_g$ of the proper group, the pair (l_0, l_1) defining this representation has the form $(0, l_1)$ or $(l_0, 0)$.

Second Case: The representation $g' \to T_{g'}$ of the proper group breaks up into two irreducible components.

Let one of these be specified by the pair (l_0, l_1). Since the second component is conjugate to it, it is specified by the pair $\pm(l_0, -l_1)$. In view of the fact that these components are not equivalent neither of the numbers l_0 and l_1 can be zero.

Thus, if the representation of the proper group induced by an irreducible representation of the complete group, has two components, then these are specified by the pairs (l_0, l_1) and $\pm(l_0, -l_1)$, where neither of the numbers l_0 and l_1 equals zero.

§ 3. The Operator of Spatial Reflection

We shall find the form of the spatial reflection operator S in terms of the basis $\{\xi_{lm}\}$ - the canonical basis of the representation $g' \to T_{g'}$ of the proper group.

I. We consider to begin with the first case (the representation $g' \to T_{g'}$ is irreducible).

We write

$$S\xi_{lm} = \sum_{l'm'} s_{lml'm'}\xi_{l'm'}. \tag{4}$$

We shall determine the general form of the numbers $s_{lml'm'}$.

We make use of the relations (1) which state that the operator S commutes with the operators of the representation $\tilde{g} \to T_{\tilde{g}}$ of the group of rotations ($\tilde{g}$ a rotation) induced by the representation of the complete group.

In paragraph 8 of the previous section we found the general form of the matrix of such an operator in terms of the canonical basis $\{\xi_{lm}\}$. According to formula (32) obtained there, the numbers $s_{lml'm'}$ have the form:

$$s_{lml'm'} = s_l \delta_{mm'} \delta_{ll'} \tag{4'}$$

(we omit the indices τ and τ' appearing in formula (32)). It remains for us to find the numbers s_l.

To this end we use the relation $SF_3 = -F_3S$ or $SF_3\xi_{lm} = -F_3S\xi_{lm}$. Substituting in this equation the expression for the operator F_3 (see Sect.2(14)) and remembering that the number A_l entering into the expression for this operator is equal to zero in our case, we obtain

$$s_l = -s_{l-1}\,^*$$

Thus we obtain

$$s_l = (-1)^{[l]} s_{l_0}$$

(where l_0 is the minimum weight participating in the representation $g' \to T_{g'}$). Since $S^2 = E$, we have $s_{l_0} = \pm 1$.

Thus we obtain two expressions for the operator S which differ only in sign

$$S\xi_{lm} = (-1)^{[l]} \xi_{lm} \tag{5}$$

and

$$S\xi_{lm} = (-1)^{[l]+1} \xi_{lm}. \tag{5'}$$

This is the final form of the operator S.

We note that the expressions (5) and (5') for the operator S lead to two non-equivalent representations of the complete group.**

* According to formula (16), Sect.2, $A_l = \frac{il_0l_1}{l(l+1)}$. Since either $l_0 = 0$ or $l_1 = 0$, then $A_l = 0$.

** We now prove that the two formulae for the operator S lead to two non-equivalent representations $g \to T_g^{(1)}$ and $g \to T_g^{(2)}$ of the complete group. Assume they are equivalent, i.e., let there be an operator B such that

$$BT_g^{(1)} = T_g^{(2)}B.$$

Since for elements of the proper group g', $T_{g'}^{(1)} = T_{g'}^{(2)}$ and the representation $g' \to T_{g'}^{(1)}$ of the proper group is irreducible, then $B = \lambda E$. Hence it follows that

$$S^{(2)} = BS^{(1)}B^{-1} = S^{(1)}$$

i.e. the operator S is the same for two equivalent representations $g \to T_g^{(1)}$ and $g \to T_g^{(2)}$ i.e. it is specified either by formula (5) or by formula (5').

Thus we have obtained the following result:

In the case when an irreducible representation $g \to T_g$ of the complete Lorentz group generates an irreducible representation $g' \to T_{g'}$, of the proper group, the latter is equivalent to its conjugate $g' \to T_{(g')^{*-1}}$ (i.e. is defined by a pair of the form $(0, l_1)$ or $(l_0, 0)$), and the operator S in terms of the canonical basis of the representation $g' \to T_{g'}$ has the form (5) or (5') (which differ from each other only in sign).

It is easily seen that we have also essentially proved the converse assertion:

Any irreducible representation $g' \to T_{g'}$ of the proper group which is equivalent to its conjugate (i.e. is defined by a pair of the form $(0, l_1)$ or $(l_0, 0)$) may be extended to a representation of the complete Lorentz group, acting in the same space, in two non-equivalent ways which differ in the sign of the operator S. The operator S itself acts either according to formula (5) or according to formula (5').

II. We pass on to the second case. The representation $g' \to T_{g'}$ of the proper group generated by an irreducible representation of the complete Lorentz group is reducible and decomposes into the sum of two representations $g' \to T^{\tau}_{g'}$ and $g' \to T^{\dot{\tau}}_{g'}$ which act in the spaces R^{τ} and $R^{\dot{\tau}}$ respectively.

In this case the spaces R^{τ} and $R^{\dot{\tau}}$ are mapped into each other by the operator S: $SR^{\tau} = R^{\dot{\tau}}$ and $SR^{\dot{\tau}} = R^{\tau}$. We choose for the space R the basis $\{\xi^{\tau}_{lm}, \xi^{\dot{\tau}}_{lm}\}$ made up of canonical bases of the representations $g' \to T^{\tau}_{g'}$ and $g' = T^{\dot{\tau}}_{g'}$ in the spaces R^{τ} and $R^{\dot{\tau}}$. We shall find the form of the operator S in the basis $\{\xi^{\tau}_{lm}, \xi^{\dot{\tau}}_{lm}\}$.

We write

$$S\xi^{\tau}_{lm} = \sum s^{\tau\dot{\tau}}_{lml'm'}\xi^{\dot{\tau}}_{l'm'} \quad \text{and} \quad S\xi^{\dot{\tau}}_{lm} = \sum s^{\dot{\tau}\tau}_{lml'm'}\xi^{\tau}_{l'm'}. \tag{6}$$

We must find the general form of the numbers $s^{\tau\dot{\tau}}_{lml'm'}$ and $s^{\dot{\tau}\tau}_{lml'm'}$. Again we make use, first of all, of the fact that the operator S is permutable with all operators of the representation of the group of rotations generated by the representation of the general Lorentz group being studied. As we recalled above, the general form of such an operator was found in a previous section (paragraph 8). Using the formula (32) derived there, we obtain

$$s^{\tau\dot{\tau}}_{lml'm'} = s^{\tau\dot{\tau}}_{l}\delta_{mm'}\delta_{ll'}; \qquad s^{\dot{\tau}\tau}_{lml'm'} = s^{\dot{\tau}\tau}_{l}\delta_{mm'}\delta_{ll'}. \tag{6'}$$

Thus the operator S has the following form in terms of the basis $\{\xi^{\tau}_{lm}, \xi^{\dot{\tau}}_{lm}\}$

$$S\xi^{\tau}_{lm} = s^{\tau\dot{\tau}}_{l}\xi^{\dot{\tau}}_{lm}; \qquad S\xi^{\dot{\tau}}_{lm} = s^{\dot{\tau}\tau}_{l}\xi^{\tau}_{lm}. \tag{6''}$$

We note, moreover, that the equation $S^2 = E$ yields

$$s_l^{\tau\dot{\tau}} s_l^{\dot{\tau}\tau} = 1.$$

It remains to determine the numbers $s_l^{\tau\dot{\tau}}$ and $s_l^{\dot{\tau}\tau}$. To this end we turn to the commutation relation $F_3 S = -S F_3$. Using the formulae (see Sect.2(6') and (13)) we obtain the following equations:

$$A_l^{\tau} s_l^{\tau\dot{\tau}} + A_l^{\dot{\tau}} s_l^{\tau\dot{\tau}} = 0,$$
$$C_l^{\tau} s_l^{\tau\dot{\tau}} + C_l^{\dot{\tau}} s_{l-1}^{\tau\dot{\tau}} = 0,$$
$$C_l^{\dot{\tau}} s_l^{\dot{\tau}\tau} + C_l^{\tau} s_{l-1}^{\dot{\tau}\tau} = 0.$$

Since for two mutually conjugate representations $g \to T_g^{\tau}$ and $g \to T_g^{\dot{\tau}}$, $A_l^{\dot{\tau}} = -A_l^{\tau}$ and $C_l^{\dot{\tau}} = C_l^{\tau}$ (see formulae (16), Sect.2), the first equation is satisfied automatically, and from the two others it follows that

$$s_l^{\tau\dot{\tau}} = -s_{l-1}^{\tau\dot{\tau}} \quad \text{and} \quad s_l^{\dot{\tau}\tau} = -s_{l-1}^{\dot{\tau}\tau}.$$

Hence
$$\left.\begin{aligned} & s_l^{\tau\dot{\tau}} = (-1)^{[l]} s_{l_0}^{\tau\dot{\tau}}, \qquad s_l^{\dot{\tau}\tau} = (-1)^{[l]} s_{l_0}^{\dot{\tau}\tau} \\ & \text{and} \qquad s_{l_0}^{\tau\dot{\tau}} s_{l_0}^{\dot{\tau}\tau} = 1. \end{aligned}\right\} \tag{7}$$

Formulae (7) for the numbers $s_l^{\tau\dot{\tau}}$ and $s_l^{\dot{\tau}\tau}$ give the general form of the operator S in terms of the basis $\{\xi_{lm}^{\tau}, \xi_{lm}^{\dot{\tau}}\}$. We note that in deriving these formulae we used only the commutation relations between the operator S and the infinitesimal operators H_3, H_+, H_-, F_3, F_+, F_- of the representation $g' \to T_{g'}$ of the proper Lorentz group and the equation $S^2 = E$.

Thus, formulae (7) give the general form of the operator S which satisfies the commutation relations (1), (2) and acts in the space where the irreducible representation of the complete Lorentz group (and the induced irreducible representation $g' \to T_{g'}$ of the proper group) is specified.

We shall make use of this observation in the next paragraph for the construction of the representations of the general Lorentz group.

The operator S defined by formulae (7) may be reduced to a simpler form by a transformation of the basis $\{\xi_{lm}^{\tau}, \xi_{lm}^{\dot{\tau}}\}$ which does not alter the form of the operators H_+, H_-, H_3, F_+, F_-, F_3. To this end we take $\xi_{lm}'^{\tau} = \alpha \xi_{lm}^{\tau}$, $\xi_{lm}'^{\dot{\tau}} = \beta \xi_{lm}^{\dot{\tau}}$. We then find that in the new basis $\{\xi_{lm}'^{\tau} \xi_{lm}'^{\dot{\tau}}\}$

$$s_{l_0}'^{\tau\dot{\tau}} = \frac{\alpha}{\beta} s_{l_0}^{\tau\dot{\tau}}, \qquad s_{l_0}'^{\dot{\tau}\tau} = \frac{\beta}{\alpha} s_{l_0}^{\dot{\tau}\tau}.$$

By choosing α and β suitably it is possible to ensure that

$$s_{l_0}'^{\tau\dot{\tau}} = s_{l_0}'^{\dot{\tau}\tau} = 1.$$

Thus for the operator S we finally obtain:

$$\left.\begin{aligned} S\xi_{lm}^{\tau} &= (-1)^{[l]}\,\xi_{lm}^{\dot{\tau}}, \\ S\xi_{lm}^{\dot{\tau}} &= (-1)^{[l]}\,\xi_{lm}^{\tau}. \end{aligned}\right\} \tag{8}$$

Hence the operator S is described in terms of the basis $\{\xi_{lm}^{\tau}, \xi_{lm}^{\dot{\tau}}\}$ by the matrix

$$S = \begin{Vmatrix} 0 & \tilde{S} \\ \tilde{S} & 0 \end{Vmatrix},$$

where the matrix $\tilde{S}$ has the form

$$\tilde{S} = \|(-1)^{l}\,\delta_{ll'}\delta_{mm'}\|.$$

We sum up:

In the case when an irreducible representation of the complete group induces a reducible representation of the proper group, the latter decomposes into a sum of two non-equivalent, mutually conjugate, irreducible representations $g' \to T_{g'}^{\tau}$ and $g' \to T_{g'}^{\dot{\tau}}$ of the proper group; in this case the pairs of numbers defining these representations have the form $\tau \sim (l_0, l_1)$ and $\dot{\tau} \sim \pm(-l_0, l_1)$ $(l_0 \neq 0,\ l_1 \neq 0)$. The operator S is specified by formula (8) in terms of the basis $\{\xi_{lm}^{\tau}, \xi_{lm}^{\dot{\tau}}\}$ made up of canonical bases of the representations $g' \to T_{g'}^{\tau}$ and $g' \to T_{g'}^{\dot{\tau}}$.

It is easily seen that the converse is also valid: two non-equivalent, mutually conjugate, irreducible representations $g \to T_{g'}^{\tau}$ and $g' \to T_{g'}^{\dot{\tau}}$ (such representations are defined by the pairs $\tau \sim (l_0, l_1)$ and $\dot{\tau} \sim (-l_0, l_1)$ $(l_0 \neq 0,\ l_1 \neq 0)$) which act in the spaces R^{τ} and $R^{\dot{\tau}}$ extend to a representation $g \to T_g$ of the complete group, which acts in the linear sum of the spaces $R^{\tau} + R^{\dot{\tau}} = R$ and is uniquely defined up to equivalence. The operator S in this case has the form (8).

Cases I and II that we have considered exhaust all possible irreducible representations of the complete group. Thus we have given a complete description of these representations.

§ 4. Irreducible Unique Representations of the General Lorentz Group

The general group is obtained from the complete Lorentz group by the inclusion of the temporal reflection t i.e. the transformation with the matrix

$$t = \begin{Vmatrix} 1 & 0 & 0 & 0 \\ 0 & 1 & 0 & 0 \\ 0 & 0 & 1 & 0 \\ 0 & 0 & 0 & -1 \end{Vmatrix}$$

and all transformations of the type tg, where g is an element of the complete group. We note that, in this case, the transformations s (spatial reflection) and t are permutable and their product is equal to the complete reflection j in four-dimensional space.

$$st = ts = j, \qquad s^2 = t^2 = j^2 = e. \tag{9}$$

The transformation j (complete reflection) is permutable with any general Lorentz transformation

$$jg = gj. \tag{9'}$$

We consider any irreducible representation of the general group $g \to T_g$. Let S, T, J be the operators corresponding to the reflections s, t and j so that

$$ST = TS = J \text{ and } J^2 = S^2 = T^2 = E. \tag{10}$$

Moreover the operator J, in virtue of relation (9), is permutable with any operator of the representation

$$JT_g = T_g J. \tag{11}$$

But an operator permutable with the operators of an irreducible representation is a multiple of the unit operator i.e. $J = \lambda E$. Since $J^2 = E$, $\lambda = \pm 1$ and either $J = E$ or $J = -E$.

Consequently, either $T = S^{-1} = S$ or $T = -S$.

Thus, an irreducible representation of the complete Lorentz group may be extended to an irreducible representation of the general group in two ways: either by the introduction of the operator $T = T_t$ according to the formula

$$T = S, \tag{12}$$

or by the introduction of the operator T according to the formula

$$T = -S. \tag{12'}$$

It is clear that all the irreducible unique representations of the general group are obtained in this manner.

§ 5. Two-valued Representations of the General Lorentz Group

As shown in Sect.1, apart from the unique representations of the general Lorentz group which are characterized by the fact that the operators S, T, J corresponding to reflections commute among themselves, the two-valued representations of the general group are also of interest. We recall that in such representations to each element $\{e, s, t, j\}$ of the group of reflections there correspond two operators $\pm E$, $\pm S$, $\pm T$, $\pm J$ differing in sign, such that the operators S, T, J anticommute with each other. We shall find the form of these operators for an irreducible two-valued representation of the general group.

It may be shown that any such representation generates a representation $g' \to T_{g'}$ of the proper group which consists of two conjugate* components $T^{\tau}_{g'}$ and $T^{\dot{\tau}}_{g'}$. In this case, the components $T^{\dot{\tau}}_{g'}$ and $T^{\tau}_{g'}$ may or may not be equivalent; in the latter case the representation of the complete group, induced by the representation of the general group, is also irreducible; in the former case the representation of the complete group is reducible but decomposes into two non-equivalent representations which differ in the form of the operator S. We shall consider both cases separately.

We consider first of all the case when T^{τ}_{g} and $T^{\dot{\tau}}_{g}$ are non-equivalent. The representation $g \to T_g$ of the general group generates in this case an irreducible representation of the complete group, in which the operator S, given in terms of the basis $\{\xi^{\tau}_{lm}, \xi^{\dot{\tau}}_{lm}\}$ is described by formula (8).

$$S\xi^{\tau}_{lm} = (-1)^{[l]}\xi^{\dot{\tau}}_{lm}, \qquad S\xi^{\dot{\tau}}_{lm} = (-1)^{[l]}\xi^{\tau}_{lm}.$$

We now find the operator T corresponding to the temporal reflection t: $t \to \pm T$. Since the elements g' of the proper group satisfy

$$tg't^{-1} = (g')^{-1},$$

the operator T satisfies a similar equation

$$TT_{g'}T^{-1} = T_{(g')^{*-1}}.$$

As we saw in paragraph 3 any such operator T for a representation $g' \to T_{g'}$ of the proper group which has two conjugate components T^{τ}_{g} and $T^{\dot{\tau}}_{g'}$, has the form (7)

$$T\xi^{\tau}_{lm} = (-1)^{[l]}t^{\tau\dot{\tau}}_{l_0}\xi^{\dot{\tau}}_{lm}, \qquad T\xi^{\dot{\tau}}_{lm} = (-1)^{[l]}t^{\dot{\tau}\tau}_{l_0}\xi^{\tau}_{lm}.$$

From the condition $ST = -TS$ and $T^2 = E$ we find that

$$t^{\tau\dot{\tau}}_{l_0}t^{\dot{\tau}\tau}_{l_0} = 1 \quad \text{and} \quad t^{\tau\dot{\tau}}_{l_0} = -t^{\dot{\tau}\tau}_{l_0} = i = \sqrt{-1}.$$

Thus the operator T is specified by the formula

$$T\xi^{\tau}_{lm} = (-1)^{[l]}i\xi^{\dot{\tau}}_{lm}, \qquad T\xi^{\dot{\tau}}_{lm} = -(-1)^{[l]}i\xi^{\tau}_{lm}.$$

---

* The proof of this fact is a refinement of the argument used in paragraph 2 of this section to prove that an irreducible representation of the complete group consists of one or two component representations of the proper group.

We now find the operator $J = TS$. We obtain

$$J\xi^{\tau}_{lm} = -i\xi^{\tau}_{lm}, \qquad J\xi^{\dot{\tau}}_{lm} = i\xi^{\dot{\tau}}_{lm}.$$

We note that the operators S, T, J correspond, in the basis $\left(\xi^{\tau}_{lm}, \xi^{\dot{\tau}}_{lm}\right)$ to the matrices

$$S = \begin{Vmatrix} 0 & \tilde{S} \\ \tilde{S} & 0 \end{Vmatrix}, \qquad T = \begin{Vmatrix} 0 & i\tilde{S} \\ -i\tilde{S} & 0 \end{Vmatrix}, \qquad J = \begin{Vmatrix} -iE & 0 \\ 0 & iE \end{Vmatrix}, \tag{13}$$

where $\tilde{S}$ is a diagonal matrix of the form $\tilde{S} = \|(-1)^{[l]}\delta_{ll}\delta_{mm}\|$, and E is the unit (identity) matrix.

It is easily verified that the operators S, T, J (13) anticommute and together with the unit operator E form a two-valued representation of the group of reflections: $s \to \pm S$; $t \to \pm T$; $j \to \pm J$; $e \to \pm E$.

Thus, in the case of a two-valued irreducible representation of the general Lorentz group, containing two non-equivalent components of the representation of the proper group, the operators S, T, J corresponding to reflections are described by the matrices (13) in terms of the canonical basis $\left\{\xi^{\tau}_{lm}, \xi^{\dot{\tau}}_{lm}\right\}$

We now consider the case when the components $T^{\tau}_{g'}$ and $T^{\dot{\tau}}_{g'}$ of the representation of the proper group induced by the two-valued representation of the general group are equivalent. In this case the representation of the complete group is reducible. It may be shown, nonetheless that the two components of the representation of the complete group are not equivalent: the operator S acts in one of them according to the formula

$$S\xi^{\tau}_{lm} = (-1)^{[l]}\xi^{\tau}_{lm}$$

and in the other according to the formula

$$S\xi^{\dot{\tau}}_{lm} = (-1)^{[l]+1}\xi^{\dot{\tau}}_{lm}.$$

The matrix of the operator S in terms of the basis $\left\{\xi^{\tau}_{lm}, \xi^{\dot{\tau}}_{lm}\right\}$ has the form

$$S = \begin{Vmatrix} \tilde{S} & 0 \\ 0 & -\tilde{S} \end{Vmatrix},$$

where again

$$\tilde{S} = \|(-1)^{[l]}\delta_{ll'}\delta_{mm'}\|.$$

We now find the form of the operator J corresponding to complete reflection. This operator commutes with the operators $T_{g'}$ of the representation $g' \to T_{g'}$ of the proper group. It is easily shown that its matrix, in the basis $\left\{\xi^{\tau}_{lm}, \xi^{\dot{\tau}}_{lm}\right\}$ is given by

$$J=\begin{Vmatrix}\lambda_{11}E & \lambda_{12}E\\ \lambda_{21}E & \lambda_{22}E\end{Vmatrix}.$$

From the conditions $SJ=-JS$ and $J^2=-E$ we have:

$$J=\begin{Vmatrix}0 & E\\ -E & 0\end{Vmatrix}$$

or

$$J\xi^{\tau}_{lm}=\xi^{\dot{\tau}}_{lm},\qquad J\xi^{\dot{\tau}}_{lm}=-\xi^{\tau}_{lm}.$$

Finally for the operator $T=JS$ we have

$$T\xi^{\tau}_{lm}=-(-1)^{[l]}\,\xi^{\dot{\tau}}_{lm},\qquad T\xi^{\dot{\tau}}_{lm}=-(-1)^{[l]}\xi^{\tau}_{lm},$$

i.e. the matrix of the operator T with respect to the basis $\{\xi^{\tau}_{lm},\xi^{\dot{\tau}}_{lm}\}$ has the form

$$T=\begin{Vmatrix}0 & -\tilde{S}\\ -\tilde{S} & 0\end{Vmatrix}.$$

It is easily verified that the operators S, T, J given above combine according to Table (13).

We note that it is possible to choose a basis in the sum of the spaces R^{τ} and $R^{\dot{\tau}}$ relative to which the operator J is diagonal; namely the basis $\{\xi_{lm},\eta_{lm}\}$ where

$$\xi_{lm}=\xi^{\tau}_{lm}+i\xi^{\dot{\tau}}_{lm},\qquad \eta_{lm}=\xi^{\tau}_{lm}-i\xi^{\dot{\tau}}_{lm}.$$

The subspaces R_{ξ} and R_{η} enveloping the bases $\{\xi_{lm}\}$, $\{\eta_{lm}\}$ are invariant and irreducible under the representation $g'\to T_{g'}$ of the proper group, and in fact the bases $\{\xi_{lm}\}$, $\{\eta_{lm}\}$ are canonical. Relative to the basis $\{\xi_{lm},\eta_{lm}\}$ the matrices of the operators S, T, J assume the form

$$S=\begin{Vmatrix}0 & \tilde{S}\\ \tilde{S} & 0\end{Vmatrix},\qquad T=\begin{Vmatrix}0 & i\tilde{S}\\ -i\tilde{S} & 0\end{Vmatrix},\qquad J=\begin{Vmatrix}-iE & 0\\ 0 & iE\end{Vmatrix}.$$

These matrices had the same form in the preceding case of two non-equivalent components $T^{\tau}_{g'}$, $T^{\dot{\tau}}_{g'}$ of the representation of the proper group.

Thus, combining these two cases we see that the irreducible two-valued representations of the general group always contain two conjugate irreducible representations of the proper group; in the space R, where our representation acts, it is always possible to select a canonical basis such that the operators S, T, J have the matrices

$$S=\begin{Vmatrix}0 & \widetilde{S} \\ \widetilde{S} & 0\end{Vmatrix}, \quad T=\begin{Vmatrix}0 & i\widetilde{S} \\ -i\widetilde{S} & 0\end{Vmatrix}, \quad J=\begin{Vmatrix}-iE & 0 \\ 0 & iE\end{Vmatrix}, \tag{14}$$

where the matrix $\widetilde{S}=\|(-1)^{[l]}\delta_{ll'}\delta_{mm'}\|$.

We note, in conclusion, that if we pass from the canonical basis $\{\xi^{\tau}_{lm}, \xi^{\dot{\tau}}_{lm}\}$ to the basis $\{\xi^{\tau}_{lm}, \sigma^{\dot{\tau}}_{lm}\}$, where

$$\xi^{\tau}_{lm}=\xi^{\tau}_{lm}, \qquad \sigma^{\dot{\tau}}_{lm}=(-1)^{l}\xi^{\dot{\tau}}_{lm},$$

then the operators S, T, J are described in the new basis by the matrices

$$J=\begin{Vmatrix}-iE & 0 \\ 0 & iE\end{Vmatrix}, \quad S=\begin{Vmatrix}0 & E \\ E & 0\end{Vmatrix}, \quad T=\begin{Vmatrix}0 & iE \\ -iE & 0\end{Vmatrix}.$$

We shall use this form of the matrices of the operators S, T, J in Sect. 5.

§ 6. The Non-degenerate Hermitian Bilinear Form, Invariant under a Representation of the Complete Lorentz Group

We shall determine the representations $g\to T_g$ of the complete Lorentz group which admit an invariant, non-degenerate Hermitian bilinear form (ψ_1,ψ_2) and we shall find the possible forms. A representation $g\to T_g$ of the complete group decomposes into irreducible components each of which contains either one irreducible component of a representation of the proper Lorentz group, or two non-equivalent, mutually conjugate components τ and $\dot{\tau}$.

We consider the representation $g'\to T_{g'}$ of the proper group induced by the representation $g\to T_g$ of the complete Lorentz group.

We recall that the representation $g'\to T_{g'}$ admits an invariant, non-degenerate Hermitian bilinear form if, and only if, the number of irreducible components of this representation defined by the pair (l_0, l_1) is equal to the number of components τ^* with the pair $(l_0, -\bar{l}_1)$. In terms of the canonical basis $\{\xi^{\tau}_{lm}\}$ of the representation $g'\to T_{g'}$ the invariant form (ψ_1, ψ_2) is then according to (39) Sect.2, given by

$$(\psi_1, \psi_2)=\sum_{\tau} a^{\tau\tau^*}s_l x^{\tau}_{lm}\bar{y}^{\tau^*}_{lm}, \tag{15}$$

where x^{τ}_{lm}, y^{τ}_{lm} are the coordinates of ψ_1 and ψ_2 in terms of the basis $\{\xi^{\tau}_{lm}\}$ and $a^{\tau\tau^*}=\bar{a}^{\tau^*\tau}$ are any complex numbers different from zero only for the components $\tau\sim(l_0, l_1)$ and $\tau^*\sim(l_0, -\bar{l}_1)$; $s_l=\pm 1$ (for a finite representation $s_l=(-1)^{[l]}$). For the invariance of the form (15) under the complete group, we must clearly also have

$$(S\psi_1, S\psi_2)=(\psi_1, \psi_2) \tag{16}$$

where S is the operator corresponding to spatial reflection. Hence

$$(S\xi_{lm}^{\tau}, S\xi_{lm}^{\tau^*}) = (\xi_{lm}^{\tau}, \xi_{lm}^{\tau^*}). \tag{17}$$

Substituting in equation (17) the expression for S (see (5), (5'), (8)), we find that the numbers specifying the invariant bilinear form must satisfy the condition

$$a^{\dot{\tau}\dot{\tau}^*} = \pm a^{\tau\tau^*}. \tag{18}$$

In this case, if $\tau = \dot{\tau}$ (i.e. the component τ is defined by the pair $\tau(0, l_1)$ or $\tau(l_0, 0)$ and consequently $\tau^* = \dot{\tau}^*$), the operator S must act identically in the spaces R^{τ} and R^{τ^*} i.e., either according to the formulae

$$S\xi_{lm}^{\tau} = (-1)^{[l]}\xi_{lm}^{\tau} \text{ and } S\xi_{lm}^{\tau^*} = (-1)^{[l]}\xi_{lm}^{\tau^*}, \tag{19}$$

or according to the formula

$$S\xi_{lm}^{\tau} = (-1)^{[l]+1}\xi_{lm}^{\tau} \text{ and } S\xi_{lm}^{\tau^*} = (-1)^{[l]+1}\xi_{lm}^{\tau^*} \tag{20}$$

(see paragraph 3, case 1).

We now formulate the result obtained.

Together with each irreducible component χ of a representation of the complete Lorentz group consisting of two component representations τ and $\dot{\tau}$ of the proper group

$$\chi \sim (\tau(l_0, l_1), \dot{\tau}(l_0, -l_1)), \quad (l_0 \neq 0, \ l_1 \neq 0)$$

we consider another irreducible component χ^*, which also consists of two component representations τ^* and $\dot{\tau}^*$ of the proper group

$$\chi^*(\tau^*(l_0, -\bar{l}_1); \ \tau^*(l_0, \bar{l}_1)).$$

In the same way for each irreducible component of a representation of the complete group χ, consisting of one component representation of the proper group

$$\chi \sim \tau(0, l_1) \text{ or } \chi \sim \tau(l_0, 0),$$

we consider an irreducible component χ^* consisting of the component τ^*,

$$\chi^* \sim \tau^*(0, -\bar{l}_1) \text{ or } \chi^* \sim \tau^*(l_0, 0),$$

such that the operator S acts identically in the components χ and χ^* (corresponding to the cases (19) or (20).

Thus we have proved that the representation $g \to T_g$ of the complete group permits an invariant, non-degenerate Hermitian bilinear form, if, and only if, the number of mutually equivalent components $\chi_1, \ldots, \chi_s, \ldots$ coincide with the number of mutually equivalent components $\overset{*}{\chi}_1, \ldots, \overset{*}{\chi}_s, \ldots$. In this case the form (ψ_1, ψ_2) itself is given by (15) with the supplementary condition (18).

We apply the result just obtained to the case of an irreducible representation of the complete group. If this representation consists of two non-equivalent components $\tau \sim (l_0, l_1)$ and $\dot{\tau} \sim (l_0, -l_1)$ of the proper group, then an invariant form (ψ_1, ψ_2) exists for it only when

1) $\tau = \tau^*$, or

2) $\tau = \dot{\tau}^*$,

i.e. when either $(l_0, l_1) = (l_0, -\bar{l}_1)$, or $(l_0, -l_1) = (l_0, -\bar{l}_1)$.

In the first case l_1 is a pure imaginary, in the second l_1 is real.

If the representation $g \to T_g$ of the complete group contains one component $\tau \sim (0, l_1)$, then an invariant form exists only for l_1 real or pure imaginary. Finally, for a representation of the complete group with component $\tau \sim (l_0\ 0)$ an invariant form (ψ_1, ψ_2) always exists.

Thus, an irreducible representation of the complete group admits an invariant non-degenerate Hermitian form if, and only if, this representation contains the components (τ and $\dot{\tau}$) with a real or pure imaginary l_1. In particular a finite irreducible representation of the complete group always admits an invariant, non-degenerate Hermitian form.

We turn to the case of a reducible representation which admits an invariant form. A procedure similar to that in paragraph 8 of the previous section which was applied to a form which is invariant under a representation of the proper Lorentz group, enables us to reduce the invariant non-degenerate bilinear form (15) to a somewhat simpler form; it is possible to choose a basis such that for each component $\tau\,(l_0, l)$ there exists only one component $\tau^*\,(l_0, -\bar{l}_1)$ such that $a^{\tau\tau^*} \neq 0$ in which case:

1) Whenever the pair of components $(\tau, \dot{\tau})$ defining an irreducible representation of the complete group coincides with the pair of components $(\tau^*, \dot{\tau}^*)$ (which will be the case, as we have seen, if l_1 is purely imaginary: $\tau = \tau^*$, or l_1 is real $\dot{\tau} = \tau^*$), it is possible to secure that

$$a^{\tau\tau^*} = \pm 1. \tag{21}$$

2) In the case when the pairs of components $(\tau, \dot{\tau})$ and $(\tau^*, \dot{\tau}^*)$ do not coincide(l_1 is not real and not a pure imaginary) it is possible to choose a basis such that

$$a^{\tau\tau^*} = 1. \tag{22}$$

We note that if among the irreducible components of the representation $g \to T_g$ there are components with l_1 real or purely imaginary (case 1) then we can construct essentially different invariant forms differing in the sign of the corresponding $a^{\tau\tau^*}$.

Section 4. Spinors and Spinor Representations of the Proper Lorentz Group

As we saw in the first part of the book, it is possible to construct all the irreducible representations of the group of rotations with the aid of spinors - quantities which are transformed under rotations of three-dimensional space.

Here we shall define the spinors for the proper Lorentz group and we shall use them to obtain all its finite irreducible representations.

§ 1. Spinors of Rank 1

In Sect.1 a two-dimensional, two-valued irreducible representation of the proper Lorentz group was constructed

$$g_a \to \pm a, \tag{1}$$

where

$$a = \begin{Vmatrix} a_{00} & a_{01} \\ a_{10} & a_{11} \end{Vmatrix} \tag{2}$$

is a complex, second-order matrix, with unit determinant, defined except for sign.

Now suppose that in each orthogonal coordinate system (x_0, x_1, x_2, x_3) * of four-dimensional space, there is specified a pair of complex numbers (a^0, a^1), defined except for sign, and that each Lorentz transformation $g \rightleftharpoons g_a$ of the coordinate system with matrix (2) effects the following transformation

$$\left.\begin{aligned} a^{0'} &= a_{00}a^0 + a_{01}a^1, \\ a^{1'} &= a_{10}a^0 + a_{11}a^1. \end{aligned}\right\} \tag{1'}$$

Such a pair of numbers is then known as an undotted spinor of the first rank with respect to the proper Lorentz group.

The two-dimensional, complex linear space $R_{(1,\,0)}$, in which the representation (1) acts, is sometimes known as the space of the undotted spinors

* We recall that an orthogonal coordinate system in four-dimensional space is a system (x_0, x_1, x_2, x_3) with regard to which the form $S^2(x)$ is given by $S^2(x) = x_0^2 - x_1^2 - x_2^2 - x_3^2$, i.e. by means of the matrix I (see Sect.1, paragraph 1).

of rank 1, and the representation (1) itself is called the undotted spinor representation of rank 1.

Together with the representation (1) we consider another two-valued irreducible representation, namely the one specified by the formula:

$$g_a \to \pm \bar{a}, \tag{3}$$

where

$$\bar{a} = \begin{Vmatrix} \bar{a}_{00} & \bar{a}_{01} \\ \bar{a}_{10} & \bar{a}_{11} \end{Vmatrix} \tag{4}$$

is the matrix whose elements are the complex conjugates of the elements of matrix (2).

Now let there be specified in each orthogonal coordinate system (x_0, x_1, x_2, x_3) of four-dimensional space, a pair of complex numbers $(a^{\dot{0}}, a^{\dot{1}})$ defined except for sign; suppose that each Lorentz transformation $g = g_a$ of the coordinate system effects a transformation with matrix (4) according to the formula

$$\left.\begin{aligned} a^{\dot{0}'} &= \bar{a}_{00}a^{\dot{0}} + \bar{a}_{01}a^{\dot{1}}, \\ a^{\dot{1}'} &= \bar{a}_{10}a^{\dot{0}} + \bar{a}_{11}a^{\dot{1}}. \end{aligned}\right\} \tag{3'}$$

Then the pair of numbers is known as a dotted spinor of the first rank with respect to the proper Lorentz group.

The two-dimensional complex space $R_{(\dot{1}, 0)}$ in which the representation (3) acts is known as the space of dotted spinors of rank 1, and the representation (3) is known as the dotted representation of rank 1.

It will be shown below that the dotted and undotted representations of rank 1 are conjugate one with the other (see also Sect.2, § 6). We proceed to the calculation of the infinitesimal operators of the spinor representations of rank 1 (dotted and undotted) and we shall evaluate the pairs (l_0, l_1) defining them.

We shall see that these pairs are as follows: for an undotted representation

$$l_0 = \frac{1}{2}, \qquad l_1 = \frac{3}{2},$$

and for a dotted one

$$l_0 = -\frac{1}{2}, \qquad l_1 = \frac{3}{2}\,^* .$$

---

* We recall that in paragraph 6, Sect.2 the numbers l_0, l_1 were evaluated except for the sign of l_0. Here we define the sign of l_0 for each of the spinor representations of rank 1.

We note that the undotted spinors $e_0=(1, 0)$ and $e_1=(0, 1)$ form a basis of the space $R_{(1,0)}$ of undotted spinors in terms of which the operators of representation (1) have the matrices a as before. Similarly, the dotted spinors $e_{\dot{0}}=(1, 0)$ and $e_{\dot{1}}=(0, 1)$ form a basis of the space of dotted spinors $R_{(\dot{0}, \dot{1})-}$ in terms of which the operators of the representation (3) have matrices a.

We shall calculate below the matrices of the infinitesimal operators H_+, H_-, H_3, F_+, F_-, F_3 of the representations (1) and (3) in terms of the bases (e_0, e_1) and $(e_{\dot{0}}, e_{\dot{1}})$. In passing we establish the connexion between the bases (e_0, e_1), $(e_{\dot{0}}, e_{\dot{1}})$ and the canonical bases $(\xi_{1/2}, \xi_{-1/2})$, $(\eta_{1/2}, \eta_{-1/2})$ of the spinor representations (1) and (3).

We start with the undotted spinor representation. We recall the relation between the matrices a and Lorentz transformations.

To each vector (x_0, x_1, x_2, x_3) from R^4 we assigned the Hermitian matrix

$$c=\left\| \begin{matrix} x_0-x_3 & x_1+ix_2 \\ x_1-ix_2 & x_0+x_3 \end{matrix} \right\|.$$

Then any transformation of the form

$$c'=aca^*,$$

where a is a complex matrix of the second order with $\det a=1$, specifies a proper Lorentz transformation in R^4.

In Sect.1 we showed that any proper Lorentz transformation g may be obtained in this way and that two matrices corresponding to one and the same transformation g differ only in sign.

We now find the matrices a which correspond to rotations in the plane (x_1, x_2) (about the x_3 -axis).

A transformation in the plane (x_1, x_2) has the form

$$\begin{aligned} x_1' &= x_1\cos\varphi + x_2\sin\varphi, \\ x_2' &= -x_1\sin\varphi + x_2\cos\varphi, \\ x_3' &= \qquad\qquad\qquad x_3, \\ x_0' &= \qquad\qquad\qquad x_0 \end{aligned}$$

Hence

$$\left\| \begin{matrix} x_0' - x_3' & x_1' + ix_2' \\ x_1' - ix_2' & x_0' + x_3' \end{matrix} \right\| = \left\| \begin{matrix} x_0 - x_3 & (x_1 + ix_2) e^{-i\varphi} \\ (x_1 - ix_2) e^{i\varphi} & x_0 + x_3. \end{matrix} \right\|.$$

It is easily verified that the required transformation is specified by the matrix

$$a = \pm \left\| \begin{matrix} e^{-\frac{i\varphi}{2}} & 0 \\ 0 & e^{\frac{i\varphi}{2}} \end{matrix} \right\|,$$ in other words

$$\left\| \begin{matrix} e^{-\frac{i\varphi}{2}} & 0 \\ 0 & e^{\frac{i\varphi}{2}} \end{matrix} \right\| \cdot \left\| \begin{matrix} x_0 - x_3 & x_1 + ix_2 \\ x_1 - ix_2 & x_0 + x_3 \end{matrix} \right\| \cdot \left\| \begin{matrix} e^{\frac{i\varphi}{2}} & 0 \\ 0 & e^{-\frac{i\varphi}{2}} \end{matrix} \right\| =$$

$$= \left\| \begin{matrix} x_0 - x_3 & (x_1 + ix_2) e^{-i\varphi} \\ (x_1 - ix_2) e^{i\varphi} & x_0 + x_3 \end{matrix} \right\|$$

In this manner the rotations in the plane (x_1, x_2) correspond to the matrices

$$a(\varphi) = \pm \left\| \begin{matrix} e^{-\frac{i\varphi}{2}} & 0 \\ 0 & e^{\frac{i\varphi}{2}} \end{matrix} \right\|,$$

and the infinitesimal operator H_3 has the form

$$H_3 = \left\| \begin{matrix} \frac{1}{2} & 0 \\ 0 & -\frac{1}{2} \end{matrix} \right\|. \tag{5}$$

Hence it is clear that the vectors of the basis (e_0, e_1) of the two-dimensional space $R_{(1,0)}$ coincide to within a multiplier, with the vectors of the canonical basis $\left(\xi_{\frac{1}{2}}, \xi_{-\frac{1}{2}}\right)$ of our representation.

If we calculate the operators H_+ and H_-* (we shall not do so) then we can convince ourselves that the bases (e_0, e_1) and $\left(\xi_{\frac{1}{2}}, \xi_{-\frac{1}{2}}\right)$ coincide precisely

$$e_1 = \xi_{\frac{1}{2}}, \qquad e_2 = \xi_{-\frac{1}{2}}.$$

---

* Such a calculation was performed in the first part of the book, Sect.2, p.33

Thus, the undotted spinors with components (1, 0) and (0, 1) form the canonical basis $\left\{\xi_{\frac{1}{2}}, \xi_{-\frac{1}{2}}\right\}$ for the undotted spinor representation.

As usual the operators H_+ and H_- have the form

$$H_+ = \begin{Vmatrix} 0 & 0 \\ 1 & 0 \end{Vmatrix}, \qquad H_- = \begin{Vmatrix} 0 & 1 \\ 0 & 0 \end{Vmatrix}. \tag{5'}$$

We now find the operator F_3 - the infinitesimal operator corresponding to transformations in the plane (x_0, x_3). These transformations are given by:

$$\begin{aligned} x_0' &= x_0 \operatorname{ch} t && + x_3 \operatorname{sh} t, \\ x_1' &= \quad x_1, \\ x_2' &= \quad x_2, \\ x_3' &= x_0 \operatorname{sh} t && + x_3 \operatorname{ch} t. \end{aligned}$$

Hence

$$\begin{Vmatrix} x_0' - x_3' & x_1' + ix_2' \\ x_1' - ix_2' & x_0' + x_3' \end{Vmatrix} = \begin{Vmatrix} (x_0 - x_3)\, e^{-t} & x_1 + ix_2 \\ x_1 - ix_2 & (x_0 + x_3)\, e^{t} \end{Vmatrix}.$$

Again it is easily verified that such a transformation is attained by means of the matrix

$$a(t) = \pm \begin{Vmatrix} e^{-\frac{t}{2}} & 0 \\ 0 & e^{\frac{t}{2}} \end{Vmatrix}. \tag{6}$$

Thus transformations in the plane (x_0, x_3) correspond to the matrices (6) and the infinitesimal operator F_3 of these transformations is equal to

$$F_3 = -\frac{i}{2}\begin{pmatrix} 1 & 0 \\ 0 & -1 \end{pmatrix}. \tag{7}$$

The operators F_+ and F_- have the form

$$F_+ = -i \begin{Vmatrix} 0 & 0 \\ 1 & 0 \end{Vmatrix}; \qquad F_- = -i \begin{Vmatrix} 0 & 1 \\ 0 & 0 \end{Vmatrix}. \tag{7'}$$

(We leave the verification to the reader).

Comparing formulae (7) and (7') with formulae (13)-(16) of Sect.2, we deduce that our representation is specified by the numbers $l_0 = \frac{1}{2}$; $l_1 = \frac{3}{2}$. We now consider the dotted spinor representation. This differs from the previous case in that the basis $(e_{\dot{0}}, e_{\dot{1}})$ of the space of dotted spinors is not canonical.

The canonical basis $\left\{\eta_{\frac{1}{2}}, \eta_{-\frac{1}{2}}\right\}$ of the dotted representation is connected with the basis $\{e_{\dot{0}}, e_{\dot{1}}\}$ by the matrix

$$\tau = \begin{pmatrix} 0 & 1 \\ -1 & 0 \end{pmatrix},$$

i.e.

$$\eta_{\frac{1}{2}} = -e_{\dot{1}}, \qquad \eta_{-\frac{1}{2}} = e_{\dot{0}}.$$

In other words, <u>the dotted spinors with the components (0,-1) and (1, 0) form the canonical basis $\left\{\eta_{\frac{1}{2}}, \eta_{-\frac{1}{2}}\right\}$ of the dotted spinor representation.</u> In order to be convinced of the fact that the basis $\left(\eta_{\frac{1}{2}}, \eta_{-\frac{1}{2}}\right)$ is canonical, we shall determine the operators of the dotted representation and find its infinitesimal operators in terms of this basis. It is clear that if the operators of representation (3) have the matrices a_g, in the basis $(e_{\dot{0}}, e_{\dot{1}})$, then in the basis $\left(\eta_{\frac{1}{2}}, \eta_{-\frac{1}{2}}\right)$ they have the matrices:

$$\tau \bar{a}_g \tau^{-1}.$$

But

$$\tau \bar{a}_g \tau^{-1} = \left(\overset{*}{a}_g\right)^{-1} \quad \text{[see Sect.1, paragraph 4, (11')]}$$

Thus, in the basis $\left\{\eta_{\frac{1}{2}}, \eta_{-\frac{1}{2}}\right\}$ the dotted representation (3) has the matrices

$$g_a \to \left(\overset{*}{a}_g\right)^{-1} \text{*)}. \tag{8}$$

If $\tilde{g}$ is a rotation then a_g is a unitary matrix: $a_g = \left(\overset{*}{a}_g\right)^{-1}$. Therefore, for the rotations $\tilde{g}$, the representation $g \to a_g$ coincides with the representation

$$g_a \to \left(\overset{*}{a}_g\right)^{-1}.$$

Consequently the operators H_+, H_-, H_3 of the dotted spinor representation of rank 1 are specified by the matrices $\tilde{H}_+$, $\tilde{H}_-$, $\tilde{H}_3$ respectively, in the basis $\left\{\eta_{\frac{1}{2}}, \eta_{-\frac{1}{2}}\right\}$. They are as follows

$$\tilde{H}_+ = \begin{Vmatrix} 0 & 0 \\ 1 & 0 \end{Vmatrix}, \qquad \tilde{H}_- = \begin{Vmatrix} 0 & 1 \\ 0 & 0 \end{Vmatrix}, \qquad \tilde{H}_3 = \begin{Vmatrix} \frac{1}{2} & 0 \\ 0 & -\frac{1}{2} \end{Vmatrix}. \tag{9}$$

---

* We note that formula (8) clearly means that the dotted spinor representation is conjugate to the undotted and is therefore defined by the pair of numbers $\left(-\frac{1}{2}, \frac{3}{2}\right)$ (see paragraph 5, Sect. 2).

The formulae just obtained mean that the basis $\left\{\eta_{\frac{1}{2}}, \eta_{-\frac{1}{2}}\right\}$ really is canonical for this representation. We now find the matrices of the operators F_{+}, F_{-}, F_3 with respect to the basis $\left\{\eta_{\frac{1}{2}}, \eta_{-\frac{1}{2}}\right\}$. We have already seen that the matrix $(a_{g_{03}})$ corresponding to a hyperbolic screw in the plane (x_0, x_3) has the form (cf (6))

$$a_{g_{03}} = \begin{Vmatrix} e^{-\frac{t}{2}} & 0 \\ 0 & e^{\frac{t}{2}} \end{Vmatrix}.$$

Hence,

$$(a^*_{g_{03}})^{-1} = \begin{Vmatrix} e^{\frac{t}{2}} & 0 \\ 0 & e^{-\frac{t}{2}} \end{Vmatrix}. \tag{8'}$$

The matrix $\tilde{F}_3$ of the infinitesimal operator F_3 relative to the basis $\left\{\eta_{\frac{1}{2}}, \eta_{-\frac{1}{2}}\right\}$ is thus given by

$$\tilde{F}_3 = -\frac{i}{2}\begin{Vmatrix} -1 & 0 \\ 0 & 1 \end{Vmatrix}. \tag{10}$$

Finally the operators F_{+}, F_{-} have the following matrices in terms of the basis $\left\{\eta_{\frac{1}{2}}, \eta_{-\frac{1}{2}}\right\}$

$$\tilde{F}_{+} = i\begin{Vmatrix} 0 & 0 \\ 1 & 0 \end{Vmatrix}, \quad \tilde{F}_{-} = i\begin{Vmatrix} 0 & 1 \\ 0 & 0 \end{Vmatrix}. \tag{10'}$$

From a comparison of formulae (13)-(16), Sect.2 and the form of the infinitesim operators (F_3, F_{-}, F_{+}) we deduce that the representation (3) is defined by the pair $\left(-\frac{1}{2}, \frac{3}{2}\right)$.

Finally we derive the matrices of the infinitesimal operators H_3, H_{+}, H_{-}, F_3, F_{+}, F_{-} in the basis $\{e_{\dot{0}}, e_{\dot{1}}\}$.

We have

$$\dot{H}_3 = \tau\tilde{H}_3\tau^{-1} = \begin{Vmatrix} -\frac{1}{2} & 0 \\ 0 & \frac{1}{2} \end{Vmatrix}.$$

similarly we obtain:

$$\left.\begin{array}{ll} \dot{H}_{+}=\begin{Vmatrix} 0 & -1 \\ 0 & 0 \end{Vmatrix}, & \dot{F}_{+}=-i\begin{Vmatrix} 0 & -1 \\ 0 & 0 \end{Vmatrix}, \\ \dot{H}_{-}=\begin{Vmatrix} 0 & 0 \\ -1 & 0 \end{Vmatrix}, & \dot{F}_{-}=-i\begin{Vmatrix} 0 & 0 \\ -1 & 0 \end{Vmatrix}, \\ \multicolumn{2}{c}{\dot{F}_{3}=-\frac{i}{2}\begin{Vmatrix} 1 & 0 \\ 0 & -1 \end{Vmatrix}.} \end{array}\right\} \quad (10'')$$

Thus we have found the infinitesimal operators, canonical bases $\left\{\xi_{\frac{1}{2}}, \xi_{-\frac{1}{2}}\right\}$ and $\left\{\eta_{\frac{1}{2}}, \eta_{-\frac{1}{2}}\right\}$ and the numbers l_0, l_1 specifying the general spinor representations of the first rank. To sum up:

The undotted spinor representation $g_a \to \pm a$ of rank 1 is specified by the pair $\left(\frac{1}{2}, \frac{3}{2}\right)$, its canonical basis $\left\{\xi_{\frac{1}{2}}, \xi_{-\frac{1}{2}}\right\}$ consists of the undotted spinors (1, 0) and (0, 1) and the operators of the representation (1) have the matrices a in this basis. The infinitesimal operators in the basis $\left\{\xi_{\frac{1}{2}}, \xi_{-\frac{1}{2}}\right\}$ have the form (5), (5'), (7) and (7'). The dotted spinor representation $g_a \to \pm \bar{a}$ of rank 1 is specified by the pair $\left(-\frac{1}{2}, \frac{3}{2}\right)$, its canonical basis $\left\{\eta_{\frac{1}{2}}, \eta_{-\frac{1}{2}}\right\}$ consists of the dotted spinors (0, -1), (1, 0) and the operators of representation (3) itself have the matrices $(a^*)^{-1}$ in this basis. The infinitesimal operators of the dotted representation have the form (9), (10) and (10') in terms of the basis $\left\{\eta_{\frac{1}{2}}, \eta_{-\frac{1}{2}}\right\}$ The canonical basis $\left\{\eta_{\frac{1}{2}}, \eta_{-\frac{1}{2}}\right\}$ in the space of the dotted spinors is connected with the basis $e_{\dot{0}}=(1, 0)$, $e_{\dot{1}}=(0, 1)$, which expresses the dotted representation by the the matrices $\bar{a}$, by the transformation τ having matrix

$$\tau=\begin{Vmatrix} 0 & 1 \\ -1 & 0 \end{Vmatrix}.$$

The infinitesimal operators H_{+}, H_{-}, H_3, F_{+}, F_{-}, F_3, of the dotted representation in the basis $\{e_{\dot{0}}, e_{\dot{1}}\}$ have the form (10").

We note that the canonical basis of a representation is in many respects very convenient. In this connexion, the components of a dotted spinor with respect to the canonical basis $\left\{\eta_{\frac{1}{2}}, \eta_{-\frac{1}{2}}\right\}$ are often considered. If we denote these components by $a_{\dot{0}}$ and $a_{\dot{1}}$, then they are clearly connected with the components $(a^{\dot{0}}, a^{\dot{1}})$ by the relations

$$a_{\dot{0}} = a^{\dot{1}},$$
$$a_{\dot{1}} = -a^{\dot{0}}.$$

Clearly the quantities $(a_{\dot{0}}, a_{\dot{1}})$ transform under proper Lorentz transformations by the matrix $(a^*)^{-1}$. We shall examine these quantities $(a_{\dot{0}}, a_{\dot{1}})$ (known as dotted spinors with lower indices) in detail in the next section.

§ 2. Lowering of the Indices of Spinors of the First Rank

We consider the bilinear (non-hermitian!) form

$$\sum_{\alpha, \beta} \tau_{\alpha\beta} a^{\alpha} b^{\beta} \qquad (\alpha = 0, 1, \quad \beta = 0, 1) \tag{11}$$

with the matrix

$$\tau = \|\tau_{\alpha\beta}\| = \begin{Vmatrix} 0 & +1 \\ -1 & 0 \end{Vmatrix}.$$

Here a^{α}, b^{β} are two undotted spinors.

As we have observed more than once, we have

$$a\tau a^{\tau} = \tau,$$

where a^{τ} is the transpose of matrix a. This equation means that the bilinear form (11) is invariant under the representation $g_a \to a$ acting in the space of the undotted spinors.

Similarly, the form

$$\sum \tau_{\dot{\alpha}\dot{\beta}} a^{\dot{\alpha}} \bar{b}^{\dot{\beta}} \tag{11'}$$

($a^{\dot{\alpha}}$, $b^{\dot{\beta}}$ are dotted spinors) is invariant under the representation $g_a \to \bar{a}$. By means of the matrix τ we use the undotted spinor a^{β} to construct the quantity

$$a_{\alpha} = \sum_{\beta = 0,1} \tau_{\alpha\beta} a^{\beta} \qquad (\alpha = 0, 1). \tag{11"}$$

The quantity (a_0, a_1) will be known as an undotted spinor of rank 1 with a lower index, and the operation (11") as the lowering of the index. It is easily verified that a Lorentz transformation g_a transforms the spinor (a_0, a_1) by the matrix $\tau a \tau^{-1} = (a^{\tau})^{-1}$.

It is clear that in the space of all spinors (a_0, a_1) there acts a representation of the proper group equivalent to the representation in the space of the spinors (a^0, a^1) with upper indices.

Similarly it is possible to lower the index of a dotted spinor:

$$a_{\dot{\alpha}} = \sum_{\dot{\beta}=0,1} \tau_{\dot{\alpha}\dot{\beta}} a^{\dot{\beta}}. \tag{11'''}$$

The quantity $(a_{\dot{0}}, a_{\dot{1}})$ is called a dotted spinor of rank 1 with a lower index.

Under a proper Lorentz transformation g_a the spinor $(a_{\dot{0}}, a_{\dot{1}})$ transforms by means of the matrix $\tau\bar{a}\tau^{-1} = (a^*)^{-1}$. Thus, in the space of all spinors $(a_{\dot{0}}, a_{\dot{1}})$ there acts a representation of the proper group $g_a \to \pm (a^*)^{-1}$, equivalent to the representation $g_a \to \pm \bar{a}$. We note that it follows from the results of the previous section that in the space of all spinors $(a_{\dot{0}}, a_{\dot{1}})$ the basis $e^{\dot{0}} = (1, 0)$, $e^{\dot{1}} = (0, 1)$, in terms of which all the matrices $(a^*)^{-1}$, are expressed, coincides with the canonical basis of the representation $g_a \to (a^*)^{-1}$.

3. Spinors of Higher Ranks

We consider a 2^k-dimensional complex space R^{2^k}, each point of which is defined by a set of 2^k numbers $a^{\alpha_1 \cdot \alpha_2 \cdots \alpha_k}$ $(\alpha_i = 0, 1)$. We specify the representation $g_a \to T_a$, in R^{2^k} according to the formula

$$a^{\alpha_1' \cdot \alpha_2' \cdots \alpha_k'} = \sum a_{\alpha_1'\alpha_1} a_{\alpha_2'\alpha_2} \cdots a_{\alpha_k'\alpha_k} a^{\alpha_1 \cdots \alpha_k}, \tag{12}$$

where the summation is carried out over the entire set $(\alpha_1 \ldots \alpha_k)$; $a = \| a_{\alpha_i'\alpha_i} \|$ is the matrix corresponding to the transformation g_a. Let there be specified for each orthogonal coordinate system (x_0, x_1, x_2, x_3) a set of 2^k complex numbers $a^{\alpha_1 \cdots \alpha_k}$ $(\alpha_i = 0, 1)$, defined except for sign, which transform by proper Lorentz transformation $g = g_a$ of coordinates according to formula (12). Such a set of numbers is known as an undotted spinor of rank k with respect to the proper Lorentz group.

Representation (12) is known as an undotted spinor representation of rank k. We note that, inasmuch as in formula (12) the matrix $\|a_{\alpha'\alpha}\|$ acts upon each index independently, representation (12) is the product of k representations of type (1), i.e. the product of k undotted spinor representations of rank 1 (equally the space R^{2^k} is the product of k two-dimensional spaces).

In the same way we define a dotted spinor representation acting in the 2^n-dimensional space of all quantities $a^{\dot{\alpha}_1 \cdots \dot{\alpha}_n}$ according to the formula

$$a^{\dot{\alpha}_1' \cdots \dot{\alpha}_n'} = \sum \bar{a}_{\dot{\alpha}_1'\dot{\alpha}_1} \bar{a}_{\dot{\alpha}_2'\dot{\alpha}_2} \cdots \bar{a}_{\dot{\alpha}_n'\dot{\alpha}_n} a^{\dot{\alpha}_1 \cdots \dot{\alpha}_n}, \tag{13}$$

the matrix $a = \| a_{\dot{\alpha}_i'\dot{\alpha}_i} \|$ is, as above, the matrix corresponding to the

Lorentz transformation $g = g_a$.

Let there be specified in each orthogonal coordinate system (x_0, x_1, x_2, x_3) a set of 2^n-complex numbers $a^{\dot{\alpha}_1\dot{\alpha}_2 \cdots \dot{\alpha}_n}$ $(\dot{\alpha}_j = 0, 1)$, defined except for sign, which transforms by a proper Lorentz transformation $g = g_a$ of coordinates according to formula (13). Such a set of numbers is known as a dotted spinor of rank n with respect to the proper Lorentz group.

The representation specified by formula (13) is known as a dotted spinor representation of rank n. Analogously to the undotted case, this is clearly the product of n dotted spinor representations of rank 1.

We consider finally the most general case.

In the 2^{k+n}-dimensional space of all quantities $a^{\alpha_1 \cdots \alpha_k \dot{\alpha}_1 \cdots \dot{\alpha}_n}$ we define a representation of the proper group by means of the formula

$$a'^{\alpha'_1 \cdots \alpha'_k \dot{\alpha}'_1 \cdots \dot{\alpha}'_n} = \\ = \sum a_{\alpha'_1\alpha_1} \cdots a_{\alpha'_k\alpha_k} \bar{a}_{\dot{\alpha}'_1\dot{\alpha}_1} \cdots \bar{a}_{\dot{\alpha}'_n\dot{\alpha}_n} a^{\alpha_1 \cdots \alpha_k \dot{\alpha}_1 \cdots \dot{\alpha}_n}. \tag{14}$$

Let there be specified in each orthogonal coordinate system (x_0, x_1, x_2, x_3) a set of 2^{k+n}- complex numbers $a^{\alpha_1 \cdots \alpha_k \dot{\alpha}_1 \cdots \dot{\alpha}_n}$, defined except for sign, which transform by a proper Lorentz transformation $g = g_a$ of coordinates according to formula (14). Such a set of numbers is known as a spinor with k undotted and n dotted indices or, more briefly, a spinor of rank (k, n) with respect to the proper Lorentz group. Representation (14) is known as the spinor representation of rank (k, n) *

Representation (14) is the product of k undotted and n dotted spinor representations of rank 1. Representation (14) of rank (k, n) may also be considered as the product of two representations: an undotted spinor representation of rank k (i.e. a representation of rank $(k, 0)$) and a dotted spinor representation of rank n (i.e. a representation of rank $(0, n)$). We note here that when g_a is a rotation, the matrix a, as we have seen, is unitary.

Thus, representation (12) (the undotted spinor representation) for the group of rotations coincides with the ordinary spinor representation which we considered in Part I (Sect. 6).

Formulae (13) and (14) specifying the spinor representations of ranks $(0, n)$ and (k, n) of the proper group, do not reduce immediately when g_a is a rotation (i.e. when a is a unitary matrix), to the formula defining a spinor representation of the group of rotations (see Part I, Sect.6, formula (3)). Nonetheless

---

* A spinor of rank $(k, 0)$ is clearly an undotted spinor of rank k, and a spinor of rank $(0, n)$ is a dotted spinor of rank n.

the representation of the group of rotations generated by formula (14) is equivalent to the spinor representation of rank $n+k$ of this group*.

Representation (14) in the space of spinors of rank (k, n) is, generally speaking, reducible, i.e. the 2^{k+n}-dimensional space of the representation (14) contains invariant subspaces.

We now select in each spinor space one such invariant subspace $R_{(k, n)}$ in which, as it happens, representation (14) is irreducible. We shall now show, moreover, that all the irreducible finite representations of the proper group are exhausted by the representations in the subspace $R_{(k, n)}$.

§ 4. Symmetric Spinors. The Realization of all the Finite Irreducible Representations of the Proper Group

We start again with particular cases.

I. An undotted spinor $a^{\alpha_1 \cdots \alpha_k}$ of rank $(k, 0)$ is called symmetric if it is unaltered by any permutation of the indices $(\alpha_1 \ldots \alpha_k)$. It is clear that the symmetric spinors of rank $(k, 0)$ form a subspace in the space of all spinors. We shall denote this subspace by $R_{(k, 0)}$ (its dimensionality is $k+1$).

In formula (12) specifying the representation (12) in the space of all spinors of rank $(k, 0)$ the matrix acts upon each index in the same way. Hence a symmetric spinor is transformed by formula (12) once again into a symmetric spinor, i.e. the subspace $R_{(k, 0)}$ in invariant under the representation (12).

We shall show that representation (12) in the space $R_{(k, 0)}$ is irreducible. In fact when g_a is a rotation, this representation coincides with the spinor representation of the group of rotations, and the latter, as was proved in the first part of the book (see Sect.6), is irreducible in the space of symmetric spinors

---

* As we have seen above, the transition from the matrix a to the matrix $\bar{a}$ in the group of unitary matrices is realized by means of the formula

$$\bar{a} = \tau a \tau^{-1}, \qquad \tau = \begin{Vmatrix} 0 & -1 \\ 1 & 0 \end{Vmatrix}. \tag{8''}$$

Since τ is a unitary matrix and $\det \tau = 1$, equation (8") is also satisfied by the operators of the representation

$$T_a = T_\tau T_{\bar{a}} T_\tau^{-1}.$$

This means that the representations $g_a \to T_a$ and $g_a \to T_{\bar{a}}$ of the group of rotations are equivalent. From what has been said it is clear that if we consider the operators which correspond to rotations in the dotted spinor representation of rank n then we obtain a representation equivalent to the spinor representation of the group of rotations of rank n.
Since the spinor representation of rank (k, n) is the product of an undotted one of rank k and a dotted one of rank n, consequently it generates a representation of the group of rotations equivalent to the product of its spinor representations of ranks k and n, i.e. equivalent to the spinor representation of rank $k+n$.

of rank k and is specified by the weight $l=\frac{k}{2}$. Hence it follows that representation (12) of the proper Lorentz group in the space $R_{(k,0)}$ is all the more irreducible and contains only one weight $l=\frac{k}{2}$. This means that $|l_0|=\frac{k}{2}$ and $|l_1|=\frac{k}{2}+1$. To determine the signs of the numbers l_1 and l_0 it is necessary to evaluate the infinitesimal operator F_3 of our representation. We shall not do this, but merely show that the signs of l_0 and l_1 are the same for all undotted symmetric spinors. (We have seen this in paragraph 1 in the example of an undotted spinor of rank 1).

Thus, representation (12) which acts in the space $R_{k,0}$ of all symmetric spinors of rank $(k, 0)$ is irreducible and defined by the pair

$$l_0=\frac{k}{2}, \qquad l_1=\frac{k}{2}+1. \tag{15}$$

We shall call such a representation a spinor irreducible representation of rank $(k, 0)$ and denote it by $T_g^{(k,0)}$.

It is clear that, any irreducible finite representation for which the pair of numbers l_0 and l_1 are of the same sign and $l_1=l_0+1$, is equivalent to some representation $T_g^{(k,0)}$.

II. A dotted spinor $a^{\dot{\alpha}_1 \dots \dot{\alpha}_n}$ of rank $(0, n)$ is called symmetric if it is unaltered by any permutation of the indices. Amongst all spinors of rank$(0, n)$ the symmetric spinors form a subspace $R_{(0,n)}$ which is invariant under the dotted representation (13).

Since for the group of rotations the representation (13) is equivalent to the ordinary spinor representation of rank n of this group (see the previous section), and the latter is irreducible in the space of spinors and has the weight $l=\frac{n}{2}$, therefore the representation of the Lorentz group is also irreducible in $R_{(0,n)}$ and contains exactly one weight i.e. $|l_0|=\frac{n}{2}$, $|l_1|=\frac{n}{2}+1$. It may be proved by evaluating the infinitesimal operator F_3 of representation (13) in $R_{(0,n)}$, that the numbers l_0 and l_1 have opposite signs for representations of the Lorentz group in the space of dotted symmetric spinors. (For a representation of rank 1 this was carried out in section 1).

It is convenient for us to take

$$l_0=-\frac{n}{2}, \qquad l_1=\frac{n}{2}+1.$$

Thus, the representation of the proper Lorentz group specified by formula (13) in the space $R_{(0,n)}$ of all symmetric spinors of rank $(0, n)$ is irreducible and defined by the pair

$$l_0 = -\frac{n}{2}, \qquad l_1 = \frac{n}{2} + 1. \tag{16}$$

We shall call such a representation an irreducible spinor representation of rank $(0, n)$ and denote it $T_g^{(0, n)}$. Any finite irreducible representation with a pair (l_0, l_1) such that $|l_1| = |l_0| + 1$ and l_0 and l_1 are of opposite sign, is equivalent to some representation $T_g^{(0, n)}$.

We have seen that the representations $T_g^{(k, 0)}$ and $T_g^{(0, \dot{n})}$ are also irreducible under the group of rotations (they contain just one weight). Conversely, any representation of the proper Lorentz group, irreducible under the group of rotations, is equivalent either to $T_g^{(k, 0)}$ or $T_g^{(0, n)}$ *

We now turn to the general case.

III. A spinor of rank (k, n) (with k undotted and n dotted indices) is called symmetric if it is unaltered under all possible permutations both of the dotted indices amongst themselves and of the undotted indices amongst themselves. The symmetric spinors of rank (k, n) form a subspace in the space of all spinors of rank (k, n) (we denote it by $R_{(k, n)}$); the dimensionality of $R_{(k, n)}$ is equal to $(k+1)(n+1)$. This subspace is invariant under representation (14) since the matrix a acts in the same way upon all the undotted indices, and the matrix $\bar{a}$, in the same way upon all the dotted indices.

It so happens that representation (14) in $R_{(k, n)}$ is irreducible. We shall not prove this **, but we shall merely find the pair (l_0, l_1) specifying this irreducible representation.

We note that the space $R_{(k, n)}$ of all symmetric spinors of rank (k, n) is the product of the spaces $R_{(k, 0)}$ and $R_{(0, n)}$ of all symmetric spinors of ranks $(k, 0)$ and $(0, n)$ respectively:

$$R_{(k, n)} = R_{(k, 0)} \times R_{(0, n)},$$

---

* For according to Sect. 2, paragraph 6 an irreducible representation of the proper group containing only one weight $l = |l_0|$, satisfies the following equation

$$|l_1| = |l_0| + 1.$$

Two cases are possible:

(1) l_1 and l_0 are of the same sign ($l_0 > 0$ and $l_1 > 0$). Such a representation is equivalent to the undotted spinor representation $T_g^{(2l_0, 0)}$.

(2) l_0 and l_1 are of opposite sign ($l_0 < 0$, $l_1 > 0$); in this case such a representation is equivalent to the dotted spinor representation $T_g^{(0, 2|l_0|)}$.

** In paragraph 5 of this section we construct a representation equivalent to representation (14) which acts in a space of forms $p(\xi, \bar{\xi})$ in the variables ξ $\bar{\xi}$. The irreducibility of the spinor representation (14) is easily verified by means of the formulae specifying the above representation.

and representation (14) acting in $R_{(k, n)}$ (we denote it by $T_g^{(k, n)}$), is the product of the representations $g \to T_g^{(k, 0)}$ and $g \to T_g^{(0, n)}$, acting in $R_{(k, 0)}$ and $R_{(0, n)}$,

$$T_g^{(k, n)} = T_g^{(k, 0)} \times T_g^{(0, n)}.$$

The representations $T_g^{(k, 0)}$ and $T_g^{(0, n)}$ in $R_{(k, 0)}$ and $R_{(0, n)}$ are irreducible under the group of rotations and their weights are $\frac{k}{2}$ and $\frac{n}{2}$. In this case the product of these representations $T_g^{(k, 0)} \times T_g^{(0, n)} = T_g^{(k, n)}$ contains all the weights from $\frac{|k-n|}{2}$ to $\frac{k+n}{2}$ once only. This means that the representation $T_g^{(k, n)}$ in the space $R_{(k, n)}$ is defined by the pair (l_0, l_1) where

$$|l_0| = \frac{k-n}{2}, \qquad |l_1| = \frac{k+n}{2} + 1.$$

If we evaluate the infinitesimal operator F_3 for the representation $T_g^{(k, n)}$ then we find that

$$l_0 = \frac{k-n}{2}, \qquad l_1 = \frac{k+n}{2} + 1.$$

Thus, the representation (14) $g \to T_g^{(k, n)}$ of the proper group in the space of all symmetric spinors of rank (k, n) is irreducible and specified by the pair (l_0, l_1) where the numbers l_0 and l_1 are given by

$$l_0 = \frac{k-n}{2}, \qquad l_1 = \frac{k+n}{2} + 1. \tag{17}$$

The representation $T_g^{(k, n)}$ in $R_{(k, n)}$ will be called the spinor irreducible representation of rank (k, n). It is clear that it is possible to choose the numbers k and n so as to obtain any pair (l_0, l_1) specifying a finite representation of the proper group (i.e. any pair such that l_0 and l_1 are simultaneously integral or half-integral and $|l_1| > |l_0|$). Thus we have proved that any finite irreducible representation of the proper group is equivalent to some irreducible spinor representation $T_g^{(k, n)}$. In other words, the spinor representations exhaust all the finite irreducible representations of the proper Lorentz group.

We note that it is clear from formula (17) that the representations $T_g^{(k, n)}$ and $T_g^{(n, k)}$ are defined by the pairs (l_0, l_1) and $(-l_0, l_1)$ respectively; and this means that they are conjugate.

Finally we make an observation which will be used in the sequel.

We consider representation (12) acting in the 2^k-dimensional space of all undotted spinors. We select in this space a basis consisting of those spinors for which one, and only one, component differs from zero *.

* In spinor notation such a spinor may be written $\xi^{\alpha_1 \cdots \alpha_k} = \delta_{\beta_1 \alpha_1} \delta_{\beta_2 \alpha_2} \cdots \delta_{\beta_k \alpha_k}$, where the component different from zero corresponds to the set $(\beta_1 \beta_2 \ldots \beta_k)$.

If we evaluate the matrices A_g of the operators of the representation $g \to T_g$ (17), in terms of this basis, then, as is clear from formula (12), the elements of this matrix are polynomials in the elements of the matrix a:

$$A_g = \| P_{\varkappa\varkappa'} (a_{00}, a_{01}, a_{10}, a_{11}) \|$$

($\varkappa$ and $\varkappa'$ are two sets of spinor indices), i.e. they are complex analytic functions of the variables $(a_{00}, a_{01}, a_{10}, a_{11})$.

If we transfer to another basis then the matrices are transformed according to the formula

$$\tilde{A} = SA_gS^{-1}$$

and the elements of the new matrix $\tilde{A}_g$ are linear combinations of the elements of the matrix A_g and so will again be complex analytic functions of the variables $(a_{00}, a_{01}, a_{10}, a_{11})$. If the space decomposes into a linear sum of invariant subspaces, then adapting the basis in the entire space to this decomposition and using the above remarks, we see that the matrices of the operators of any representation $g_a \to T_a$ in any invariant subspace of the space of all spinors of rank $(k, 0)$, in particular in the subspace $R_{(k, 0)}$ of all symmetric spinors have elements which are complex analytic functions of $(a_{00}\ a_{10}, a_{01}, a_{11})$. It is clear that this property also holds for the matrices of any finite representation equivalent to the representation $T_g^{(k, 0)}$.

It is clear from formula (13) that the elements of the matrices of the operators T_g, of the representation (12) in the space of all dotted spinors are polynomials in $(\bar{a}_{00}, \bar{a}_{01}, \bar{a}_{10}, \bar{a}_{11})$, i.e. anti-analytic functions of the variables $(a_{00}, a_{01}, a_{10}, a_{11})$. This property of the matrix elements holds both for an irreducible spinor representation of rank $(0, n)$ and for any representation equivalent to it.

Finally, the matrix elements for the operators of representation (14) in the space of all symmetric spinors of rank (k, n) are neither analytic nor anti-analytic functions of the variables $(a_{00}, a_{01}, a_{10}, a_{11})$. It follows from what has been said that if the matrix elements of the operators in some basis of a finite representation are analytic (anti-analytic) functions of the variables $(a_{00}, a_{10}, a_{01}, a_{11})$, then the irreducible components of the representation are equivalent only to undotted (dotted) spinor representations, i.e. to representations of ranks $(k, 0)$($(0, n)$). We shall make use of this observation in Sect.6.

In concluding this section we consider by way of an example how a vector representation of the proper Lorentz group acting in the space $R^{(1)}$ is realized by means of spinors. In paragraph 7, Sect.2 we found that the representation $g \to g$ is defined by the pair $(l_0, l_1) = (0, 2)$. It follows that this representation is equivalent to a spinor representation $T_g^{(1,1)}$ of rank (1, 1) and is realized in the space $R_{(1, 1)}$ of all spinors with one dotted and one undotted index. We now find an explicit expression for the coordinates of the vector x_0, x_1, x_2, x_3 in terms of the components of the spinor $(a^{0\dot{0}}, a^{0\dot{1}}, a^{1\dot{0}}, a^{1\dot{1}})$. First of all we note that the space $(R_{1,1})$ is a four-dimensional complex space. In this connexion, the representation $g \to g$, which is equivalent to the representation $T_g^{(1,1)}$ has, up to now, been deemed to act in a four-dimensional real space; naturally it is now considered to act in a four-dimensional complex space; so that the coordinates x_0, x_1, x_2, x_3 may be complex. We now recall that the fundamental correspondence $g_a \to a$ between proper Lorentz transformations and the matrices a ($\det a = 1$) was constructed as follows: if to the vector (x_0, x_1, x_2, x_3) is assigned the matrix c

$$c = \begin{Vmatrix} x_0 - x_3 & x_2 - ix_1 \\ x_2 + ix_1 & x_0 + x_3 \end{Vmatrix},$$

then every transformation of the matrices c of the form

$$c' = aca^*$$

generates in the space $R^{(4)}$ a proper Lorentz transformation g_a, which we put in correspondence with the matrix a: $a \sim g_a$. Whereas, in the case of real coordinates x_0, x_1, x_2, x_3 the matrix c had to be Hermitian, in the present case of complex coordinates x_0, x_1, x_2, x_3 the matrix c may be any arbitrary complex second-order matrix.

Thus any proper Lorentz transformation upon the vectors x: $x' = g_a x$ has the following expression by means of the matrices c:

$$\begin{Vmatrix} x'_0 - x'_3 & x'_2 - ix'_1 \\ x'_2 + ix'_1 & x'_0 + x'_3 \end{Vmatrix} = a \begin{Vmatrix} x_0 - x_3 & x_2 - ix_1 \\ x_2 + ix_1 & x_0 + x_3 \end{Vmatrix} a^*. \tag{17'}$$

The spinor $(a^{0\dot{0}}, a^{0\dot{1}}, a^{1\dot{0}}, a^{1\dot{1}})$ of rank (1, 1) may also be written in the form of a matrix

$$c = \begin{Vmatrix} a^{0\dot{0}} & a^{0\dot{1}} \\ a^{1\dot{0}} & a^{1\dot{1}} \end{Vmatrix}.$$

Whereupon it is easily verified that the representation $T_g^{(1,1)}$, acts upon the matrix c according to the formula

$$c' = aca^*$$

or, in full

$$\begin{Vmatrix} a'^{0\dot{0}} & a'^{0\dot{1}} \\ a'^{1\dot{0}} & a'^{1\dot{1}} \end{Vmatrix} = a \begin{Vmatrix} a^{0\dot{0}} & a^{0\dot{1}} \\ a^{1\dot{0}} & a^{1\dot{1}} \end{Vmatrix} a^*.$$

Comparing the formula just obtained with formula (17') we see that

$$a^{0\dot{0}} = x_0 - x_3, \qquad a^{0\dot{1}} = x_2 - ix_1,$$
$$a^{1\dot{1}} = x_0 + x_3, \qquad a^{1\dot{0}} = x_2 + ix_1.$$

Hence

$$x_0 = \frac{a^{0\dot{0}} + a^{1\dot{1}}}{2}, \qquad x_1 = \frac{a^{1\dot{0}} - a^{0\dot{1}}}{2i},$$

$$x_2 = \frac{a^{1\dot{0}} + a^{0\dot{1}}}{2}, \qquad x_3 = \frac{a^{1\dot{1}} - a^{0\dot{0}}}{2}.$$

These formulae may be written more concisely thus:

$$x_k = \frac{1}{2} \sum_{\alpha\dot{\alpha}} \sigma^{(k)}_{\alpha\dot{\alpha}} a^{\alpha\dot{\alpha}}, \tag{17"}$$

where the $\|\sigma^{(k)}_{\alpha\dot{\alpha}}\| = \sigma^{(k)}$ are the so-called Pauli matrices

$$\sigma^0 = \begin{Vmatrix} 1 & 0 \\ 0 & 1 \end{Vmatrix}, \quad \sigma^{(1)} = \begin{Vmatrix} 0 & -i \\ i & 0 \end{Vmatrix}, \quad \sigma^{(2)} = \begin{Vmatrix} 0 & 1 \\ 1 & 0 \end{Vmatrix}, \quad \sigma^{(3)} = \begin{Vmatrix} -1 & 0 \\ 0 & 1 \end{Vmatrix}.$$

We now make a further observation. The representation $T_g^{(1,\,1)}$ in the space of all spinors of rank (1, 1) is the product of the representations $T_g^{(1,\,0)}$ and $T_g^{(0,\,1)}$ acting in the spaces of all dotted and undotted spinors respectively.

In connexion with this the coordinates of the vector x_0, x_1, x_2, x_3 may be expressed in terms of the components of the undotted spinor (a^0, a^1) and the dotted spinor $(a^{\dot{0}}, a^{\dot{1}})$.

In fact, if we put

$$a^0 a^{\dot{0}} = a^{0\dot{0}}, \qquad a^0 a^{\dot{1}} = a^{0\dot{1}}, \qquad a^1 a^{\dot{0}} = a^{1\dot{0}}, \qquad a^1 a^{\dot{1}} = a^{1\dot{1}},$$

then the numbers $(a^{0\dot{0}}, a^{0\dot{1}}, a^{1\dot{0}}, a^{1\dot{1}})$ form a spinor of rank (1, 1). Hence we obtain

$$x_k = \frac{1}{2} \sum \sigma^{(k)}_{\alpha\dot{\alpha}} a^{\alpha} a^{\dot{\alpha}}.$$

This is also the expression for the vector in terms of dotted and undotted spinors.

§ 5. Lowering the Index of Spinors of Higher Rank

We recall that from the spinors (a^0, a^1) and $(a^{\dot{0}}, a^{\dot{1}})$ of the first rank we obtained, by means of the matrix $\|\tau_{\beta\alpha}\| = \tau$ spinors with lower indices

$$a_\beta = \sum \tau_{\beta\alpha} a^\alpha \text{ and } a_{\dot{\beta}} = \sum \tau_{\dot{\beta}\dot{\alpha}} a^{\dot{\alpha}}. \tag{18}$$

Transformation (18) was called lowering the index.

The spinors (a_0, a_1) and $(a_{\dot{0}}, a_{\dot{1}})$ with a lower index are transformed under a proper Lorentz transformation g_a by means of the matrices $(a^\tau)^{-1}$ and $(a^*)^{-1}$ respectively.

We may lower the indices for spinors of any rank in a similar manner. Let $a^{\alpha_1 \cdots \alpha_k \dot{\alpha}_1 \cdots \dot{\alpha}_n}$ be a spinor of rank (k, n). The quantity

$$a^{\alpha_{i_1} \cdots \alpha_{i_s} \dot{\alpha}_{j_1} \cdots \dot{\alpha}_{j_p}}_{\beta_{k_1} \cdots \beta_{k_l} \dot{\beta}_{r_1} \cdots \dot{\beta}_{r_m}} = \sum \tau_{\beta_{k_1} \alpha_{k_1}} \cdots \tau_{\dot{\beta}_{r_1} \dot{\alpha}_{r_1}} \cdots a^{\alpha_1 \cdots \alpha_k \dot{\alpha}_1 \cdots \dot{\alpha}_n} \quad (19)$$

is called a spinor of rank (k, n) with (s, p) upper indices and (l, m) lower indices $(l + s = k;\ p + m = n)$ while operation (19) is described as lowering some of the indices. We now examine how the spinor $a^{\alpha_{i_1} \cdots \alpha_{i_s} \dot{\alpha}_{j_1} \cdots \dot{\alpha}_{j_p}}_{\beta_{k_1} \cdots \beta_{k_l} \dot{\beta}_{r_1} \cdots \dot{\beta}_{r_m}}$ is transformed under proper Lorentz transformations. The spinor $a^{\alpha_1 \cdots \alpha_k \dot{\alpha}_1 \cdots \dot{\alpha}_n}$ is transformed according to formula (14), where the matrix a acts upon each undotted spinor and the matrix $\bar{a}$ upon each dotted spinor. In the example of a dotted spinor of the first rank we saw that upon lowering the undotted index the matrix a is replaced by the matrix $(a^*)^{-1}$. It is easily seen that this will also be the case for a spinor of any rank, i.e. the transformation formula of the spinor

$$a^{\alpha_{i_1} \cdots \alpha_{i_s} \dot{\alpha}_{j_1} \cdots \dot{\alpha}_{j_p}}_{\beta_{k_1} \cdots \beta_{k_l} \dot{\beta}_{r_1} \cdots \dot{\beta}_{r_m}}$$

has the form

$$a'^{\alpha'_{i_1} \cdots \alpha'_{i_s} \dot{\alpha}'_{j_1} \cdots \dot{\alpha}'_{j_p}}_{\beta'_{k_1} \cdots \beta'_{k_l} \dot{\beta}'_{r_1} \cdots \dot{\beta}'_{r_m}} = \sum a_{\alpha'_{i_1} \alpha_{i_1}} \cdots \bar{a}_{\dot{\alpha}'_{j_1} \dot{\alpha}_{j_1}} \cdots \bar{a}^{\beta'_{k_1} \beta_{k_1}} \cdots a^{\dot{\beta}'_{r_1} \dot{\beta}_{r_m}} \cdots$$
$$\cdots a^{\alpha_{i_1} \cdots \alpha_{i_s} \dot{\alpha}_{j_1} \cdots \dot{\alpha}_{j_p}}_{\beta_{k_1} \cdots \beta_{k_l} \dot{\beta}_{r_1} \cdots \dot{\beta}_{r_m}}, \quad (20)$$

where the upper undotted indices are acted upon by the matrix a and the lower dotted indices by the matrix $(a^*)^{-1} = \left\| a^{\beta'_i \beta_i} \right\|$; similarly for the other indices.

A representation of the proper Lorentz group specified by formula (20) in the space of spinors $a^{\alpha_{i_1} \cdots \alpha_{i_s} \dot{\alpha}_{j_1} \cdots \dot{\alpha}_{j_p}}_{\beta_{k_1} \cdots \beta_{k_l} \dot{\beta}_{r_1} \cdots \dot{\beta}_{r_m}}$, is clearly equivalent to representation (14) in the space of spinors of rank (k, n) $a^{\alpha_1 \cdots \alpha_k \dot{\alpha}_1 \cdots \dot{\alpha}_n}$ with upper indices.

We consider the special case of spinors $a^{\alpha_1 \cdots \alpha_k}_{\dot{\beta}_1 \cdots \dot{\beta}_n}$ with k upper undotted indices and n lower dotted indices.

The representation of the proper group in the space of such spinors acts according to the formula

$$a^{\alpha_1 \ldots \alpha_k}_{\dot\beta_1 \ldots \dot\beta_n} = \sum a_{\alpha_1 \alpha_1'} \ldots a_{\alpha_k \alpha_k'} a^{\dot\beta_1 \dot\beta_1'} \ldots a^{\dot\beta_n \dot\beta_n'} a^{\alpha_1' \ldots \alpha_k'}_{\dot\beta_1' \ldots \dot\beta_n'}. \tag{21}$$

It is clear that the symmtric spinors $a^{\alpha_1 \ldots \alpha_k}_{\dot\beta_1 \ldots \dot\beta_n}$ form a subspace R_n^k which is invariant and irreducible under representation (21). In this case, the representation (21) specified in R_n^k, is equivalent to an irreducible spinor representation $T_g^{(k,\,n)}$ of rank $(k,\,n)$, i.e. is defined by the pair

$$\left(l_0 = \frac{k-n}{2};\; l_1 = \frac{k+n}{2} + 1\right).$$

We note that the representation conjugate to $T_g^{(k,\,n)}$, which is defined by the pair $\left(\frac{n-k}{2},\; \frac{n+k}{2} + 1\right)$, may be realized in the space R_k^n of spinors with n upper undotted indices and k lower dotted indices according to the formula

$$b^{\alpha_1' \ldots \alpha_n'}_{\dot\beta_1' \ldots \dot\beta_k'} = \sum a_{\alpha_1' \alpha_1} \ldots a_{\alpha_n' \alpha_n} a^{\dot\beta_1' \dot\beta_1} \ldots a^{\dot\beta_k' \dot\beta_k} b^{\alpha_1 \ldots \alpha_n}_{\dot\beta_1 \ldots \dot\beta_k}. \tag{22}$$

There is a natural correspondence between the spinors $a^{\alpha_1 \ldots \alpha_k}_{\dot\beta_1 \ldots \dot\beta_n}$ from R_n^k and $b^{\alpha_1 \ldots \alpha_n}_{\dot\beta_1 \ldots \dot\beta_k}$ from R_k^n: each spinor $a^{\alpha_1 \ldots \alpha_k}_{\dot\beta_1 \ldots \dot\beta_n}$ from R_n^k yields on interchanging the upper indices with the lower, the corresponding spinor $b^{\alpha_1' \ldots \alpha_n'}_{\dot\beta_1' \ldots \dot\beta_k'}$ from R_k^n; and vice-versa

$$a^{\alpha_1 \ldots \alpha_k}_{\dot\beta_1 \ldots \dot\beta_n} \longleftrightarrow b^{\alpha_1' \ldots \alpha_n'}_{\dot\beta_1' \ldots \dot\beta_k'}, \quad \alpha_i' = \dot\beta_i, \quad \dot\beta_i' = \alpha_i. \tag{23}$$

Correspondence (23) has the following obvious property. Let $e_1 \ldots e_N$ and $f_1 \ldots f_N$ be the bases of R_n^k and R_k^n respectively whose elements correspond according to (23)

$$f_i \longleftrightarrow e_i.$$

Let A_g be the matrix of representation (21) in terms of the basis $e_1 \ldots e_N$, and $\tilde{A}_g$ the matrix of representation (22) in terms of the basis $f_1 \ldots f_N$.

This being the case, we have

$$\tilde{A}_{g_a} = A_{(g_a)^{*-1}} \quad \text{or} \quad \tilde{A}_{g_a} = A_{g_{(a^*)^{-1}}}. \tag{24}$$

In fact the interchanging of the upper and lower indices clearly involves changing the matrix a in formulae (21) and (22) into the matrix $(a^*)^{-1'}$. This is precisely the meaning of equation (24).

Correspondence (23) between the spinors $a^{\alpha_1 \ldots \alpha_k}_{\dot\beta_1 \ldots \dot\beta_n}$ and $b^{\dot\alpha_1 \ldots \dot\alpha_n}_{\beta_1 \ldots \beta_k}$ will be used in the next section for the construction of the finite irreducible representations of the complete Lorentz group.

§ 6. Another Description of Spinor Representation

We consider a set of homogeneous polynomials $p(z_0, z_1, \bar z_0, \bar z_1)$ of degree k in z_0 and z_1 and degree n in $\bar z_0$ and $\bar z_1$, i.e. polynomials of the form

$$p(z_0, z_1, \bar z_0, \bar z_1) = \sum_{\substack{p+r=k \\ s+t=n}} b_{p,r,s,t} z_0^p z_1^r \bar z_0^s \bar z_1^t. \tag{25}$$

We denote the space of all such polynomials by $\tilde R_{(k,n)}$. Each polynomial from $\tilde R_{(k,n)}$ may be written in the following form:

$$p(z_0, z_1, \bar z_0, \bar z_1) = \sum_{\substack{(\alpha_1 \ldots \alpha_k) \\ (\dot\alpha_1 \ldots \dot\alpha_n)}} \frac{1}{k!\,n!} a^{\alpha_1 \ldots \alpha_k \dot\alpha_1 \ldots \dot\alpha_n} z_{\alpha_1} \ldots z_{\alpha_k} \bar z_{\dot\alpha_1} \ldots \bar z_{\dot\alpha_n} \tag{26}$$

$$(\alpha_i = 0, 1; \ \dot\alpha_i = 0, 1),$$

where the numbers $a^{\alpha_1 \ldots \alpha_k \dot\alpha_1 \ldots \dot\alpha_n}$ are unaltered by any permutation of the dotted indices amongst themselves, and also of the undotted indices amongst themselves. Thus, to each polynomial of the form (25) there corresponds a set of numbers $a^{\alpha_1 \ldots \alpha_k \dot\alpha_1 \ldots \dot\alpha_n}$, which are unaltered by permutations of dotted and undotted indices separately.

We specify in the space $\tilde R_{(k,n)}$ a representation of the group $\mathfrak{A}$ in the following manner. Let the variables z_0, z_1 be transformed according to

$$\begin{aligned} z_0 &= a_{00} z_0' + a_{01} z_1', \\ z_1 &= a_{10} z_0' + a_{10} z_1', \end{aligned} \tag{27}$$

and $\bar z_0$, $\bar z_1$ correspondingly

$$\begin{aligned} \bar z_0 &= \bar a_{00} \bar z_0' + \bar a_{01} \bar z_1', \\ \bar z_1 &= \bar a_{10} \bar z_0' + \bar a_{10} \bar z_1'. \end{aligned} \tag{27'}$$

After such a change of variables the polynomial $p(z_0, z_1, \bar z_0, \bar z_1)$ becomes the polynomial $\tilde p(z_0', z_1', \bar z_0', \bar z_1')$ of the variables $z_0', z_1', \bar z_0', \bar z_1'$, which again belongs to $\tilde R_{(k,n)}$. The transformation $p \to \tilde p$ in $\tilde R_{(k,n)}$ is linear and specifies a representation of the group $\mathfrak{A}$:

$$T_a p = \tilde p.$$

We write down the manner in which the coefficients $a^{\alpha_1 \dots \alpha_k \dot{\alpha}_1 \dots \dot{\alpha}_n}$ are transformed under this transformation T_a:

$$T_a p = T_a \left(\frac{1}{k!\, n!} \sum a^{\alpha_1 \dots \alpha_k \dot{\alpha}_1 \dots \dot{\alpha}_n} z_{\alpha_1} \dots z_{\alpha_k} \bar{z}_{\dot{\alpha}_1} \dots \bar{z}_{\dot{\alpha}_n} \right) =$$

$$= \frac{1}{k!\, n!} \sum a^{\alpha_1 \dots \alpha_k \dot{\alpha}_1 \dots \dot{\alpha}_n} \left(\sum a_{\alpha_1' \alpha_1} z'_{\alpha_1} \right) \times \left(\sum a_{\alpha_2' \alpha_2} z'_{\alpha_2} \right) \times \dots$$

$$\dots \times \left(\sum a_{\alpha_k' \alpha_k} z'_{\alpha_k} \right) \times \left(\sum \bar{a}_{\dot{\alpha}_1' \dot{\alpha}_1} \bar{z}'_{\dot{\alpha}_1} \right) \times \dots \times \left(\sum \bar{a}_{\dot{\alpha}_n' \dot{\alpha}_n} \bar{z}'_{\dot{\alpha}_n} \right) =$$

$$= \frac{1}{k!\, n!} \sum a^{\alpha_1' \dots \alpha_k' \dot{\alpha}_1' \dots \dot{\alpha}_n'} z'_{\alpha_1} \dots z'_{\alpha_k} \bar{z}'_{\dot{\alpha}_1} \dots \bar{z}'_{\dot{\alpha}_n}, \tag{28}$$

where

$$a^{\alpha_1' \dots \alpha_k' \dot{\alpha}_1' \dots \dot{\alpha}_n'} = \sum a_{\alpha_1' \alpha_1} \dots a_{\alpha_k' \alpha_k} \bar{a}_{\dot{\alpha}_1' \dot{\alpha}_1} \dots \bar{a}_{\dot{\alpha}_n' \dot{\alpha}_n} a^{\alpha_1 \dots \alpha_k \dot{\alpha}_1 \dots \dot{\alpha}_n}. \tag{29}$$

The formula just obtained coincides with formula (14). This means that the coefficients of the polynomial $p(z_0, z_1, \bar{z}_0, \bar{z}_1)$ constitute a symmetric spinor of rank (k, n). Consequently, representation (28) in the space $\tilde{R}_{(k, n)}$ of all the polynomials $p(z_0, z_1, \bar{z}_0, \bar{z}_1)$ is equivalent to representation (14) in the space $R_{(k, n)}$ of symmetric spinors of rank (k, n). The latter is irreducible and, as we have seen, is defined by the pair:

$$\left(l_0 = \frac{k-n}{2},\ l_1 = \frac{k+n}{2} + 1 \right).$$

We now transform the above formulae somewhat. Each homogeneous polynomial $p(z_0, z_1, \bar{z}_0, \bar{z}_1)$ of degree k in z_0 and z_1 and degree n in $\bar{z}_0$ and $\bar{z}_1$ may be written in the form

$$p(z_0, z_1, \bar{z}_0, \bar{z}_1) = z_1^k \bar{z}_1^n p\left(\frac{z_0}{z_1},\ 1,\ \frac{\bar{z}_0}{\bar{z}_1},\ 1 \right). \tag{30}$$

If we put

$$\frac{z_0}{z_1} = \xi, \quad \frac{\bar{z}_0}{\bar{z}_1} = \bar{\xi}.$$

Then we may write:

$$p(z_0, z_1, \bar{z}_0, \bar{z}_1) = z_1^k \bar{z}_1^n q(\xi, \bar{\xi}),$$

where $q(\xi, \bar{\xi}) = p(\xi, 1, \bar{\xi}, 1)$ is a polynomial of degree k in ξ and degree n in $\bar{\xi}$. We shall denote the space of all polynomials $q(\xi, \bar{\xi})$ by $\tilde{\tilde{R}}_{(k, n)}$.

It is clear that representation (28) acting in the space $\tilde{R}_{(k, n)}$ of homogeneous polynomials may be considered as acting in the space $\tilde{\tilde{R}}_{(k,n)}$ of polynomials $q(\xi, \bar{\xi})$.

We now find the formulae specifying this representation. We have

$$q(\xi, \bar{\xi}) = \frac{1}{z_1^k \bar{z}_1^n} p(z_0, z_1, \bar{z}_0, \bar{z}_1),$$

$$T_g q = \frac{1}{z_1^k \bar{z}_1^n} T_g p.$$

Using this, we find

$$T_g q(\xi, \bar{\xi}) = \frac{1}{z_1^k \bar{z}_1^n} T_g \left[z_1^k \bar{z}_1^n q\left(\frac{z_0}{z_1}, \frac{\bar{z}_0}{\bar{z}_1}\right)\right] =$$

$$= \frac{1}{z_1^k \bar{z}_1^n} (a_{10} z_0 + a_{11} z_1)^k (\bar{a}_{10} \bar{z}_0 + \bar{a}_{11} \bar{z}_1)^n q\left(\frac{a_{00} z_0 + a_{01} z_1}{a_{10} z_0 + a_{11} z_1}, \frac{\bar{a}_{00} \bar{z}_0 + \bar{a}_{01} \bar{z}_1}{\bar{a}_{10} \bar{z}_0 + \bar{a}_{11} \bar{z}_1}\right) =$$

$$= (a_{10}\xi + a_{11})^k (\bar{a}_{10}\bar{\xi} + \bar{a}_{11})^n q\left(\frac{a_{00}\xi + a_{01}}{a_{10}\xi + a_{11}}, \frac{\bar{a}_{00}\bar{\xi} + \bar{a}_{01}}{\bar{a}_{10}\bar{\xi} + \bar{a}_{11}}\right).$$

Thus, we see finally that an irreducible representation of the proper group acting in the space of polynomials $q(\xi, \bar{\xi})$ of degree k in ξ and degree n in $\bar{\xi}$ is specified by the formula

$$T_g q(\xi, \bar{\xi}) = (a_{10}\xi + a_{11})^k (\bar{a}_{10}\bar{\xi} + \bar{a}_{11})^n q\left(\frac{a_{00}\xi + a_{01}}{a_{10}\xi + a_{11}}; \frac{\bar{a}_{00}\bar{\xi} + \bar{a}_{01}}{\bar{a}_{10}\bar{\xi} + \bar{a}_{11}}\right). \quad (31)$$

From this formula it is clear that the effect of the operators of the representation T_g amounts to a change of variables in the polynomial $q(\xi, \bar{\xi})$ by means of a bilinear transformation, and a multiplication of $q(\xi, \bar{\xi})$ by some expressions which depend upon the coefficients of this transformation.

§ 7. Unitary Representations of the Proper Lorentz Group

In the previous section the representation $g_a \to T_{g_a}$:

$$T_{g_a} q(\xi, \bar{\xi}) = (a_{10}\xi + a_{11})^k (\bar{a}_{10}\bar{\xi} + \bar{a}_{11})^n q\left(\frac{a_{00}\xi + a_{01}}{a_{10}\xi + a_{11}}, \frac{\bar{a}_{00}\bar{\xi} + \bar{a}_{01}}{\bar{a}_{10}\bar{\xi} + \bar{a}_{11}}\right), \quad (31')$$

acting in the space $\tilde{\tilde{R}}_{kn}$ of polynomials $q(\xi, \bar{\xi})$ was examined. We recall that this representation is defined by the pair

$$(l_0, l_1) = \left(\frac{k-n}{2}, \frac{k+n}{2} + 1\right).$$

Replacing k and n by their expressions in terms of l_0 and l_1: $k = l_0 + l - 1$, $n = l_1 - l_0 - 1$, we may rewrite formula (31) in the form

$$T_{g_a} q(\xi, \bar{\xi}) =$$

$$= (a_{10}\xi + a_{11})^{l_0 + l_1 - 1} (\bar{a}_{10}\bar{\xi} + \bar{a}_{11})^{l_1 - l_0 - 1} q\left(\frac{a_{00}\xi + a_{01}}{a_{10}\xi + a_{11}}; \frac{\bar{a}_{00}\bar{\xi} + \bar{a}_{01}}{\bar{a}_{10}\bar{\xi} + \bar{a}_{11}}\right). \quad (32)$$

It is obvious that this formula remains meaningful for arbitrary complex l_1 and half integral l_0 if only we replace the polynomials $q(\xi, \bar{\xi})$ by a suitable family of functions. Whereupon it appears that the following singular theorem holds:

Any irreducible representation of the proper Lorentz group is equivalent to a representation specified by formula (32) in a suitably selected function space.

The proof of this theorem proceeds according to the following scheme: in Sect.2 we found all possible irreducible sextets of operators $H_+, H_-, H_3, F_+, F_-, F_3$, satisfying the commutation relations (I'-XV'). It so happens that if we evaluate the infinitesimal operators of the representation specified by formula (32) for various values of (l_0, l_1) then all the sextets of the irreducible operators $H_+, H_-, H_3, F_+, F_-, F_3$ described in Sect.2 are obtained. Firstly we prove the theorem formulated above and then we deduce that each sextet of operators $\{H_{+,-,3}, F_{+,-,3}\}$ constructed in Sect.2 in fact serves as infinitesimal operators of some irreducible representation of the Lorentz group. In other words, to every admissible pair (l_0, l_1) there corresponds an irreducible representation of the proper Lorentz group.*

We construct the unitary representations of the proper Lorentz groups by means of formula (32).

A. The fundamental series of unitary representations. The numbers (l_0, l_1) defining a unitary representation belonging to the fundamental series have the form (see Sect.2) $l_0 = \frac{m}{2}$, $l_1 = i\rho$, where m is an arbitrary integer and ρ is an arbitrary real number.

As carrier space for the representation we consider the Hilbert space of functions $f(\xi)$ with the scalar product

$$(f_1, f_2) = \int f_1(\xi)\bar{f}_2(\xi)\,dx\,dy \qquad (\xi = x + iy).$$

By a simple calculation, which we shall omit here, it is easy to convince ourselves of the fact that the operators

$$T_{g_a} f(\xi) = (a_{10}\xi + a_{11})^{\frac{m}{2}-1+i\rho}\,\overline{(a_{10}\xi + a_{11})}^{-\frac{m}{2}-1+i\rho}\, f\left(\frac{a_{00}\xi + a_{01}}{a_{10}\xi + a_{11}}\right)$$

are unitary in this Hilbert space.

It is easily proved that this representation is in fact irreducible and belongs to the fundamental series, the numbers defining it being $l_0 = \frac{m}{2}$, $l_1 = i\rho$**

B. The supplementary series of unitary representations: $l_0 = 0$ $|l_1| = \rho < 1$. The numbers defining a representation of this series are (see Sect.2) $l_0 = 0$, $l_1 = \rho$,

---

* For a detailed proof according to this scheme see the paper of M.A.Naimark, Uspekhi Matem. Nauk. No. 4 (1955), 89-90.

** See the paper cited above by M.A.Naimark, Uspekhi Matem.Nauk.IX, No.4 (1955), 68-78.

where ρ is real, $0 < \rho < 1$. In order to realize representations of the supplementary series with the aid of formula (32) we consider the Hilbert space of functions $f(\xi)$ with the scalar product

$$(f_1 f_2) = \int |\xi_1 - \xi_2|^{-2-2\rho} f_1(\xi_1) \overline{f_2(\xi_2)}\, dx_1\, dy_1\, dx_2\, dy_2$$

$$(\xi_1 = x_1 + iy_1, \quad \xi_2 = x_2 + iy_2).$$

In accordance with formula (32) the operations of the representation are specified by:

$$T_{g_a} f = |a_{10}\xi + a_{11}|^{2\rho - 2} f\left(\frac{a_{00}\xi + a_{01}}{a_{10}\xi + a_{11}}\right).$$

Just as in the previous case, it is easily proved that this representation is unitary and irreducible and that the number pair defining it is $(l_0, l_1) = (0, \rho)$.

§ 8. An Observation on Tensors

Together with the realization of the finite representations of the proper Lorentz group by means of spinors, another - a tensor realization of these representations is often used.

Definition of a tensor. Let there be specified in each orthogonal coordinate system a set of 4^n numbers $t_{k_1 k_2 \ldots k_n}$ $(k_i = 0, 1, 2, 3)$, which, under a proper Lorentz transformation $g = \| g_{k'k} \|$ from one coordinate system to another transform according to the formula

$$t'_{k'_1 k'_2 \ldots k'_n} = \sum g_{k'_1 k_1} g_{k'_2 k_2} \cdots g_{k'_n k_n} t_{k_1 k_2 \ldots k_n} \quad (k'_i, k_i = 0, 1, 2, 3). \tag{33}$$

Then the set of numbers $t_{k_1 \ldots k_n}$ is called a tensor of the n-th rank with respect to the proper Lorentz group *.

A representation acting in the 4^n-space of tensors will be called the tensor representation of the proper Lorentz group of rank n.

For example the quantities

$$t_{k_1 k_2 \ldots k_n} = x_{k_1} x_{k_2} x_{k_3} \ldots x_{k_n}$$

---

* Together with tensors $t_{k_1 \ldots k_n}$ with lower indices, tensors of the form $t^{j_1 \ldots j_s}_{k_1 \ldots k_n}$ are also considered; under a proper Lorentz transformation, the matrix $\| g^{j'j} \| = (\| g_{k'k} \|^{T})^{-1}$ acts upon the upper indices:

$$t'^{j'_1 \ldots j'_s}_{k'_1 \ldots k'_n} = \sum g^{j'_1 j_1} \cdots g^{j'_s j_s} g_{k'_1 k_1} \cdots g_{k'_n k_n} t^{j_1 \ldots j_n}_{k_1 \ldots k_n}. \tag{34}$$

However, representation (34) is equivalent to representation (33) (see Sect. 1, paragraph 1) and so we shall not consider it specially.

(x_0, x_1, x_2. x_3 are the coordinates of a point in a four-dimensional space) constitute a tensor of the rank n; the quantities x_k ($k = 0, 1, 2, 3$) thus constitute a tensor of the first rank, otherwise known as a vector. We note that the tensor representation (33) of rank n is equivalent to the n-fold product of the identity representation of the Lorentz group with itself, the latter representation acting in the space $R^{(4)}$ $(x_0x_1x_2x_3)$ according to the formula

$$x'_{k'} = \sum g_{k'k} x_k.$$

Representation (33) in tensor space is, generally speaking, reducible. We now consider its possible irreducible components.

Since the representation $g \to g$ of the proper group acting in four-dimensional space, contains only integral weights l ($l = 0, 1$), the tensor representation (33), equivalent to the n-fold product of the representation $g \to g$, also contains only integral weights. This means that the numbers l_0, l_1, defining irreducible components of the tensor representation are integral. In Sect.6 the converse proposition will be proved, namely: any finite irreducible representation of the proper group whose defining numbers l_0, l_1 are integral, is equivalent to an irreducible component of some tensor representation. We note also that using formula (33) it is possible to specify a representation of the complete (and also the general) Lorentz group in tensor space, if the matrix s, corresponding to spatial reflection, is taken as a matrix of type $\|g_{k'k}\|$. Thus it is possible to extend a representation of the proper Lorentz group to a representation of the complete (and the general) Lorentz group.

According to the results of Sect.3 it follows from this, in particular, that the tensor representation contains, together with each irreducible component of the representation (33) defined by the pair (l_0, l_1), its conjugate component, defined by the pair $(-l_0, l_1)$.

In Sect.5 we shall revert once more to the representation of the complete group in tensor space.

In conclusion we consider in greater detail the tensor representation of the second rank

$$t'_{k'_1k'_2} = \sum g_{k'_1k_1} g_{k'_2k_2} t_{k_1k_2}, \tag{35}$$

acting in the 16-dimensional space of $t_{k_1k_2}$. This representation is reducible. We shall find its irreducible subspaces.

We note that the symmetric tensors $(t_{k_1k_2} = t_{k_2k_1})$ form a subspace $R^{(10)}$ of dimensionality (10) which is invariant under representation (35). Similarly, the skew symmetric tensors $(t_{k_1k_2} = -t_{k_2k_1})$ form a subspace $R^{(6)}$ (of dimensionality 6) which is invariant under the tensor representation (35). The subspaces $R^{(6)}$ and $R^{(10)}$ have no common elements different from zero, and their sum is the entire space $R^{(16)}$ of tensors of rank 2.

$$R^{(16)} = R^{(10)} \dot{+} R^{(6)}.$$

However the representations in the invariant subspaces $R^{(6)}$ and $R^{(10)}$ are still reducible, and we must further decompose these subspaces.

Symmetric tensors. We note that the symmetric tensor $\delta_{k_1k_2}$ with components $-\delta_{00} = \delta_{11} = \delta_{22} = \delta_{33} = 1$ and $\delta_{k_1k_2} = 0$ for $k_1 \neq k_2$ is unaltered by transformation (35). Thus tensors of the form $\alpha\delta_{k_1k_2}$ form a one-dimensional subspace $R^{(1)}$ (the representation in $R^{(1)}$ is defined by the pair 0, 1). Before taking the decomposition of the space $R^{(10)}$ any further we note that the expression $t_{11} + t_{22} + t_{33} - t_{00}$ (which will be called the trace of the tensor $t_{k_1k_2}$) is not altered by transformation of the form (35). Consequently, the subspace $R^{(9)}$ of symmetric tensors with trace equal to zero is invariant under our representation. It happens, moreover, than R^{9} is irreducible. (We shall convince ourselves of this later; see Sect. 6).

We now find the pair (l_0, l_1) defining the irreducible representation in $R^{(9)}$. To this end we note that rotations $\|g_{ik}\|$ in no way affect the indices $k_i = 0$ in formula (35); i.e. the following groups of components of the tensor $t_{k_1k_2}$ are transformed independently under rotations: $\{t_{00}\}$ $\{t_{01} = t_{10};\ t_{02} = t_{20};\ t_{03} = t_{30}\}$; $\{t_{11},\ t_{22},\ t_{33},\ t_{12} = t_{21},\ t_{31} = t_{13},\ t_{23} = t_{32}\}$. Hence the subspace $R^{(5)}$ of symmetric tensors $t_{k_1k_2}$, for which $t_{00} = t_{01} = t_{02} = t_{03} = 0$ is invariant under the group of rotations. The non-zero components of such a tensor $t_{11}, t_{22}, t_{33}, t_{12}, t_{13}, t_{23}$ form a symmetric tensor of the second rank with respect to the group of rotations (with trace equal to zero). Thus, the representation of the group of rotations generated in $R^{(5)}$ by formula (35) coincides with the representation of this group acting in the space of symmetric tensors of the second rank with trace equal to zero. The latter as we know from the first part of the book (see Sect.5, p.64) is irreducible and has weight $l = 2$. In similar fashion we may convince ourselves that the space $R^{(3)}$ of tensors for which only the components $t_{01} = t_{10},\ t_{02} = t_{20},\ t_{03} = t_{30}$ differ from zero, is irreducible under the group of rotations with weight $l = 1$.

Finally, a tensor from $R^{(9)}$ with components $t_{00} = 1,\ t_{11} = t_{22} = t_{33} = \frac{1}{3}$, $t_{k_1k_2} = 0$ for $k_1 \neq k_2$ forms a scalar with respect to the group of rotations (a representation with weight $l = 0$).

Thus, the irreducible representation of the proper Lorentz group acting in the space $R^{(9)}$ contains the weights $l = 2,\ 1,\ 0$ and is consequently defined by the pair $(l_0, l_1) = (0, 3)$

We have analysed the entire subspace $R^{(10)}$ of symmetric tensors of rank 2 into two irreducible subspaces: the scalar $\delta_{k_1k_2}$ $(R^{(1)})$ (a representation with the pair (0, 1)); and $R^{(9)}$ consisting of all symmetric tensors with zero trace (a representation with the pair (0, 3)).

We note that both these subspaces are invariant under the representation of the complete Lorentz group, specified by formula (35).

Antisymmetric tensors: $t_{ik} = -t_{ki}$. We again note that the following two groups of non-zero components of such a tensor $t_{01} = -t_{10}$, $t_{02} = -t_{20}$, $t_{03} = -t_{30}$ and $t_{12} = -t_{21}$, $t_{13} = -t_{31}$, $t_{23} = -t_{32}$ are transformed independently under rotations. Consequently the three-dimensional subspaces of tensors, for which either the first or the second group of components vanish, are invariant and irreducible under the group of rotations (and have the weight $l = 1$). It follows immediately from this that a representation of the proper Lorentz group in $R^{(6)}$ is reducible (an irreducible representation of the proper group does not contain two equal weights) and consists of two irreducible representations acting in the spaces $R^{(3)}$ and $\dot{R}^{(3)}$ and containing just one weight $l = 1$. The pairs defining such representations have the form $(\pm 1, 2)$. As was noted on p. 249 a tensor representation contains the conjugate of each of its irreducible components. Therefore the representations specified in $R^{(3)}$ and $\dot{R}^{(3)}$ are not equivalent and one of them is specified by the pair (1, 2) and the other by the pair (-1, 2).

The space $R^{(6)}$ of antisymmetric tensors $t_{k_1k_2}$ is, according to formula (35), invariant under spatial reflection. Since $R^{(6)}$ consists of two conjugate components, the results of Sect. 3 then imply that the representation of the complete group specified in $R^{(6)}$ by formula (35) is irreducible.

Thus the space $R^{(6)}$ of antisymmetric tensors of rank 2 decomposes into two subspaces $R^{(3)}$ and $\dot{R}^{(3)}$ irreducible under the representation of the proper Lorentz group. The pairs defining these representations are (-1, 2) and (1, 2). The representation of the complete Lorentz group in the space $R^{(6)}$ is irreducible.

We now sum up:

The space of tensors $t_{k_1k_2}$ of rank 2 decomposes into the sum of four subspaces which are irreducible under representation (35) of the proper Lorentz group:

1. $R^{(1)}\delta_{k_1k_2}$ - a scalar (pair (0, 1));
2. $R^{(9)}$ - the space of symmetric tensors with trace $t_{11} + t_{22} + t_{33} - t_{00}$, equal to zero (pair (0, 3));
3. and 3'. the two spaces $R^{(3)}$ and $\dot{R}^{(3)}$ of antisymmetric tensors (pairs (-1, 2) and (1, 2)). The sum of the spaces $R^{(3)} + \dot{R}^{(3)}$ is the entire space $R^{(6)}$ of antisymmetric tensors of rank 2.

The spaces $R^{(1)}, R^{(9)}$ and $R^{(6)}$ are invariant and irreducible under the representation of the complete Lorentz group specified by formula (35). In concluding this paragraph we make the following observation. As we know, any finite representation, including a tensor representation, is equivalent to some spinor representation. Here we shall express the components of the tensor $t_{k_1 \dots k_n}$ of rank n explicitly in terms of the components of the spinor $a^{\alpha_1 \dots \alpha_n \dot{\alpha}_1 \dots \dot{\alpha}_n}$ with n dotted and n undotted indices. To this end we consider a tensor of the form

$$t_{k_1 \dots k_n} = x_{k_1} x_{k_2} \dots x_{k_n} \qquad (k_i = 0, 1, 2, 3), \tag{36}$$

where x_{k_i} are the coordinates of a vector. As we have already seen in (17"), paragraph 4, Sect.4).

$$x_{k_i} = \sum_{\alpha_i \dot{\alpha}_i} \sigma^{(k_i)} a^{\alpha_i \dot{\alpha}_i},$$

where $a^{\alpha_1 \dot{\alpha}_1}$ are the components of a spinor of rank (1, 1). Hence

$$x_{k_1} x_{k_2} \ldots x_{k_n} = \sum \sigma^{(k_1)}_{\alpha_1 \dot{\alpha}_1} \sigma^{(k_2)}_{\alpha_2 \dot{\alpha}_2} \ldots \sigma^{(k_n)}_{\alpha_n \dot{\alpha}_n} a^{\alpha_1 \dot{\alpha}_1} a^{\alpha_2 \dot{\alpha}_2} \ldots a^{\alpha_n \dot{\alpha}_n}.$$

It is clear that the product

$$a^{\alpha_1 \dot{\alpha}_1} a^{\alpha_2 \dot{\alpha}_2} \ldots a^{\alpha_n \dot{\alpha}_n} = a^{\alpha_1 \alpha_2 \ldots \alpha_n \dot{\alpha}_1 \dot{\alpha}_2 \ldots \dot{\alpha}_n}$$

is a spinor with n dotted and n undotted indices (generally speaking, non-symmetric). Thus a tensor of the form (36), and consequently also any tensor of rank n, may be written in the form

$$t_{k_1 \ldots k_n} = \sum \sigma^{(k_1)}_{\alpha_1 \dot{\alpha}_1} \sigma^{(k_2)}_{\alpha_2 \dot{\alpha}_2} \ldots \sigma^{(k_n)}_{\alpha_n \dot{\alpha}_n} a^{\alpha_1 \ldots \alpha_n \dot{\alpha}_1 \ldots \dot{\alpha}_n}$$

(the summation is taken over all sets $(\alpha_1 \ldots \alpha_n \dot{\alpha}_1 \ldots \dot{\alpha}_n)$).

§ 9. The Difference Between Spinor and Tensor Representations of the Lorentz Group

We note that the tensor representation of the Lorentz group specified by formula (35) may be extended to a representation of the entire group of non-degenerate linear transformations a in four-dimensional space. In order to do this it suffices to substitute into this formula the matrix of the linear transformation a. Thus, the tensors $t_{k_1 k_2 \ldots k_n}$, which we considered above only in orthogonal coordinate systems (i.e. those systems which transform into one another by Lorentz transformations) may be written down in any coordinate system.

The situation is different for the spinor representations $T_g^{(k,n)}$, which are not equivalent to tensor representations (in particular, a dotted or undotted spinor representation of rank 1). These representations of the Lorentz group cannot be extended to a representation of the entire group of linear transformations of four-dimensional space. Thus spinors which we defined only in an orthogonal coordinate system cannot be defined in a natural fashion in an oblique coordinate system.

Section 5. Finite Representations of the Complete and General Lorentz Groups Bispinors

In the previous section we realized all the finite irreducible representations of the proper Lorentz group (or, more precisely, the group $\mathfrak{A}$ of complex, second-order matrices with determinant equal to one) in the spaces $R_{(k,n)}$ of symmetric spinors of rank (k, n) (k is the number of undotted, n the number of dotted indices).

In this section we shall use spinors to realize the finite irreducible representations of the complete group.

We recall first of all the fundamental results of Sect.3, where all the irreducible representations of the complete group were described. Any such representation $g \to T_g$ generates a representation $g' \to T_{g'}$ of the proper group consisting either of one component or of two non-equivalent components τ and $\dot{\tau}$.

In the first case the representation $g' \to T_{g'}$ is equivalent to its conjugate representation and is consequently defined by the pair $(0, l_1)$ or $(l_0, 0)$. Conversely, any irreducible representation of the proper group which is equivalent to its conjugate can be extended to a representation of the complete group - this can be done in two ways differing in the choice of sign for the operator S.

In the second case when two non-equivalent irreducible components τ and $\dot{\tau}$ of the representation $g' \to T_{g'}$ of the proper group participate in the representation of the complete group, these components are conjugate one to the other and are defined by the pairs $\tau \sim (l_0, l_1)$, $\dot{\tau} \sim (l_0, -l_1)$, $l_0 \neq 0$, $l_1 \neq 0$. Conversely, a representation of the proper Lorentz group which decomposes into two mutually conjugate, non-equivalent, irreducible representations τ and $\dot{\tau}$, may be extended in unique fashion (to within equivalence) to an irreducible representation of the complete Lorentz group.

We start with the simplest case.

§ 1. A Bispinor of the First Rank

We recall that in Sect.4 we constructed an irreducible representation of the proper Lorentz group in the two-dimensional complex space $R_{(1,0)}$, acting according to the formula

$$\left.\begin{aligned} a^{0'} &= a_{00}a^0 + a_{01}a^1, \\ a^{1'} &= a_{10}a^0 + a_{11}a^1, \end{aligned}\right\} \tag{1}$$

where $a = \|a_{\alpha'\alpha}\|$ is the matrix corresponding to the transformation g_a of the proper group (the correspondence $g_a \sim a$ was described at length in Sect.1, paragraph 2).

The quantity (a^0, a^1) which is transformed according to formula (1) was called an undotted spinor of rank 1 with respect to the proper Lorentz group. This representation, as we saw, is specified by the pair $\left(\frac{1}{2}, \frac{3}{2}\right)$ and its canonical basis consists of the spinors $\xi_{\frac{1}{2}} = (1, 0)$, $\xi_{-\frac{1}{2}} = (0, 1)$.

From the results of Sect.3 it follows that in the space $(R_{1,0})$ of undotted spinors of rank 1, representation (1) of the proper group cannot be extended to a representation of the complete Lorentz group. In order to do this it is necessary to specify one further irreducible representation of the proper group, the conjugate to the representation $g_a \to \pm a$, i.e. the representation defined by the pair $\left(-\frac{1}{2}, \frac{3}{2}\right)$.

The representation conjugate to the representation $g_a \to \pm a$ acts, as we already know, in the space of dotted spinors with lower indices according to the formula

$$\left.\begin{aligned} a'_{\dot{0}} &= a^{\dot{0}\dot{0}} a_{\dot{0}} + a^{\dot{0}\dot{1}} a_{\dot{1}}, \\ a'_{\dot{1}} &= a^{\dot{1}\dot{0}} a_{\dot{0}} + a^{\dot{1}\dot{1}} a_{\dot{1}}, \end{aligned}\right\} \tag{2}$$

where $\|a^{\dot{\alpha}\dot{\beta}}\| = (a^*)^{-1}$. The basis consisting of the vectors $\{e^{\dot{0}}, e^{\dot{1}}\}$, in terms of which all the matrices $(a^*)^{-1}$ are written, coincides with the canonical basis $\left\{\eta_{\frac{1}{2}}, \eta_{-\frac{1}{2}}\right\}$ of the representation $g_a \to \pm (a^*)^{-1}$.

If we now consider the representation of the proper group consisting of the components $g_a \to \pm a$ and $g_a \to \pm a^{*-1}$, then the operators of this representation are expressible by fourth-order matrices relative to the canonical basis $\left\{\xi_{\frac{1}{2}}, \xi_{-\frac{1}{2}}, \eta_{\frac{1}{2}}, \eta_{-\frac{1}{2}}\right\}$:

$$T_{g_a} = \pm \begin{Vmatrix} a & 0 \\ 0 & a^{*-1} \end{Vmatrix}. \tag{3}$$

According to the results of Sect.3, the operator S has relative to this basis, the matrix

$$S = \begin{Vmatrix} 0 & E \\ E & 0 \end{Vmatrix}, \quad E = \begin{Vmatrix} 1 & 0 \\ 0 & 1 \end{Vmatrix}. \tag{4}$$

We leave it to the reader to verify directly, without using the results of Sect.3, that the operators T_g and S satisfy the necessary relations

$$ST_gS^{-1} = T_{g^{*-1}}, \quad S^2 = E.$$

We denote the space in which the above representation of the complete group acts by R_1^1; the coordinates of a vector of R_1^1 in the canonical basis are $a^0, a^1, a_{\dot{0}}, a_{\dot{1}}$. Formulae (3) and (4) clearly mean that, upon transforming the orthogonal coordinate system by a transformation from the proper Lorentz group, the numbers $a^0, a^1, a_{\dot{0}}, a_{\dot{1}}$ are transformed according to the formulae

$$\left.\begin{aligned} a^{0'} &= a_{00}a^0 + a_{01}a', \\ a^{1'} &= a_{10}a^0 + a_{11}a', \end{aligned}\right\}$$

$$\left.\begin{aligned} a'_{\dot{0}} &= a^{\dot{0}\dot{0}}a_{\dot{0}} + a^{\dot{0}\dot{1}}a_{\dot{1}}, \\ a'_{\dot{1}} &= a^{\dot{1}\dot{0}}a_{\dot{0}} + a^{\dot{1}\dot{1}}a_{\dot{1}}, \end{aligned}\right\} \quad (5)$$

and, under spatial reflection s - according to the formulae

$$\left.\begin{aligned} a'^{0} &= a_{\dot{0}}, \\ a'^{1} &= a_{\dot{1}}, \\ a'_{\dot{0}} &= a^{0}, \\ a'_{\dot{1}} &= a^{1}. \end{aligned}\right\} \quad (6)$$

Suppose that a quartet of numbers $(a^0, a^1, a_{\dot{0}}, a_{\dot{1}})$ is specified in each orthogonal coordinate system of the four-dimensional space R^4, which is uniquely determined except for sign. Suppose that a transformation of the coordinate system by a proper Lorentz transformation $g = g_a$ gives rise to a transformation of the quartet with matrix (3) according to formula (5), and that the corresponding effect of a spatial reflection of the coordinate system is given by formula (6). Then we say that the quartet of numbers is a bispinor of the first rank.

The first two components (a^0, a^1) of the bispinor transform under proper Lorentz transformations as an undotted spinor of the first rank with upper indices, and the last two components $(a_{\dot{0}}, a_{\dot{1}})$ transform as a dotted spinor of the first rank with lower indices. The spatial reflection interchanges these two pairs.

We also note that rotations which correspond to unitary matrices $a = (a^*)^{-1}$, transform both pairs of components in the same way.

In the space R_1^1 of bispinors of the first rank the representation $g \to T_g$ of the complete group can be extended in two distinct ways to a unique representation of the general group, according as the temporal reflection t is related to the operator $T = +S$ or $T = -S$.

We note that it is also possible to specify in the space R_1^1 a two-valued representation of the general group (see Sects.1 and 3), if temporal reflection is given the matrix T.

$$T = \begin{Vmatrix} 0 & iE \\ -iE & 0 \end{Vmatrix},$$

and total reflection the matrix J

$$J = \begin{Vmatrix} -iE & 0 \\ 0 & iE \end{Vmatrix}$$

(this representation of the general group is closely connected with the so-called Dirac equation (see Sect.9)). Such a two-valued representation of the general group in the space R_1^1 may be specified in a unique manner (to within equivalence).

§ 2. The General Case. Bispinors of rank (k, n)

As we proved in Sect.4, every irreducible finite representation of the proper group may be realized in the space R_n^k of symmetric spinors with k upper undotted and n lower dotted indices according to the formula,

$$a^{\alpha_1' \dots \alpha_k'}_{\dot\beta_1' \dots \dot\beta_n'} = \sum a_{\alpha_1'\alpha_1} \dots a_{\alpha_k'\alpha_k} a^{\dot\beta_1'\dot\beta_1} \dots a^{\dot\beta_n'\dot\beta_n} a^{\alpha_1 \dots \alpha_k}_{\dot\beta_1 \dots \dot\beta_n}, \qquad (7)$$

$$\left\| a^{\dot\beta_i'\dot\beta_i} \right\| = \left\| a_{\alpha_i'\alpha_i} \right\|^{*-1}.$$

The pair defining this representation has the form $\left(l_0 = \frac{k-n}{2};\ l_1 = \frac{k+n}{2} + 1\right)$. The representation conjugate to representation (7) acts in the space R_k^n of spinors $b^{\alpha_1 \dots \alpha_n}_{\dot\beta_1 \dots \dot\beta_k}$ with n upper undotted indices and k lower dotted indices according to the formula

$$b^{\alpha_1' \dots \alpha_n'}_{\beta_1' \dots \beta_k'} = \sum a_{\alpha_1'\alpha_1} \dots a_{\alpha_n'\alpha_n} a^{\dot\beta_1'\dot\beta_1} \dots a^{\dot\beta_k'\dot\beta_k} b^{\alpha_1 \dots \alpha_n}_{\dot\beta_1 \dots \dot\beta_k}. \qquad (8)$$

After this recapitulation we pass on to the construction of the irreducible finite representations of the complete group.

As usual, we distinguish two cases:

I. $k \neq n$ or $l_0 \neq 0$. In this case, as we know from Sect.3, an irreducible representation of the complete group can be constructed in a unique manner, to within equivalence, in a linear sum of spaces

$$R(k, n) = R_n^k \dot{+} R_k^n.$$

We shall write this representation of the complete group explicitly. Each vector from the space $R(k, n)$ is specified by a set of numbers

$$\left\{ a^{\alpha_1 \dots \alpha_k}_{\dot\beta_1 \dots \dot\beta_n}, \quad b^{\alpha_1 \dots \alpha_n}_{\dot\beta_1 \dots \dot\beta_k} \right\}. \qquad (9)$$

In the space $R(k, n)$ the representation of the complete group acts in the following way. Under proper Lorentz transformations g_a the set $\left\{ a^{\alpha_1 \dots \alpha_k}_{\dot\beta_1 \dots \dot\beta_n}, b^{\alpha_1 \dots \alpha_n}_{\dot\beta_1 \dots \dot\beta_k} \right\}$ is transformed according to the formulae

$$\left.\begin{aligned} a^{\alpha_1' \dots \alpha_k'}_{\dot\beta_1' \dots \dot\beta_n'} &= \sum a_{\alpha_1'\alpha_1} \dots a_{\alpha_k'\alpha_k} a^{\dot\beta_1'\dot\beta_1} \dots a^{\dot\beta_k'\dot\beta_k} a^{\alpha_1 \dots \alpha_n}_{\beta_1 \dots \beta_k}, \\ b^{\alpha_1' \dots \alpha_n'}_{\dot\beta_1' \dots \dot\beta_k'} &= \sum a_{\alpha_1'\alpha_1} \dots a_{\alpha_n'\alpha_n} a^{\dot\beta_1'\dot\beta_1} \dots a^{\dot\beta_k'\dot\beta_k} b^{\alpha_1 \dots \alpha_n}_{\dot\beta_1 \dots \dot\beta_k}, \end{aligned}\right\} \qquad (10)$$

i.e. the spaces R_n^k and R_k^n are invariant and irreducible subspaces under representation (10) of the proper group and in these spaces representation (10) coincides with representations (7) and (8) respectively.

The operator S corresponding to the reflection s is given by

$$\left.\begin{aligned} a^{\alpha_1' \cdots \alpha_k'}_{\dot\beta_1' \cdots \dot\beta_n'} &= \sum \delta^{\alpha_1' \dot\beta_1} \cdots \delta^{\alpha_k' \dot\beta_k} \delta_{\alpha_1 \dot\beta_1'} \cdots \delta_{\alpha_n \dot\beta_n'} b^{\alpha_1 \cdots \alpha_n}_{\dot\beta_1 \cdots \dot\beta_k}, \\ b^{\alpha_1' \cdots \alpha_n'}_{\dot\beta_1' \cdots \dot\beta_k'} &= \sum \delta^{\alpha_1' \dot\beta_1} \cdots \delta^{\alpha_n' \dot\beta_n} \delta_{\alpha_1 \dot\beta_1'} \cdots \delta_{\alpha_k \dot\beta_k'} a^{\alpha_1 \cdots \alpha_k}_{\dot\beta_1 \cdots \dot\beta_n} \end{aligned}\right\} \tag{11}$$

i.e. the numbers $a^{\alpha_1 \cdots \alpha_n}_{\dot\beta_1 \cdots \dot\beta_k}$ and $a^{\alpha_1 \cdots \alpha_k}_{\dot\beta_1 \cdots \dot\beta_n}$ are interchanged and simultaneously their upper indices are interchanged with the lower. We recall that in paragraph 5, Sect. 4 we constructed a correspondence between the spinors $a^{\alpha_1 \cdots \alpha_k}_{\dot\beta_1 \cdots \dot\beta_n}$ of R_n^k and $b^{\alpha_1 \cdots \alpha_n}_{\dot\beta_1 \cdots \dot\beta_k}$ of R_k^n

$$a^{\alpha_1 \cdots \alpha_k}_{\dot\beta_1 \cdots \dot\beta_n} \longleftrightarrow b^{\alpha_1' \cdots \alpha_n'}_{\dot\beta_1' \cdots \dot\beta_k'}, \qquad \begin{matrix} \alpha_i' = \dot\beta_i \\ \dot\beta_i' = \alpha_i \end{matrix}. \tag{12}$$

It is clear that the operator S specifies precisely this correspondence. We must convince ourselves that formulae (10) and (11) actually define a representation of the complete group. To do so it suffices to verify that for any operator T_{g_a} of the representation (11) of the proper group the equation

$$ST_{g_a}S = T_{(g_a^*)^{-1}} = T_{g(a^*)^{-1}}$$

is satisfied (see Sect. 3).

Let $\{e_1, \ldots, e_N\}$ and $\{f_1, \ldots, f_N\}$ be bases of R_n^k and R_k^n, which are carried to one another by the operator S,

$$Se_i = f_i, \quad Sf_i = e_i \qquad (i = 1, \ldots, N).$$

It is clear that the operator S is described in terms of the basis $(e_1, \ldots, e_N, f_1, \ldots, f_N)$ by the matrix

$$S = \begin{Vmatrix} 0 & E \\ E & 0 \end{Vmatrix}, \qquad E = \begin{Vmatrix} 1 & 0 & 0 \ldots 0 \\ 0 & 1 & 0 \ldots 0 \\ 0 & 0 & 0 \ldots 1 \end{Vmatrix},$$

and the operators of representation (10) of the proper group by matrices of the form

$$T_{g_a} = \begin{Vmatrix} A_{g_a} & 0 \\ 0 & A_{g(a^*)^{-1}} \end{Vmatrix}.$$

this follows from the observation at the end of paragraph 5, Sect. 4, concerning the correspondence (12) between the spinors $a^{\alpha_1 \cdots \alpha_k}_{\dot\beta_1 \cdots \dot\beta_n}$ and $b^{\alpha_1 \cdots \alpha_n}_{\dot\beta_1 \cdots \dot\beta_k}$. The equation $ST_{g_a}S^{-1} = T_{g(a^*)^{-1}}$ is now verified directly. Thus, we have in fact obtained a representation of the complete Lorentz group, acting in the space $R(k, n)$ of all quantities $\left(a^{\alpha_1 \cdots \alpha_k}_{\dot\beta_1 \cdots \dot\beta_n},\ b^{\alpha_1 \cdots \alpha_n}_{\dot\beta_1 \cdots \dot\beta_k}\right)$. It follows from the results of Sect.3 that the representation of the complete group in the space $R(k, n)$ $(k \neq n)$ is irreducible.

We now give the general definition of quantities

$$\left\{ a^{\alpha_1 \cdots \alpha_k}_{\dot\beta_1 \cdots \dot\beta_n}, \quad b^{\alpha_1 \cdots \alpha_n}_{\dot\beta_1 \cdots \dot\beta_k} \right\}.$$

Let there be specified in each orthogonal coordinate system a set of $2 \times 2^{k+n}$ numbers $\left\{ a^{\alpha_1 \cdots \alpha_k}_{\dot\beta_1 \cdots \dot\beta_n}, b^{\alpha_1 \cdots \alpha_n}_{\dot\beta_1 \cdots \dot\beta_k} \right\}$ uniquely defined except for sign, which transform under a proper Lorentz transformation $g = g_a$ of the coordinate system according to formulae (10). Suppose further that the numbers $\left\{ a^{\alpha_1 \cdots \alpha_k}_{\dot\beta_1 \cdots \dot\beta_n}, b^{\alpha_1 \cdots \alpha_n}_{\dot\beta_1 \cdots \dot\beta_k} \right\}$ transform according to formula (11) under a spatial reflection of the coordinate system. Then this set of numbers is known as a bispinor of rank (k, n).

The first set of numbers of the bispinor transform under proper Lorentz transformations as a spinor of rank (k, n) with k upper undotted indices and n lower dotted indices, and the second set of numbers as a spinor of rank (n, k) with n upper undotted indices and k lower dotted indices. Under rotations $a = (a^*)^{-1}$ the first and second sets of numbers of the bispinor are transformed in the same way.

The bispinors of rank (k, n) form a $2 \times 2^{k+n}$ - dimensional linear space. Formulae (10) and (11) specify a representation in this space of the complete Lorentz group which is, generally speaking, reducible.

The space $R(k, n)$ which we have just constructed is clearly a subspace of the space of all bispinors which are invariant under representation (10) and (11).

We shall call the bispinors belonging to the space $R(k, n)$ symmetric bispinors of rank (k, n) (they are composed of two symmetric spinors). Thus representation (10) constructed above is the representation acting in the space $R(k, n)$ of symmetric bispinors of rank (k, n). In the case when $k \neq n$, it is irreducible.

II. We now consider the case $n = k$. All the constructions of the previous section obviously hold true here, the sole exception being that the representation of the complete group constructed in the space of symmetric bispinors of rank (k, k) is (by Sect.3) reducible. However, when $n = k$ i.e. $l_0 = 0$) an irreducible representation of the complete group may be specified in two non equivalent ways in the space R^k_k of symmetric spinors of rank (k, k).

The operator S is defined by the formula

$$a^{\alpha'_1 \dots \alpha'_k}_{\dot\beta'_1 \dots \dot\beta'_k} = \pm \sum \delta^{\alpha'_1 \dot\beta_1} \dots \delta^{\alpha'_k \dot\beta_k} \delta_{\alpha_1 \dot\beta'_1} \dots \delta_{\alpha_k \dot\beta'_k} a^{\alpha_1 \dots \alpha_k}_{\dot\beta_1 \dots \dot\beta_k}. \quad (13)$$

§ 3. Irreducible Representations of the General Group

Unique representations. As we know from the results of Sect.3, any irreducible representation of the complete group extends to a unique representation of the general group, if the temporal reflection t is assigned to the operator $T = \pm S$, and the complete reflection j to the operator $J = \pm E$.

In this manner it is possible to construct an irreducible unique representation of the general group in the space of bispinors of any rank. For this, in case I $(k \neq n)$ the operators T and J have the form

$$\left.\begin{aligned} T: &\left\{\begin{aligned} a^{\alpha'_1 \dots \alpha'_k}_{\dot\beta'_1 \dots \dot\beta'_n} &= \pm \sum \delta^{\alpha'_1 \dot\beta_1} \dots \delta^{\alpha'_k \dot\beta_k} \delta_{\alpha_1 \dot\beta'_1} \dots \delta_{\alpha_n \dot\beta'_n} b^{\alpha_1 \dots \alpha_n}_{\dot\beta_1 \dots \dot\beta_k}, \\ b^{\alpha'_1 \dots \alpha'_n}_{\dot\beta'_1 \dots \dot\beta'_k} &= \pm \sum \delta^{\alpha'_1 \dot\beta_1} \dots \delta^{\alpha'_n \dot\beta_n} \delta_{\alpha_1 \dot\beta'_1} \dots \delta_{\alpha_k \dot\beta'_k} a^{\alpha_1 \dots \alpha_k}_{\dot\beta_1 \dots \dot\beta_k}, \end{aligned}\right. \\ J: &\left\{\begin{aligned} a^{\alpha'_1 \dots \alpha'_k}_{\dot\beta'_1 \dots \dot\beta'_n} &= \pm a^{\alpha_1 \dots \alpha_k}_{\dot\beta_1 \dots \dot\beta_n}, \\ b^{\alpha'_1 \dots \alpha'_n}_{\dot\beta'_1 \dots \dot\beta'_k} &= \pm b^{\alpha_1 \dots \alpha_n}_{\dot\beta_1 \dots \dot\beta_k}, \end{aligned}\right. \qquad \begin{aligned} \alpha'_i &= \alpha_i, \\ \dot\beta'_i &= \dot\beta_i, \end{aligned} \end{aligned}\right\} \quad (14)$$

and in case II $(k = n)$ the operators T and J, like the operator S, act in the space of spinors of rank (k, k) according to the formulae

$$\left.\begin{aligned} T:\ & a^{\alpha'_1 \dots \alpha'_k}_{\dot\beta'_k \dots \dot\beta'_k} = \pm \sum \delta^{\alpha'_1 \dot\beta_1} \dots \delta^{\alpha'_k \dot\beta_k} \delta_{\alpha_1 \dot\beta'_1} \dots \delta_{\alpha_k \dot\beta'_k} a^{\alpha_1 \dots \alpha_k}_{\dot\beta_1 \dots \dot\beta_k}, \\ J:\ & a^{\alpha'_1 \dots \alpha'_k}_{\dot\beta'_1 \dots \dot\beta'_k} = \pm a^{\alpha_1 \dots \alpha_k}_{\dot\beta_1 \dots \dot\beta_k}, \qquad \alpha'_i = \alpha_i, \quad \dot\beta'_i = \dot\beta_i. \end{aligned}\right\} \quad (14')$$

Two-valued representations of the general group. As we have seen in Sect.3 any such representation generates a representation of the complete group consisting of two conjugate irreducible components of the proper group. In this case it is possible to choose a basis (see paragraph 5, Sect.3) so that the operators S, T, J have the matrices

$$S = \begin{Vmatrix} 0 & E \\ E & 0 \end{Vmatrix}, \quad T = \begin{Vmatrix} 0 & iE \\ -iE & 0 \end{Vmatrix}, \quad J = \begin{Vmatrix} -iE & 0 \\ 0 & iE \end{Vmatrix}.$$

Conversely, any representation of the proper group consisting of two conjugate components may be extended to an irreducible two-valued representation of the general group.

A representation of the complete group, acting in the space of symmetric bispinors of rank (k, n) contains two conjugate components of the proper group. In this manner, representation (10), (11) of the complete group in the space of bispinors can be extended to an irreducible, two-valued representation of the general group.

In this case the operators T and J act according to the formulae

$$T: \left\{ \begin{aligned} a^{\alpha_1' \ldots \alpha_k'}_{\dot\beta_1' \ldots \dot\beta_n'} &= i \sum \delta^{\alpha_1' \dot\beta_1} \ldots \delta^{\alpha_k' \dot\beta_k} \delta_{\alpha_1 \dot\beta_1'} \ldots \delta_{\alpha_n \dot\beta_n'} b^{\alpha_1 \ldots \alpha_n}_{\dot\beta_1 \ldots \dot\beta_k}, \\ b^{\alpha_1' \ldots \alpha_n'}_{\dot\beta_1' \ldots \dot\beta_k'} &= -i \sum \delta^{\alpha_1' \dot\beta_1} \ldots \delta^{\alpha_n' \dot\beta_n} \delta_{\alpha_1 \dot\beta_1'} \ldots \delta_{\alpha_k \dot\beta_k'} a^{\alpha_1 \ldots \alpha_k}_{\dot\beta_1 \ldots \dot\beta_n}, \end{aligned} \right\} \tag{15}$$

$$J: \left\{ \begin{aligned} a^{\alpha_1' \ldots \alpha_k'}_{\dot\beta_1' \ldots \dot\beta_n'} &= -i a^{\alpha_1 \ldots \alpha_k}_{\dot\beta_1 \ldots \dot\beta_n}, \qquad \alpha_i' = \alpha_i, \\ b^{\alpha_1' \ldots \alpha_n'}_{\dot\beta_1' \ldots \dot\beta_k'} &= i b^{\alpha_1 \ldots \alpha_k}_{\dot\beta_1 \ldots \dot\beta_n}, \qquad \dot\beta_i' = \dot\beta_i, \end{aligned} \right\} \tag{15'}$$

and the operator S according to formula (11).

It is easily seen that if, in the space of symmetric bispinors $R(k, n)$, we select a basis such that operator S (11) be written down by the matrix

$$S = \begin{Vmatrix} 0 & E \\ E & 0 \end{Vmatrix},$$

then the operators, T, J (15) and (15') have in this basis the form

$$J = \begin{Vmatrix} -iE & 0 \\ 0 & iE \end{Vmatrix}; \quad T = \begin{Vmatrix} 0 & iE \\ -iE & 0 \end{Vmatrix}.$$

The operators S, T, J anticommute and, together with E, form a two-valued representation of the group of reflections: $e \to \pm E$; $s \to \pm S$; $t \to \pm T$; $j = \pm J$.

§ 4. Tensor Representations of the Complete and General Lorentz Groups

We recall that the tensor representation of the proper group, i.e. the representation specified by the formula

$$t'_{k_1' k_2' \ldots k_n'} = \sum g_{k_1' k_1} g_{k_2' k_2} \ldots g_{k_n' k_n} t_{k_1 k_2 \ldots k_n} \tag{16}$$

(the indices k_i assume the values $k_i = 0, 1, 2, 3$; $\| g_{k_i' k_i} \|$ is the matrix of a proper Lorentz transformation) may be naturally extended to a representation of the complete Lorentz group by permitting the matrix $\| g_{ik} \|$ in formula (16) to be the matrix of the spatial reflection s

$$s = \begin{Vmatrix} -1 & 0 & 0 & 0 \\ 0 & -1 & 0 & 0 \\ 0 & 0 & -1 & 0 \\ 0 & 0 & 0 & 1 \end{Vmatrix}$$

or the matrix $-s$.

The operator S is defined in this case by the formula

$$t'_{k'_1 \dots k'_n} = \sum \left(\pm \delta_{k'_1 k_1}\right)\left(\pm \delta_{k'_2 k_2}\right) \cdots \left(\pm \delta_{k'_n k_n}\right) t_{k_1 \dots k_n}, \tag{16'}$$

where

$$-\delta_{00} = \delta_{11} = \delta_{22} = \delta_{33} = 1 \quad \text{for } k'_i = k_i$$

and

$$\delta_{k'_i k_i} = 0 \quad \text{for } k'_i \neq k_i.$$

If in formula (16') we choose a plus sign in an even number of brackets then the corresponding quantity is called a proper tensor. The quantities which are transformed under reflection by formula (16') in which the plus sign is chosen in an odd number of brackets, are called pseudotensors (pseudo-scalar, pseudovector).

Analogous to the operator S, it is possible to specify by formula (16) the operators T and J corresponding to temporal and spatial (sic) reflections t, j.

In this case we obtain a unique tensor representation of the general group. A tensor representation of the proper Lorentz group may be extended to a two-valued representation of the general group. According to the general construction, described in Sect.3, this representation is specified thus: we consider a pair of tensors $t_{k_1 \dots k_n}$ and $t_{k_1 \dots k_n}$, which are transformed independently under proper Lorentz transformations and under the reflection s according to formula (16) and (16'). In this case the tensor $t_{k_1 \dots k_n}$ is transformed by formula (16') with a plus sign, and for the tensor $\tilde{t}_{k_1 \dots k_n}$ there is a minus sign in front of this formula. The operators T and J act upon $t_{k_1 \dots k_n}$ and $\tilde{t}_{k_1 \dots k_n}$ as follows

$$T: \begin{cases} t'_{k'_1 \dots k'_n} = i \sum \left(\pm \delta_{k'_1 k_1}\right)\left(\pm \delta_{k'_2 k_2}\right) \cdots \left(\pm \delta_{k'_n k_n}\right) \tilde{t}_{k_1 \dots k_n}, \\ \tilde{t}'_{k'_1 \dots k'_n} = -i \sum \left(\pm \delta_{k'_1 k_1}\right)\left(\pm \delta_{k'_2 k_2}\right) \cdots \left(\pm \delta_{k'_n k_n}\right) t_{k_1 \dots k_n}, \end{cases}$$

$$J: \begin{cases} t'_{k_1 \dots k_n} = \tilde{t}_{k_1 \dots k_n}, \\ \tilde{t}'_{k_1 \dots k_n} = -t_{k_1 \dots k_n} \end{cases}$$

Section 6. The product of Two irreducible Finite Representations of the Proper Lorentz Group*

§ 1. The Decomposition of the Kronecker Product of Two Irreducible Representations of the Proper Lorentz Group into Irreducible Constituents

As in the case of the group of rotations, the following problem is of interest: Given two irreducible representations of the proper group: one $g \to T_g^{(1)}$ acting in space R_1 and the other $g \to T_g^{(2)}$, acting in space R_2. Is the representation $\tau_g = T_g^{(1)} \times T_g^{(2)}$ acting in $R_1 \times R_2$ generally speaking, reducible? What irreducible components does it contain?

Here we shall solve this problem for the case of finite, irreducible representations $g \to T_g^{(1)}$, $g \to T_g^{(2)}$. In this case it is convenient to consider these representations realized in spinor spaces (see Sect.4). We recall that any finite representation of the proper Lorentz group may be realized in a space of symmetric spinors of rank (k, n), where k is the number of undotted indices and n is the number of dotted indices, the number pair defining this representation being

$$l_0 = \frac{k-n}{2}, \qquad l_1 = \frac{k+n}{2} + 1 .$$

We pass on to the solution of our problem. We shall start with particular cases.

1. In Sect.4 it was shown that an irreducible representation in the space of symmetric spinors of rank (k, n) is the product of two irreducible representations $T_g^{(k,0)}$ and $T_g^{(0,n)}$:

$$T_g^{(k,n)} = T_g^{(k,0)} \times T_g^{(0,n)}.$$

The first of these representations is realized in the space of undotted spinors, and below we shall call it simply the undotted representation; the second acts in the space of dotted spinors and is called simply the dotted representation.

2. Let there be specified the product of two undotted representations

$$T_g^{(k_1,0)} \times T_g^{(k_2,0)}.$$

We shall show that the irreducible components of this product are again undotted representations. To this end we make use of our observation in Sect.4, paragraph 3 concerning the fact that finite representations $g \to T_g$, for which the matrix elements of the operators T_g relative to some basis are analytical functions of the elements of the second order matrix a: $(a_{00}, a_{01}, a_{10}, a_{11})$, decompose only into undotted spinor representations. We called such representations $g \to T_g$ analytic. It is clear that the irreducible undotted

* For the definition of the product of representations see Part I, Sect.4.

representation $g \to T_g^{(k_1, 0)}$ is itself analytic. Moreover it is obvious that the product of two analytic representations is analytic.

Thus, the representation $T_g^{(k_1,0)} \times T_g^{(k_2, 0)}$ is analytic and therefore, contains only undotted components $T_g^{(k', 0)}$.

We now determine the possible ranks k' of such components. To this end, we recall that the representations $T_g^{(k_1,0)}$ and $T_g^{(k_2, 0)}$ are irreducible under the group of rotations and have weights $\frac{k_1}{2}$ and $\frac{k_2}{2}$ respectively.

Since each representation $T_g^{(k'_1, 0)}$ is also irreducible under the group of rotations and has the weight $\frac{k'}{2}$, $\frac{k'}{2}$ assumes all the values from $\frac{|k_1 - k_2|}{2}$ to $\frac{k_1 + k_2}{2}$ (once only).

Thus, the ranks k' of the undotted components $T_g^{(k', 0)}$ which enter into the decomposition of $T_g^{(k_1, 0)} \times T_g^{(k_2, 0)}$ assume (once only) the values

$$k' = |k_1 - k_2|, \quad |k_1 - k_2| + 2, \ \ldots, \quad k_1 + k_2 - 2, \quad k_1 + k_2.$$

3. Similarly, the product of two dotted representations

$$T_g^{(0, n_1)} \times T_g^{(0, n_2)}$$

decomposes only into dotted components $T_g^{(0, n')}$, the ranks n' of which can assume the values

$$n' = |n_1 - n_2|, \quad |n_1 - n_2| + 2; \quad n_1 + n_2 - 2, \quad n_1 + n_2$$

(each once only).

4. The general case. As was shown initially the product of the undotted representation $T_g^{(k, 0)}$ by the dotted one $T_g^{(0, n)}$

$$T_g^{(k, 0)} \times T_g^{(0, n)}$$

is irreducible and equivalent to the representation $T_g^{(k, n)}$ in the space of symmetric spinors of rank (k, n):

$$T_g^{(k, 0)} \times T_g^{(0, n)} = T_g^{(k, n)}.$$

We shall use this fact for the decomposition into irreducible components of the representation

$$\tau_g = T_g^{(k_1, n_1)} \times T_g^{(k_2, n_2)},$$

i.e. for the solution of the general problem of the decomposition of the product of irreducible finite representations of the proper Lorentz group. We make use of the equations

$$T_g^{(k_1, n_1)} = T_g^{(k_1, 0)} \times T_g^{(0, n_1)}$$

and

$$T_g^{(k_2, n_2)} = T_g^{(k_2, 0)} \times T_g^{(0, n_2)}.$$

Then for τ_g we obtain

$$\tau_g = (T_g^{(k_1, 0)} \times T_g^{(0, n_1)}) \times (T_g^{(k_2, 0)} \times T_g^{(0, n_2)}).$$

Interchanging the factors and using the associative property of the product of two representations, we obtain *

$$\tau_g = (T_g^{(k_1, 0)} \times T_g^{(k_2, 0)}) \times (T_g^{(0, n_1)} \times T_g^{(0, n_2)}).$$

The first bracket in this product gives a representation containing only undotted components $T_g^{(k', 0)}, k' = |k_1 - k_2|, |k_1 - k_2| + 2, \ldots, k_1 + k_2$. The second bracket decomposes into dotted components $T_g^{(0, n')}$ $n' = |n_1 - n_2|$, $|n_1 - n_2| + 2, \ldots, n_1 + n_2 - 2, n_1 + n_2$. It is clear that the representation τ contains all possible representations of the form $T_g^{(k', 0)} \times T_g^{(0, n')}$ (precisely once). But each such representation is irreducible and is realized in the space of spinors of rank (k', n') i.e.

$$T_g^{(k', 0)} \times T_g^{(0, n')} = T_g^{(k' n')}.$$

Thus, we finally conclude that the product of two finite irreducible representations of the proper Lorentz group $T_g^{(k_1, n_1)}$ and $T_g^{(k_2, n_2)}$ (realized by spinors of rank (k_1, n_1) and (k_2, n_2) respectively) decomposes into the irreducible components $T_g^{(k', n')}$, where k' and n' independently assume the values

$$k' = |k_1 - k_2|, \quad |k_1 - k_2| + 2, \ldots, \quad k_1 + k_2 - 2, \quad k_1 + k_2,$$
$$n' = |n_1 - n_2|, \quad |n_1 - n_2| + 2, \ldots, \quad n_1 + n_2 - 2, \quad n_1 + n_2,$$

and each such component enters only once into the analysis.

* The possibility of interchanging the factors and the associative property of representation multiplication follows easily from the definition of the product of representations (see Part I, Sect.4).

We recall once again that the numbers (k, n) are related to the numbers (l_0, l_1) by which an irreducible representation is usually specified in this book, by the formulae

$$l_0 = \frac{k-n}{2}, \qquad l_1 = \frac{k+n}{2} + 1. \tag{1}$$

We now consider an example: The product of two identity representations $g \to g$ of the Lorentz group.

As we saw in Sect.4, § 7, such a product is equivalent to a tensor representation of rank 2. In that paragraph we decomposed this representation into irreducible components. We shall find this analysis once again, starting from the results just obtained.

The representation $g \to g$ is defined by the pair (0, 2) the space $R^{(4)}$ contains two subspaces which are invariant under the group of rotations, namely: one-dimensional - the axis x_0 ($l = l_0 = 0$), and three-dimensional: $x_0 = 0$ ($l = 1$, $l_1 = 2$).) From formula (1) it follows that the representation $g \to g$ is realized by the spinors of rank (1, 1) (i.e., is equivalent to the representation $T_g^{(1,1)}$). Thus, we must decompose into its irreducible components the product

$$T_g^{(1,1)} \times T_g^{(1,1)}.$$

We have

$$T_g^{(1,1)} \times T_g^{(1,1)} = \sum T_g^{(k',n')},$$

where

$$k' = 0,2; \quad n' = 0,2$$

(in all, four components are obtained).

We draw up a table (see Sect. 4, paragraph 7)

n' \ k'	0	2
0	scalar $l_0 = 0$, $l_1 = 1$	$l_0 = 1$, $l_1 = 2$ antisymmetric tensor
2	antisymmetric tensor $l_0 = -1$, $l_1 = 2$	symmetric tensor of rank 2 with trace equal to zero ($t_{11} + t_{22} + t_{33} - t_{00} = 0$), $l_0 = 0$, $l_1 = 3$

Obviously we have again obtained the same four components as in Sect.4, paragraph 7. From this table it is clear, in particular, that the symmetric tensors with zero trace (the component $T_g^{(2,2)}$) form an irreducible subspace in the space of tensors of the first rank; in Sec. 4 we assumed this fact without proof.

§ 2. The Clebsch-Gordan Coefficients *

In this paragraph we shall find how the vectors of the canonical bases of the irreducible components of the product of the two representations $T_g^{(k_1, n_1)}$ and $T_g^{(k_2, n_2)}$ are expressed in terms of the product of the canonical bases $\{\xi_{lm}^{k_1 n_1}\}$ and $\{\xi_{lm}^{k_2 n_2}\}$ in the spaces R_{k_1, n_1} and R_{k_2, n_2}, where the representations $T_g^{(k_1, n_1)}$ and $T_g^{(k_2, n_2)}$ themselves act.

Let $\{\xi_{lm}^{k' n'}\}$ be the canonical basis for the component $T_g^{(k', n')}$.

We have

$$\xi_{lm}^{k' n'} = \sum H_{k_1 n_1\, l_1 m_1;\, k_2 n_2 l_2 m_2}^{k' n'\, lm} \xi_{l_1 m_1}^{k_1 n_1} \xi_{l_2 m_2}^{k_2 n_2}. \tag{2}$$

We express the coefficients $H_{k_1 n_1 l_1 m_1;\, k_2 n_2 l_2 m_2}^{k' n' lm}$ in terms of the Clebsch-Gordan coefficients for the group of rotations (see Part I, Sect. 10).

Above we have seen that

$$T_g^{(k_1, n_1)} = T_g^{(k_1, 0)} \times T_g^{(0, n_1)}.$$

Let $\left\{\eta_{\frac{k_1}{2}, m_1'}\right\}$ and $\left\{\bar{\eta}_{\frac{n_1}{2}, m_1''}\right\}$ be the canonical bases of the representations $T_g^{(k_1, 0)}$ and $T_g^{(0, n_1)}$. In the space R_1 where the representation $T_g^{(k_1, n_1)}$ acts, the vectors $\left\{\eta_{\frac{k_1}{2}, m_1'}, \bar{\eta}_{\frac{n_1}{2}, m_1''}\right\}$ constitute a basis. It is clear that

$$\xi_{l_1 m_1}^{k_1 n_1} = \sum_{m_1' + m_1'' = m_1} B_{\frac{k_1}{2}, m_1';\, \frac{n_1}{2}, m_1''}^{l_1 m_1} \eta_{\frac{k_1}{2}, m_1'} \bar{\eta}_{\frac{n_1}{2}, m_1''}. \tag{3}$$

On the other hand (see Sect. 10, Part I)

$$\eta_{\frac{k_1}{2}, m_1'} \bar{\eta}_{\frac{n_1}{2}, m_1''} = \sum_{\substack{|m_1| \leqslant l_1 \\ m_1' + m_1'' = m_1}} B_{\frac{k_1}{2}, l_1 - \frac{n_1}{2};\, \frac{n_1}{2}, m_1 + \frac{n_1}{2} - l_1}^{\frac{n_1}{2} + m_1', m_1} \xi_{l_1 m_1}^{k_1 n_1}. \tag{4}$$

Similarly for

$$T_g^{(k_2 n_2)} = T_g^{(k_2, 0)} \times T_g^{(0, n_2)}$$

---

* In analogy with the case of the group of rotations, we call the elements of the matrix carrying one basis into another the Clebsch-Gordan coefficients.

we have:

$$\eta_{\frac{k_2}{2},\, m_2'}\bar{\eta}_{\frac{n_2}{2},\, m_2''} = \sum_{\substack{|m_2|\leqslant l_2\\ (m_2'+m_2''=m_2)}} B^{\frac{n_2}{2}+m_2',\, m_2}_{\frac{k_2}{2},\, l_2-\frac{n_2}{2};\, \frac{n_2}{2},\, m_2+\frac{n_2}{2}-l_2}\, \xi^{k_2 k_2}_{l_2 m_2}, \tag{5}$$

where $\left\{\eta_{\frac{k_2}{2},\, m_2'}\right\}$ and $\left\{\bar{\eta}_{\frac{n_2}{2},\, m_2''}\right\}$ are the canonical bases of the representations $T_g^{(k_2,\, 0)}$ and $T_g^{(0,\, n_2)}$.

We consider the representation

$$T_g^{(k_1,\, 0)} \times T_g^{(k_2,\, 0)}.$$

Let $\left\{\xi_{\frac{k'}{2},\, \mu}\right\}$ be the canonical basis of the component $T_g^{(k',\, 0)}$ of this product $(|k_1 - k_2| \leqslant k' \leqslant k_1 + k_2)$.

It is clear that

$$\xi_{\frac{k'}{2},\, \mu'} = \sum_{\mu' = m_1' + m_2'} B^{\frac{k'}{2},\, \mu'}_{\frac{k_1}{2},\, m_1';\, \frac{k_2}{2},\, m_2'}\, \eta_{\frac{k_1}{2},\, m_1'}\, \eta_{\frac{k_2}{2},\, m_2'}. \tag{6}$$

Similarly for the canonical basis $\left\{\bar{\xi}_{\frac{n'}{2},\, \mu''}\right\}$ of the component $T_g^{(0,\, n')}$ of the product $T_g^{(0,\, n_1)} \times T_g^{(0,\, n_2)}$ we have:

$$\bar{\xi}_{\frac{n'}{2},\, \mu''} = \sum_{\mu'' = m_1'' + m_2''} B^{\frac{n'}{2},\, \mu''}_{\frac{n_1}{2},\, m_1'';\, \frac{n_2}{2},\, m_2''}\, \bar{\eta}_{\frac{n_1}{2},\, m_1''}\, \bar{\eta}_{\frac{n_2}{2},\, m_2''}. \tag{7}$$

The product $T_g^{(k',\, 0)} \times T_g^{(0,\, n')}$ forms an irreducible representation $T_g^{(k',\, n')}$ in the canonical basis $\left\{\xi^{k'n'}_{lm}\right\}$

$$\xi^{k'n'}_{lm} = \sum_{m=\mu'+\mu''} B^{lm}_{\frac{k'}{2},\, \mu';\, \frac{n'}{2},\, \mu''}\, \xi_{\frac{k'}{2},\, \mu'}\, \bar{\xi}_{\frac{n'}{2},\, \mu''} \tag{8}$$

or (see (6) and (7))

$$\xi^{k'n'}_{lm} = \sum_{m_1,\, m_1',\, m_2'} B^{lm}_{\frac{k'}{2},\, \mu';\, \frac{n'}{2},\, \mu''}\, B^{\frac{k'}{2},\, \mu'}_{\frac{k_1}{2},\, m_1';\, \frac{k_2}{2},\, m_2'} \times$$

$$\times B^{\frac{n'}{2},\, \mu''}_{\frac{n_1}{2},\, m_1'';\, \frac{n_2}{2},\, m_2''}\, \eta_{\frac{k_1}{2},\, m_1'}\, \eta_{\frac{k_2}{2},\, m_2'}\, \bar{\eta}_{\frac{n_1}{2},\, m_1''}\, \bar{\eta}_{\frac{n_2}{2},\, m_2''}.$$

or finally (see (4) and (5)),

$$\xi_{lm}^{k'n'} = \sum_{\substack{m_1', m_2', l_1 \\ l_2, m_1}} B^{lm}_{\frac{k'}{2}, \mu'; \frac{n'}{2}, \mu''} B^{\frac{k'}{2}, \mu'}_{\frac{k_1}{2}, m_1'; \frac{k_2}{2}, m_2'} B^{\frac{n'}{2}, \mu''}_{\frac{n_1}{2}, m_1''; \frac{n_2}{2}, m_2''} \times$$

$$\times B^{\frac{n_1}{2}+m_1', m_1}_{\frac{k_1}{2}, l_1-\frac{n_1}{2}; \frac{n_1}{2}, m_1+\frac{n_1}{2}-l_1} B^{\frac{n_2}{2}+m_2', m_2}_{\frac{k_2}{2}, l_2-\frac{n_2}{2}; \frac{n_2}{2}, m_2+\frac{n_2}{2}-l_2} \xi^{k_1 n_1}_{l_1 m_1} \xi^{k_2 n_2}_{l_2 m_2}.$$

We have:

$$m_1'' = m_1 - m_1', \quad m_2'' = m_2 - m_2',$$
$$\mu' = m_1' + m_2', \quad \mu'' = m - m_1' - m_2'$$

(the summation is carried out over all repeated indices).

Hence

$$\xi_{lm}^{k'n'} = \sum B^{lm}_{\frac{k'}{2}, m_1'+m_2'; \frac{n'}{2}, m-m_1'-m_2'} B^{\frac{k'}{2}, m_1'+m_2'}_{\frac{k_1}{2}, m_1'; \frac{k_2}{2}, m_2'} \times$$

$$\times B^{\frac{n'}{2}, m-m_1'-m_2'}_{\frac{n_1}{2}, m_1-m_1'; \frac{n_2}{2}, m_2-m_2'} B^{\frac{n_1}{2}+m_1', m_1}_{\frac{k_1}{2}, l_1-\frac{n_1}{2}; \frac{n_1}{2}, m_1+\frac{n_1}{2}-l_1} \times$$

$$\times B^{\frac{n_2}{2}+m_2', m_2}_{\frac{k_2}{2}, l_2-\frac{n_2}{2}; \frac{n_2}{2}, m_2+\frac{n_2}{2}-l_2} \xi^{k_1 n_1}_{l_1 m_1} \xi^{k_2 n_2}_{l_2 m_2}.$$

The summation is carried out over the following range of indices

$$-\frac{k_1}{2} \leqslant m_1' \leqslant \frac{k_1}{2}, \quad -\frac{k_2}{2} \leqslant m_2' \leqslant \frac{k_2}{2}, \quad \frac{|k_1 - n_1|}{2} \leqslant l_1 \leqslant \frac{k_1 + n_1}{2},$$

$$\frac{|k_2 - n_2|}{2} \leqslant l_2 \leqslant \frac{k_2 + n_2}{2} \text{ and } -l_1 < m_1 \leqslant l_1 \quad (m = m_1 + m_2).$$

Comparing with (2) we obtain:

$$H^{k'n'lm}_{k_1 n_1 l_1 m_1 k_2 n_2 l_2 m_2} = \sum_{m_1', m_2'} B^{lm}_{\frac{k'}{2}, m_1'+m_2'; \frac{n'}{2}, m-m_1'-m_2'} \times$$

$$\times B^{\frac{k'}{2}, m_1'+m_2'}_{\frac{k_1}{2}, m_1'; \frac{k_2}{2}, m_2'} B^{\frac{n'}{2}, m-m_1'-m_2'}_{\frac{n_1}{2}, m_1-m_1'; \frac{n_2}{2}, m_2-m_2'} B^{\frac{n_1}{2}+m_1', m_1}_{\frac{k_1}{2}, l_1-\frac{n_1}{2}; \frac{n_1}{2}, m_1+\frac{n_1}{2}-l_1} \times$$

$$\times B^{\frac{n_2}{2}+m_2', m_2}_{\frac{k_2}{2}, l_2-\frac{n_2}{2}; \frac{n_2}{2}, m_2+\frac{n_2}{2}-l_2}.$$

CHAPTER II

RELATIVISTIC-INVARIANT EQUATIONS

Section 1. General Relativistic-Invariant Equations

§ 1. Definition of Relativistic-Invariant Equations

In the quantum theory of fields the state of a particle is described by a function $\psi(x_0, x_1, x_2, x_3)$. The quantity $\psi(x_0, x_1, x_2, x_3)$ may consist of one or more components (and also, possibly, of an infinite number). The function $\psi(x_0, x_1, x_2, x_3)$ must satisfy some homogeneous, linear, differential equation *.

The form of this equation is determined by the physical properties of the particle.

We consider for example the simplest equation - the so-called Clebsch-Gordan equation:

$$\frac{\partial^2\psi}{\partial x_0^2} - \frac{\partial^2\psi}{\partial x_1^2} - \frac{\partial^2\psi}{\partial x_2^2} - \frac{\partial^2\psi}{\partial x_3^2} + \varkappa^2\psi = 0, \tag{1}$$

where the function $\psi(x_0, x_1, x_2, x_3)$ consists of one component (a scalar function). It is easily seen that if we subject the variables $(x_0, x_1, x_2, x_3) = \bar{x}$ to a Lorentz transformation g: $\bar{x}' = g\bar{x}$, then the form of equation (1) is unchanged and the function $\psi'(\bar{x}') = \psi(\bar{x})$ is the solution of equation (1) as before. For this reason, it is said that equation (1) is invariant under Lorentz transformations, or briefly, relativistic-invariant. This circumstance expresses the fact that the properties of the particle described by equation (1) (in particular, this equation itself) must be independent of the choice of a system of reference (x_0, x_1, x_2, x_3). Two distinct orthogonal ** reference

---

* We are considering the case of a free field. For interacting fields the equations are non-homogeneous.

** We recall that the coordinate system (x_0, x_1, x_2, x_3) is said to be orthogonal if the quadratic form $S^2(x^2)$ has in this system the form

$$S^2 = x_0^2 - x_1^2 - x_2^2 - x_3^2.$$

In physics this is known as an inertial reference system.

systems are related by a Lorentz transformation:

$$x'_i = \sum g_{ik} x_k, \tag{1'}$$

where $\|g_{ik}\|$ is the matrix of the Lorentz transformation.

Thus the invariance of equation (1) under Lorentz transformations is equivalent to its invariance with respect to the choice of reference system. It is clear that the latter property - the independence of the choice of reference system - is essential for all differential equations or systems of differential equations which describe real particles.

We shall consider systems of first-order equations.*

$$L_0 \frac{\partial \psi}{\partial x_0} + L_1 \frac{\partial \psi}{\partial x_1} + L_2 \frac{\partial \psi}{\partial x_2} + L_3 \frac{\partial \psi}{\partial x_3} + i\varkappa\psi = 0 \tag{2}$$

---

* It may be shown that any equation of higher order may be reduced to a system of first-order equations. As an example we consider equation (1)

$$\square \psi + \chi^2 \psi = 0.$$

We put

$$\varkappa \psi_i = \frac{\partial \psi}{\partial x_i}.$$

Equation (1) is then equivalent to the system

$$\sum_{i=1,2,3} \frac{\partial \psi_i}{\partial x_i} - \frac{\partial \psi_0}{\partial x_0} - \chi \psi = 0, \qquad \frac{\partial \psi}{\partial x_i} - \chi \psi_i = 0.$$

This system may be written as follows: Let $\Phi = \begin{Vmatrix} \psi \\ \psi_0 \\ \psi_1 \\ \psi_2 \\ \psi_3 \end{Vmatrix}$.

Then

$$L_0 \frac{\partial \Phi}{\partial x_0} + L_1 \frac{\partial \Phi}{\partial x_1} + L_2 \frac{\partial \Phi}{\partial x_2} + L_3 \frac{\partial \Phi}{\partial x_3} + i\varkappa\Phi = 0,$$

where

$$L_0 = \begin{Vmatrix} 0 & i & 0 & 0 & 0 \\ -i & 0 & 0 & 0 & 0 \\ 0 & 0 & 0 & 0 & 0 \\ 0 & 0 & 0 & 0 & 0 \\ 0 & 0 & 0 & 0 & 0 \end{Vmatrix}, \qquad L_1 = \begin{Vmatrix} 0 & 0 & -i & 0 & 0 \\ 0 & 0 & 0 & 0 & 0 \\ -i & 0 & 0 & 0 & 0 \\ 0 & 0 & 0 & 0 & 0 \\ 0 & 0 & 0 & 0 & 0 \end{Vmatrix},$$

$$L_2 = \begin{Vmatrix} 0 & 0 & 0 & -i & 0 \\ 0 & 0 & 0 & 0 & 0 \\ 0 & 0 & 0 & 0 & 0 \\ -i & 0 & 0 & 0 & 0 \\ 0 & 0 & 0 & 0 & 0 \end{Vmatrix}, \qquad L_3 = \begin{Vmatrix} 0 & 0 & 0 & 0 & -i \\ 0 & 0 & 0 & 0 & 0 \\ 0 & 0 & 0 & 0 & 0 \\ 0 & 0 & 0 & 0 & 0 \\ -i & 0 & 0 & 0 & 0 \end{Vmatrix}.$$

($\varkappa$ a real constant) where the wave function $\psi(x_0\ x_1, x_2, x_3)$ assumes values in some linear space R (in other words, has a finite or infinite number of components), and L_0, L_1, L_2, L_3 are matrices which act in this space.

We now describe precisely what is meant by a relativistic-invariant equation of the form (2).

Any Lorentz transformation g of the coordinates x_0, x_1, x_2, x_3 (a transfer from one reference system to another) must be accompanied, generally speaking, by some transformation T_g of quantity ψ

$$\psi'(x') = T_g\psi(x) \qquad (x' = gx). \tag{2'}$$

The correspondence $g \to T_g$ specifies a representation of the Lorentz group in the space R.

An equation of type (2) is said to be relativistic-invariant if its form is unchanged by a simultaneous transformation of coordinates according to formula (1) and of the function ψ according to formula (2').

We stress once again that for a complete description of a particle there must also be specified, besides a relativistic-invariant equation of form (2), the transformation law of the quantities ψ under a Lorentz transformation, i.e. the representation $g \to T_g$ in the space R must be specified.

§ 2. The Conditions for Relativistic-invariance of an Equation for the Case When $\varkappa \neq 0$.

We now derive the conditions which the matrices L_0, L_1, L_2, L_3 must satisfy in a relativistic-invariant equation with $\varkappa \neq 0$. We subject $x(x_0, x_1, x_2, x_3)$ to some Lorentz transformation g: $x' = gx$.

Whereupon ψ is transformed by the operator T_g

$$\psi'(x') = T_g\psi(x).$$

We consider how equation (2) is transformed. We have:

$$\frac{\partial\psi'}{\partial x_i} = \sum_k \frac{\partial\psi'}{\partial x'_k} g_{ki}, \qquad \psi = T_g^{-1}\psi'.$$

Substituting this expression in (2), we obtain

$$\sum_{i,\,k} L_i T_g^{-1} \frac{\partial\psi'}{\partial x'_k} g_{ki} + i\varkappa T_g^{-1}\psi' = 0.$$

Left-multiplying by the operator T_g we obtain

$$\sum_{k,\, i} T_g L_i T_g^{-1} g_{ki} \frac{\partial \psi'}{\partial x'_k} + i\varkappa\psi' = 0.$$

Comparing this equation with equation (2), we see that we must have for the invariance of equation (2)

$$\sum_i T_g L_i T_g^{-1} g_{ki} = L_k \qquad (k = 1,\ 2,\ 3,\ 0). \tag{3}$$

This is then the condition that must be satisfied by the matrices L_i

It is clearly sufficient to require the fulfillment of equation (3) for infinitely small Lorentz transformations.

Let, for example, $\|g_{ki}\|$ be the matrix of the screw $g_{12}(\varphi)$ in the plane $(x_1,\ x_2)$,

$$g_{12}(\varphi) = \begin{Vmatrix} \cos\varphi & \sin\varphi & 0 & 0 \\ -\sin\varphi & \cos\varphi & 0 & 0 \\ 0 & 0 & 1 & 0 \\ 0 & 0 & 0 & 1 \end{Vmatrix}$$

or for small φ,

$$g_{12}(\varphi) = E + \varphi \begin{Vmatrix} 0 & 1 & 0 & 0 \\ -1 & 0 & 0 & 0 \\ 0 & 0 & 0 & 0 \\ 0 & 0 & 0 & 0 \end{Vmatrix} + o(\varphi).$$

Whence for the elements of the matrix $g_{12}(\varphi)$ for small φ we obtain the expression

$$[g_{12}(\varphi)]_{ki} = \delta_{ki} - \varphi(\delta_{k2}\delta_{i1} - \delta_{k1}\delta_{i2}) + o(\varphi)$$

$$(\delta_{ki} = 1 \text{ for } k = i;\ \delta_{ki} = 0 \text{ for } k \neq i).$$

The operator $T_{g_{12}}(\varphi)$ for small φ is clearly given by

$$T_{g_{12}(\varphi)} = E + \varphi A_{12} + o(\varphi).$$

Substituting $[g_{12}(\varphi)]_{ki}$ and $T_{g_{12}(\varphi)}$ in the expression (3), we have:

$$\sum_i \{(E + \varphi A_{12}) L_i (E - \varphi A_{12}) (\delta_{ki} - \varphi[\delta_{k2}\delta_{i1} - \delta_{k1}\delta_{i2}]) + o(\varphi)\} = L_k.$$

Collecting terms of the first order in φ we obtain:

$$\sum_i \{(A_{12}L_i - L_iA_{12})\,\delta_{ki} - (L_i\delta_{k2}\delta_{i1} - L_i\delta_{k1}\delta_{i2})\} = 0$$

or

$$[A_{12}, L_k] - L_1\delta_{k2} + L_2\delta_{k1} = 0 \qquad (k = 1, 2, 3, 0).$$

Whence we obtain the following relations:

$$\left.\begin{array}{l} [A_{12}, L_1] = -L_2, \ [A_{12}, L_2] = L_1, \\ [A_{12}, L_3] = [A_{12}, L_0] = 0. \end{array}\right\} \qquad (4)$$

Similarly for A_{13} and A_{23}

$$\left.\begin{array}{l} [A_{13}, L_k] - L_1\delta_{k3} + L_3\delta_{k1} = 0, \\ [A_{23}, L_k] - L_2\delta_{k3} + L_3\delta_{k2} = 0. \end{array}\right\} \qquad (5)$$

For B_1, B_2, B_3 we obtain

$$\left.\begin{array}{l} [B_1, L_k] + L_1\delta_{k0} + L_0\delta_{k1} = 0, \\ [B_2, L_k] + L_2\delta_{k0} + L_0\delta_{k2} = 0, \\ [B_3, L_k] + L_3\delta_{k0} + L_0\delta_{k3} = 0, \end{array}\right\} \quad k = 1, 2, 3, 0. \qquad (6)$$

Thus the problem of finding all the relativistic-invariant equations reduces to finding the matrices L_1, L_2, L_3, L_0, which satisfy relations (4)-(6). From relations (6) it is clear that

$$L_k = -[B_k, L_0] \qquad (k = 1, 2, 3), \qquad (6')$$

i.e. it suffices to find the matrix L_0 only, as the others are (then) uniquely determined.

For the matrix L_0 we have the following relations:

$$[L_0, A_{12}] = [L_0, A_{13}] = [L_0, A_{23}] = 0,$$
$$[B_3, [B_3, L_0]] = L_0.$$

The remaining relations follow from these.

If we introduce the operators

$$\begin{array}{ll} H_+ = iA_{23} - A_{13}, & F_+ = iB_1 - B_2, \\ H_- = iA_{23} + A_{13}, & F_- = iB_1 + B_2, \\ H_3 = iA_{12}, & F_3 = iB_3, \end{array}$$

then we obtain

$$[L_0, H_+] = [L_0, H_-] = [L_0, H_3] = 0, \qquad (7)$$

$$[[F_3, L_0], F_3] = L_0. \qquad (8)$$

Thus the problem of determining the four matrices L_1, L_2, L_3, L_0 has been reduced to that of finding a matrix L_0 which satisfies relations (7)-(8).

If equation (2) is invariant under the complete Lorentz group, i.e. if it is unaltered by the reflection s

$$x_1' = -x_1, \quad x_2' = -x_2, \quad x_3' = -x_3, \quad x_0' = x_0,$$

of three-dimensional space, then the matrix L_0 commutes with the operator S ($S = T_s$). In fact from relation (3) we have:

$$SL_0S^{-1} = L_0 \quad \text{or} \quad [S, L_0] = 0. \tag{9}$$

Thus we conclude finally that the matrices L_1, L_2, L_3 in a relativistic-invariant equation (2) are expressed in terms of the matrix L_0 by formula (6'):

$$L_i = -[B_i L_0], \quad i = 1, 2, 3; \tag{6'}$$

the matrix L_0 is determined from the conditions (7)-(8)

$$[L_0, H_+] = [L_0, H_-] = [L_0, H_3] = 0, \qquad [[F_3, L_0], F_3] = L_0. \tag{7-8}$$

When equation (2) is invariant under the complete Lorentz group, condition (9) is added to the conditions (7)-(8).

§ 3. The Determination of the Matrices L_0, L_1, L_2, L_3.

We now find the matrix L_0 in an equation invariant under the proper Lorentz group i.e., in an equation which is unaltered by proper Lorentz transformations. As was proved in the last section, such a matrix L_0, satisfies the conditions (7)-(8).

Let the space R in which the representation $g \to T_g$ acts be decomposed into a linear sum of subspaces R^τ in each of which there acts an irreducible representation τ of the proper Lorentz group, defined by the pair

$$\tau \sim (l_0, l_1).$$

In each subspace R^τ we select a canonical basis $\{\xi_{lm}^\tau\}$, namely, the basis consisting of eigenvectors of the operator H_3. The vectors $\{\xi_{lm}^\tau\}$ clearly form a basis of the entire space R.

Let $\|c_{lml'm'}^{\tau\tau'}\|$ be the matrix of the operator L_0, in terms of this basis, i.e.

$$L_0 \xi_{lm}^\tau = \sum c_{lml'm'}^{\tau\tau'} \xi_{l'm'}^{\tau'}. \tag{10}$$

We shall find the general form of the numbers $c_{lml'm'}^{\tau\tau'}$.

From relations (7), which mean that the matrix L_0 commutes with the operators of the representation of the group of rotations, we obtain according to (32) Sect. 2:

$$c_{lml'm'}^{\tau\tau'} = c_l^{\tau\tau'} \delta_{ll'} \delta_{mm'}. \tag{11}$$

We now find the numbers $c_l^{\tau\tau'}$. We make use of the relation

$$[[F_3, L_0], F_3] \doteq L_0.$$

Applying the operators of both sides of this equation to the vector ξ_{lm}^{τ}, we obtain:

$$[[F_3, L_0], F_3]\, \xi_{lm}^{\tau} = L_0 \xi_{lm}^{\tau}.$$

Expanding this equation with the aid of relations (10) and (11) and the expression (13), Sect. 2, for F_3, we arrive at the following system of equations:

$$\left.\begin{aligned}
&C_l^{\tau'} C_{l+1}^{\tau'} c_{l+1}^{\tau\tau'} - 2C_l^{\tau'} C_{l+1}^{\tau} c_l^{\tau\tau'} + C_l^{\tau} C_{l+1}^{\tau} c_{l-1}^{\tau\tau'} = 0,\\
&C_l^{\tau} C_{l+1}^{\tau} c_{l+1}^{\tau\tau'} - 2C_l^{\tau} C_{l+1}^{\tau'} c_l^{\tau\tau'} + C_l^{\tau'} C_{l+1}^{\tau'} c_{l-1}^{\tau\tau'} = 0,\\
&C_l^{\tau'} [A_{l-1}^{\tau'} + A_l^{\tau'} - 2A_l^{\tau}]\, c_l^{\tau\tau'} = C_l^{\tau} [2A_{l-1}^{\tau'} - A_{l-1}^{\tau} - A_l^{\tau}]\, c_{l-1}^{\tau\tau'},\\
&C_l^{\tau'} [A_{l-1}^{\tau} + A_l^{\tau} - 2A_l^{\tau'}]\, c_l^{\tau\tau'} = C_l^{\tau'} [2A_{l-1}^{\tau} - A_{l-1}^{\tau'} - A_l^{\tau'}]\, c_{l-1}^{\tau\tau'},\\
&2C_{l+1}^{\tau'} C_{l+1}^{\tau} c_{l+1}^{\tau\tau'} - \{(C_{l+1}^{\tau'})^2 + (C_{l+1}^{\tau})^2 + (C_l^{\tau'})^2 + (C_l^{\tau})^2 +\\
&\qquad + (A_l^{\tau'} - A_l^{\tau})^2\}\, c_l^{\tau\tau'} + 2C_l^{\tau'} C_l^{\tau} c_{l-1}^{\tau\tau'} = 0,\\
&2(l+1)^2 C_{l+1}^{\tau'} C_{l+1}^{\tau} c_{l+1}^{\tau\tau'} - \{(l+1)^2 (C_{l+1}^{\tau'})^2 + (l+1)^2 (C_{l+1}^{\tau})^2 +\\
&\qquad + l^2 (C_l^{\tau'})^2 + l^2 (C_l^{\tau})^2\}\, c_l^{\tau\tau'} + 2l^2 C_l^{\tau'} C_l^{\tau} c_{l-1}^{\tau\tau'} = 4c_l^{\tau\tau'}.
\end{aligned}\right\} \quad (12)$$

Here A_l^{τ} and C_l^{τ} denote the quantities defined by formula (16) Sect. 2 in the irreducible representation $\tau(l_0, l_1)$. Our problem reduces to the investigation and solution of the above system of linear equations. Solving any three of equations (12) for $c_{l-1}^{\tau\tau'}$, $c_l^{\tau\tau'}$, $c_{l+1}^{\tau\tau'}$ and substituting the values obtained in the remaining equations, we convince ourselves that $c_l^{\tau\tau'}$ can differ from zero only if the components $\tau(l_0, l_1)$ and $\tau'(l_0', l_1')$ are such that

1) either

$$(l_0', l_1') = (l_0 \pm 1, l_1), \tag{13}$$

2) or

$$(l_0', l_1') = (l_0, l_1 \pm 1). \tag{14}$$

If the pairs (l_0, l_1) and (l_0', l_1) of the two components τ and τ' are connected by one or other of these two relations then we shall say that the components τ and τ' are interlocking.

In this case the numbers $c_l^{\tau\tau'}$ have the following form:

1) for $(l_0', l_1') = (l_0 + 1, l_1)$

$$\left.\begin{aligned} c_l^{\tau\tau'} &= c^{\tau\tau'}\sqrt{(l+l_0+1)(l-l_0)}, \\ c_l^{\tau'\tau} &= c^{\tau'\tau}\sqrt{(l+l_0+1)(l-l_0)}; \end{aligned}\right\} \tag{15}$$

2) for $(l_0', l_1') = (l_0, l_1+1)$

$$\left.\begin{aligned} c_l^{\tau\tau'} &= c^{\tau\tau'}\sqrt{(l+l_1+1)(l-l_1)}, \\ c_l^{\tau'\tau} &= c^{\tau'\tau}\sqrt{(l+l_1+1)(l-l_1)}, \end{aligned}\right\} \tag{16}$$

where $c^{\tau\tau'}$ and $c^{\tau'\tau}$ are arbitrary complex numbers. We stress once again that the numbers $c^{\tau\tau'}$ and $c^{\tau'\tau}$ differ from zero only for interlocking components τ and τ'.

In the remaining cases $c^{\tau\tau'} = c^{\tau'\tau} = 0$.

Thus, finally, the elements of the matrix L_0 have the form

$$c_{lm;\,l'm'}^{\tau\tau'} = c_l^{\tau\tau'}\delta_{ll'}\delta_{mm'}, \tag{17}$$

where the numbers $c_l^{\tau\tau'}$ are defined by formula (14) and (15) and are different from zero only in the case when τ and τ' are interlocking.

Using formulae (6) we obtain the following expressions for the matrices L_1, L_2, L_3.

(a) For L_3. We denote the matrix elements of L_3 by $a_{lm;\,l'm'}^{\tau\tau'}$: $L_3 = \| a_{lm;\,l'm'}^{\tau\tau'} \|$. Then

$$\left.\begin{aligned} a_{lm;\,l-1,\,m}^{\tau\tau'} &= i\sqrt{l^2-m^2}\,[C_l^{\tau}c_{l-1}^{\tau\tau'} - C_l^{\tau'}c_l^{\tau\tau'}], \\ a_{lm;\,lm}^{\tau\tau'} &= -imc_l^{\tau\tau'}\,[A_l^{\tau} - A_l^{\tau'}], \\ a_{lm;\,l+1,\,m}^{\tau\tau'} &= -i\sqrt{(l+1)^2-m^2}\,[C_{l+1}^{\tau}c_{l+1}^{\tau\tau'} - C_{l+1}^{\tau'}c_l^{\tau\tau'}]. \end{aligned}\right\} \tag{18}$$

(b) For L_1 and L_2. We denote the elements of these matrices by $b_{lm;\,l'm'}^{\tau\tau'}$ and $d_{lm;\,l'm'}^{\tau\tau'}$ respectively: $L_1 = \| b_{lm;\,l'm'}^{\tau\tau'} \|$, $L_2 = \| d_{lm;\,l'm'}^{\tau\tau'} \|$. The following equations arise:

$$\left.\begin{aligned} b_{lm;\,l-1,\,m-1}^{\tau\tau'} &= -id_{lm;\,l-1,\,m-1} = \\ &= +\frac{i}{2}\sqrt{(l+m)(l+m-1)}\,[C_l^{\tau}c_{l-1}^{\tau\tau'} - C_l^{\tau'}c_l^{\tau\tau'}], \\ b_{lm;\,l-1;\,m+1}^{\tau\tau'} &= id_{lm;\,l-1,\,m+1} = \\ &= -\frac{i}{2}\sqrt{(l-m)(l-m-1)}\,[C_l^{\tau}c_{l-1}^{\tau\tau'} - C_l^{\tau'}c_l^{\tau\tau'}], \\ b_{lm;\,l,\,m-1}^{\tau\tau'} &= -id_{lm;\,l,\,m-1} = \\ &= \frac{i}{2}\sqrt{(l+m)(l-m+1)}\,c_l^{\tau\tau'}\,[A_l^{\tau} - A_l^{\tau'}], \end{aligned}\right.$$

$$\left.\begin{aligned}
b^{\tau\tau'}_{lm;\, l,\, m+1} &= id_{lm;\, l,\, m+1} = \\
&= \frac{i}{2}\sqrt{(l-m)(l+m+1)}\, c_l^{\tau\tau'}\left[A_l^{\tau} - A_l^{\tau'}\right], \\
b^{\tau\tau'}_{lm;\, l+1,\, m-1} &= -id_{lm;\, l+1,\, m-1} = \\
&= \frac{i}{2}\sqrt{(l-m+1)(l-m+2)}\left[C^{\tau}_{l+1}c^{\tau\tau'}_{l+1} - C^{\tau'}_{l+1}c^{\tau\tau'}_{l}\right], \\
b^{\tau\tau'}_{lm;\, l+1,\, m+1} &= id_{lm;\, l+1,\, m+1} = \\
&= -\frac{i}{2}\sqrt{(l+m+1)(l+m+2)}\left[C^{\tau}_{l+1}c^{\tau\tau'}_{l+1} - C^{\tau'}_{l+1}c^{\tau\tau'}_{l}\right].
\end{aligned}\right\} \tag{19}$$

The components τ and τ' in these formulae are assumed to be interlocking. The numbers $c_l^{\tau\tau'}$ are defined according to formulae (15) and (16) and the numbers C_l^{τ} and A_l^{τ} according to formulae (16), Sect.2. Formulae (13) to (19) contain the complete solution of the problem of seeking all the equations that are invariant under proper Lorentz transformations. We see that such equations are entirely defined by the set of numbers $c^{\tau\tau'}$ and $c^{\tau'\tau}$ corresponding to the various pairs τ and τ' of interlocking components of the representation $g \to T_g$.

§ 4. Relativistic-invariant Equations with $\varkappa = 0$

All the definitions and results of the previous section have referred to relativistic equations (2) with $\varkappa \neq 0$. It is clear that these results apply also to the case $\varkappa = 0$. It happens, however, that in the latter case the conditions under which an equation is relativistic-invariant are somewhat broader.

We consider the equation

$$L_0 \frac{\partial\psi}{\partial x_0} + L_1 \frac{\partial\psi}{\partial x_1} + L_2 \frac{\partial\psi}{\partial x_2} + L_3 \frac{\partial\psi}{\partial x_3} = 0 \tag{20}$$

(The quantity ψ is as usual transformed, according to some representation $g \to T_g$ of the proper or complete Lorentz group).

We examine the condition for equation (20) to be relativistic-invariant.

As usual, we make the substitution

$$x' = gx, \qquad \psi'(x') = T_g\psi(x). \tag{21}$$

Then

$$\frac{\partial\psi(x)}{\partial x_i} = \sum T_g^{-1}\frac{\partial\psi'(x')}{\partial x'_k}\, g_{ki};$$

consequently the function $\psi'(x')$ satisfies the equation

$$\sum_{k,\,i} L_i T_g^{-1} g_{ki} \frac{\partial \psi'(x')}{\partial x'_k} = 0. \tag{22}$$

We assume that there exists a non-degenerate transformation V_g, such that

$$\sum_{k,\,i} V_g L_i T_g^{-1} g_{ki} \frac{\partial \psi'(x')}{\partial x'_k} \equiv \sum_k L_k \frac{\partial \psi'(x')}{\partial x'_k} \quad (\text{ for all } \quad \psi'(x')),$$

i.e.

$$\sum_i V_g L_i T_g^{-1} g_{ki} = L_k. \tag{23}$$

If such a transformation V_g exists it will follow that equations (20) and (22) are equivalent, i.e. that equation (2) is essentially unaltered after the substitution (21).

Thus, equation (20) is relativistic-invariant if after the simultaneous substitutions $x' = gx$ and $\psi' = T_g\psi$ the resulting equation coincides with the initial one to within a non-degenerate transformation V_g.

We note in the case $\varkappa \neq 0$ the transformation V_g reducing the equation to the initial form must coincide with the operator T_g: $V_g = T_g$. In the case $\varkappa = 0$, the transformation V_g may be different from T_g. For the matrices L_0, L_1, L_2, L_3 in the relativistic-invariant equation of the form (20) we obtain relation (23) $\sum_i V_g L_i T_g^{-1} g_{ki} = L_k$, where T_g is the matrix of the representation and V_g is the matrix of an unspecified transformation.

It is easily verified that the correspondence $g \to V_g$ specifies a representation of the proper (or complete) Lorentz group (acting in the same space as the representation $g \to T_g$).

Thus, in order that equation (20) be relativistic-invariant it is necessary that there should act in the space of quantities ψ, as well as the representation $g \to T_g$ a representation $g \to V_g$ such that equation (23) is satisfied:

$$\sum_i g_{ki} V_g L_i T_g^{-1} = L_k.$$

This is then the condition for the relativistic-invariance of equation (20).

We now find the general form of the matrices L_0, L_1, L_2, L_3 in the relativistic-invariant equation (20).

The vector ψ from R is carried by the matrix L_k into some vector ξ,

$$\xi = L_k \psi. \tag{24}$$

Now let $\{\sigma^{\tau}_{lm}\}$ be the canonical basis of the representation $g \to V_g$ and $\{\xi^{\tau'}_{l'm'}\}$ be the canonical basis of the representation $g \to T_g$. We require that the vector ξ be referred to the basis $\{\sigma^{\tau}_{lm}\}$ and the vector ψ to the basis $\{\xi^{\tau'}_{l'm'}\}$.

In this case equation (24) may be written:

$$x^{\tau}_{lm} = \sum c^{\tau\tau'\,(k)}_{lm;\,l'm'} y^{\tau'}_{l'm'},$$

where x^{τ}_{lm} are the coordinates of ξ in the basis $\{\sigma^{\tau}_{lm}\}$ and $y^{\tau'}_{l'm'}$ are the coordinates of ψ in the basis $\{\xi^{\tau'}_{l'm'}\}$. The numbers $c^{\tau\tau'\,(k)}_{lm;\,l'm'}$ are the elements of the matrix L_k; they are more precisely the elements of the matrix of L_k with respect to the two bases $\{\sigma^{\tau}_{lm}\}$ and $\{\xi^{\tau'}_{l'm'}\}$.

With the aid of reasoning analogous to that conducted in paragraph 3 we may easily convince ourselves that the numbers $c^{\tau\tau'\,(k)}_{lm;\,l'm'}$ have the same form as for matrices L_k in equation (2) with $\varkappa \neq 0$ i.e. they are specified by formulae (13)-(19) in which the quantities with the symbol τ now refer to the representation $g \to V_g$, and the quantities with the symbol τ' refer to the representation $g \to T_g$.

In particular, for the matrix L_0

$$c^{\tau\tau'}_{lm;\,l'm'} = c^{\tau\tau'}_{l} \delta_{mm'}\delta_{ll'}, \tag{25}$$

where the numbers $c^{\tau\tau'}_{l} \neq 0$ only for interlocking components τ and τ' of the two representations $g \to V_g(\tau)$ and $g \to T_g(\tau')$ and are obtained according to formulae (15) and (16).

For the sake of simplicity we have confined our attention for the time being to the case of square matrices L_0, L_1, L_2, L_3, i.e., we have assumed that the number of equations in system (20) coincides with the number of components of the wave function ψ. In this case we are able to consider the representation $g \to V_g$ according to which system (20) is transformed, as acting in the very space R in which the values of the wave function ψ lie, and where, consequently, the representation $g \to T_g$ acts.

However systems of equation of the form (20) are encountered in which the number of equations does not coincide with the number of components of ψ, i.e. the matrices L_0, L_1, L_2, L_3 are no longer square but are arbitrary rectangular matrices. In such a case the representation $g \to V_g$, which transforms the system must be considered as acting in some space $\tilde{R}$ distinct from the space R where the representation $g \to T_g$ acts. In order that such a system of equations be relativistic-invariant, the matrices L_0, L_1, L_2, L_3 must satisfy the condition

$$\sum V_g L_k g_{ik} T_g^{-1} = L_i.$$

We now find the general form of the matrices L_0, L_1, L_2, L_3 in this case. Clearly we must consider that a vector ψ of the space R is carried by the matrix L_k into a vector ξ of the space $\tilde{R}$,

$$\xi = L_k \psi. \tag{26}$$

If, as above, we denote the canonical basis of the representation $g \to V_g$ in space $\tilde{R}$ by $\{\sigma^{\tau}_{lm}\}$, the canonical basis of the representation $g \to T_g$ in space R by $\{\xi^{\tau'}_{l'm'}\}$ and the coordinates of the vectors ξ and ψ relative to the bases $\{\sigma^{\tau}_{lm}\}$ and $\{\xi^{\tau'}_{l'm'}\}$ by x^{τ}_{lm} and $y^{\tau'}_{l'm'}$ then equation (26) becomes

$$x^{\tau}_{lm} = \sum c^{\tau\tau'\,(k)}_{lm;\,l'm'} y^{\tau'}_{l'm'}.$$

The numbers $c^{\tau\tau'\,(k)}_{lm;\,l'm'}$ are the matrix elements of the matrices L_k.

The form of these numbers, as above, is given by formulae (13)-(19) where the symbol τ refers to the representation $g \to V_g$, and the symbol τ' to the representation $g \to T_g$, the components τ and τ' being interlocking.

Below we shall examine in detail two examples of relativistic-invariant equations with $\varkappa = 0$: the equation for the so-called two-component neutrino and Maxwell's equations for an electromagnetic field in vacuum.

§ 5. Equations Invariant with Respect to the Complete Lorentz Group

As we have seen (paragraph 2, equation (9)) we have such equations as $SL_0 = L_0 S$ (where S is the operator corresponding to the spatial reflection s). We examine the conditions which this relation imposes upon the matrix L_0 (i.e. upon the numbers $c^{\tau\tau'}$).

We recall that an irreducible representation of the complete Lorentz group contains either one or two irreducible conjugate representations τ and $\dot{\tau}$ of the proper Lorentz group. The pairs defining them

$$\tau \sim (l_0, l_1) \text{ and } \dot{\tau} \sim (\dot{l}_0, \dot{l}_1)$$

are connected by the relation

$$(\dot{l}_0, \dot{l}_1) = \pm (l_0 - l_1).$$

Whence, in the first case, when $\tau = \dot{\tau}$, either $l_1 = 0$, or $l_0 = 0$.

As before, let the space R, where the representation $g \to T_g$ of the complete group acts, be decomposed into a sum of subspaces R^{τ} in which an irreducible representation of the proper group is induced. It is clear that, together with each subspace R^{τ}, the space R also contains the subspace $R^{\dot{\tau}}$. We note that if the components τ and τ' "interlock", then $\dot{\tau}$ and $\dot{\tau}'$ also "interlock".

I. We consider first of all the case when $\tau \neq \dot{\tau}$ and $\tau' \neq \dot{\tau}'$. The operator S is in this case given by (see formulae (7), Sect. 3):

$$\left.\begin{aligned} S\xi^{\tau}_{lm} &= (-1)^{[l]}\xi^{\dot{\tau}}_{lm}, \\ S\xi^{\tau'}_{lm} &= (-1)^{[l]}\xi^{\dot{\tau}'}_{lm}. \end{aligned}\right\} \quad (27)$$

Substituting this expression in the relation

$$L_0 S\xi^{\tau}_{lm} = SL_0\xi^{\tau}_{lm}, \quad (27')$$

we obtain $c_l^{\dot{\tau}\dot{\tau}'} = c_l^{\tau\tau'}$ or using (15) and (16),

$$c^{\tau\tau'} = c^{\dot{\tau}\dot{\tau}'}.$$

II. Let $\tau = \dot{\tau}$ and $\tau' \neq \dot{\tau}'$ (or vice versa $\tau \neq \dot{\tau}$, and $\tau' = \dot{\tau}'$). In this case the operator S has the following form (see (5) and (6), Sect. 3)

$$S\xi^{\tau}_{lm} = (-1)^{[l]}\xi^{\tau}_{lm} \quad (28)$$

or

$$S\xi^{\tau}_{lm} = (-1)^{[l]+1}\xi^{\tau}_{lm} \quad (28')$$

and

$$S\xi^{\tau'}_{lm} = (-1)^{[l]}\xi^{\dot{\tau}'}_{lm}. \quad (28'')$$

Again substituting in the expression (27) for S and for L_0 we obtain:

$c^{\tau\tau'} = c^{\dot{\tau}\dot{\tau}'}$ (if the operator S has the form (28)),

or

$c^{\tau\tau'} = -c^{\dot{\tau}\dot{\tau}'}$ (if the operator S is specified by (28')).

III. Finally if $\tau = \dot{\tau}$ and $\tau' = \dot{\tau}'$ then it is easily seen that $c_l^{\tau\tau'} \neq 0$ only when the operator S acts in the same way in spaces R^{τ} and $R^{\tau'}$, i.e. if either (see section 3 (5) and (6))

$$S\xi^{\tau}_{lm} = (-1)^{[l]}\xi^{\tau}_{lm} \text{ and } S\xi^{\tau'}_{lm} = (-1)^{[l]}\xi^{\tau'}_{lm}, \quad (29)$$

or

$$S\xi^{\tau}_{lm} = (-1)^{[l]+1}\xi^{\tau}_{lm} \text{ and } S\xi^{\tau'}_{lm} = (-1)^{[l]+1}\xi^{\tau'}_{lm} \quad (29')$$

No other restrictions are imposed upon the numbers $c^{\tau\tau'}$ in this case.*

* We note that in the case of the presence of such interlocking elements it is often possible to define the operator S in several different ways without violating the invariance of the equation or altering the matrices L_k. Since the invariant equation is defined not only by the matrices L_k, but also by

Thus, for the matrix L_0 of an equation invariant under the complete group we have the conditions

1. $c^{\tau\tau'} = c^{\dot{\tau}\dot{\tau}'}$ $\quad (\tau \neq \dot{\tau}, \quad \tau' \neq \dot{\tau}').$ (30)

2. $c^{\tau\tau'} = \pm c^{\dot{\tau}\dot{\tau}'}$ $\quad (\tau \neq \dot{\tau}, \quad \tau' = \dot{\tau}'$ or $\tau = \dot{\tau}, \quad \tau' \neq \dot{\tau}').$ (31)

Here the plus sign applies when the operator S acts according to formulae (28) and (28") and the minus sign applies when S acts according to formulae (28') and (28").

3. For $\tau = \dot{\tau}$ and $\tau' = \dot{\tau}'$ we have

$$c^{\tau\tau'} \neq 0 \tag{32}$$

only when S acts in the same manner in R^{τ} and $R^{\tau'}$, i.e. either according to formula (29) or according to formulae (29'). In this case no other restrictions are imposed upon the numbers $c^{\tau\tau'}$.

In the case of equation (20) with $\varkappa = 0$, invariant under the complete Lorentz group, the conditions imposed upon the numbers $c^{\tau\tau'}$ remain the same as for $\varkappa \neq 0$ (see (30)-(32)). We merely recall that the components τ and τ' belong to different representations: τ' belongs to the representation $g \to T_g$ according to which the wave functions are transformed and τ to the representation $g \to V_g$ with the aid of which (20) itself is transformed.

§ 6. An Observation Concerning the Operators T_g. The Case of the General Lorentz Group

At the very beginning of this section we assumed that upon passing from one orthogonal coordinate system to another with the aid of a Lorentz transformation g, the wave functions $\psi(x)$ of a particle are transformed according to some representation $g \to T_g$ of the Lorentz group.

We note however, that inasmuch as the wave functions $\psi(x)$ themselves are defined to within a multiplier, the transformations T_g to which they are subjected must also be defined only to within a multiplier. In other words, to each g there corresponds a family of operators λT_g, differing by a (scalar) factor.

In the case of an irreducible representation it can be shown that for each transformation g it is possible to select a factor in front of the operator T_g such that to each transformation there will correspond either one operator T_g, or two such operators which differ in sign: T_g and $-T_g$, so that the following conditions are satisfied:

* (Cont'd). the transformation law for the quantities ψ (i.e. by the representation $g \to T_g$) we will have in these cases several essentially distinct equations. Scalar and pseudoscalar, vector and pseudovector equations (see Sec. 9) may serve as examples of such equations differing only in the form of the transformation S.

$$
\begin{aligned}
&1) \quad e \to \pm E, \\
&2) \quad T_{g_1} T_{g_2} = \pm T_{g_1 g_2}.
\end{aligned}
\tag{33}
$$

In the case of the proper (or complete) Lorentz group we arrive, in this manner, at a unique representation of this group when each g corresponds only to one operator T_g, and at a two-valued representation when each transformation g corresponds to two operators $g \to \pm T_g$, differing in sign, it being impossible to eliminate this indeterminacy in sign.

We now turn to the case of the general Lorentz group. Here, as we know, two cases may present themselves :-

1. The operators S, T, J, corresponding to the elements of the group of reflections s, t, j, commute. In this case the indeterminacy in sign for the operators S, T, J arises only because of the two-valued representation of the proper group. We called such representations unique representations of the general group.

2. The operators S, T, J anticommute. In this case the representations

$$
e \to \pm E, \qquad s \to \pm S, \qquad t \to \pm T, \qquad j \to \pm J
$$

themselves are already essentially two-valued and it is impossible to eliminate this lack of uniqueness (even if the representation of the proper group is also unique). We called such representations of the general group two-valued representations.

Thus, in the case of the proper or complete Lorentz group the operators according to which the wave function $\psi(x)$ are transformed specify a unique or two-valued representation of the proper (complete) group. Also in the case of the general group, the operators T_g constitute either a unique representation (S, T, J commute) or a two-valued representation S, T, J anticommute).

We shall see below that, for example, the wave functions satisfying the Dirac equations are transformed expressly according to a two-valued representation of the general group.

We note, in conclusion, that no equations invariant under the general Lorentz group (apart from the example of the Dirac equations) are considered in the remainder of this book.

Section 8. Equations Arising from Invariant Lagrangian Functions

Of special interest for physical applications of the above theory are those relativistic-invariant equations which may be obtained by the variation of some invariant Lagrangian functions. With such equations it is possible to connect invariantly a series of physical quantities, for instance, charge, energy, momentum, moment of inertia.

§ 1. The Invariant Lagrangian Function

Let a particle be described by a wave function $\psi(x)$, which transforms under a coordinate transformation from the Lorentz group according to the representation $g \to T_g$:

$$\tilde{\psi}(x') = T_g \psi(x), \qquad x' = gx.$$

Moreover, let there be specified in the space R, wherein the values of the wave function $\psi(x)$ lie, some non-degenerate, bilinear, Hermitian form (ψ_1, ψ_2).

With the aid of such a bilinear Hermitian form it is possible to construct an expression known as the Lagrangian function and a bilinear functional known as the action.

The expression

$$\mathscr{L}[\psi(x)] = \frac{1}{2i} \sum_{k=0,1,2,3} \left\{ \left(\Lambda_k \frac{\partial \psi}{\partial x_k}, \psi \right) - \left(\psi, \Lambda_k \frac{\partial \psi}{\partial x_k} \right) \right\} + \varkappa(\psi, \psi) =$$

$$= \operatorname{Im} \sum_{k=0,1,2,3} \left(\Lambda_k \frac{\partial \psi}{\partial x_k}, \psi \right) + \varkappa(\psi, \psi), \tag{1}$$

where $\Lambda_0, \Lambda_1, \Lambda_2, \Lambda_3$ are given matrices, is known as the Lagrangian function. The bilinear functional

$$S(\psi) = \int_{-\infty}^{\infty} \int_{-\infty}^{\infty} \int_{-\infty}^{\infty} \int_{-\infty}^{\infty} \mathscr{L}[\psi(x)]\, dx_0 dx_1 dx_2 dx_3. \tag{2}$$

is known as the action.

We now give a definition of an invariant Langrangian function and of the invariant action.

The Lagrangian function is called invariant if it is unaltered by any Lorentz transformation. In other words, if it remains unchanged by the simultaneous substitutions $x' = gx$ and $\tilde{\psi}(x') = T_g \psi(x)$

$$\mathscr{L}[\tilde{\psi}(x')] = \mathscr{L}[\psi(x)]. \tag{3}$$

Similarly the action $S(\psi)$ is called invariant if

$$S(\tilde{\psi}) = S(\psi). \tag{3'}$$

It is clear that to each invariant Lagrangian function there corresponds an invariant action and that every invariant action is obtained from some invariant Lagrangian function.

We shall find necessary and sufficient conditions in order that a Lagrangian function be invariant. Here, two essentially different cases arise: when $\varkappa \neq 0$ and when $\varkappa = 0$. We shall consider them separately.

First Case: $\varkappa \neq 0$. We shall show that in this case the Lagrangian function is invariant if, and only if, firstly, the Hermitian form (ψ_1, ψ_2) is invariant and, secondly, the matrices Λ_k satisfy the conditions

$$\sum g_{ik} T_g \Lambda_k T_g^{-1} = \Lambda_i, \tag{4}$$

i.e. if it is possible to construct an invariant differential equation by means of them (see Sect. 7, (3)).

Suppose the Lagrangian function is invariant. We take for $\psi(x)$ the constant function $\psi(x) \equiv \psi^0$.

Since $\frac{\partial \psi}{\partial x_k} = 0$, then $\mathscr{L}[\psi(x)] = \varkappa(\psi^0, \psi^0)$. The invariance condition (3) clearly reduces in this case to the equation

$$(\psi^0, \psi^0) = (T_g \psi^0, T_g \psi^0),$$

which means that the form (ψ_1, ψ_2) is invariant. *

We now prove that the matrices Λ_k in an invariant Langrangian function $\mathscr{L}[\psi(x)]$ satisfy condition (4). To this end we write condition (3) at greater length:

$$\psi(x) = T_g^{-1} \tilde{\psi}(x'), \qquad x' = gx,$$

$$\mathscr{L}[\psi(x)] = \operatorname{Im} \sum \left(\psi, \Lambda_k \frac{\partial \psi}{\partial x_k}\right) + \varkappa(\psi, \psi) =$$

$$= \operatorname{Im} \sum_k \left(T_g^{-1} \tilde{\psi}(x'), \Lambda_k T_g^{-1} \sum_i \frac{\partial \tilde{\psi}}{\partial x'_i} \frac{\partial x'_i}{\partial x_k}\right) + \varkappa(T_g \tilde{\psi}, T_g \tilde{\psi}) =$$

$$= \operatorname{Im} \sum_k \left(\tilde{\psi}(x') T_g \Lambda_k T_g^{-1} \sum_k g_{ik} \frac{\partial \tilde{\psi}}{\partial x'_i}\right) + \varkappa(\tilde{\psi}, \tilde{\psi}) \tag{5}$$

(in the last equation the invariance of the form (ψ_1, ψ_2) was used).

On the other hand

$$\mathscr{L}(\tilde{\psi}(x')) = \operatorname{Im} \sum_i \left(\tilde{\psi}(x'), \Lambda_i \frac{\partial \psi}{\partial x'_i}\right) + \varkappa(\tilde{\psi}, \tilde{\psi}). \tag{5'}$$

In virtue of the invariance of the Lagrangian function $\mathscr{L}[\psi(x)] = \mathscr{L}[\tilde{\psi}(x')]$.

---

* It can be proved that the bilinear form (ψ_1, ψ_2) is invariant provided that the corresponding quadratic form (ψ, ψ) is invariant.

Hence we obtain finally:

$$\operatorname{Im}\sum_i\left(\tilde{\psi}(x'),\ \Lambda_i\frac{\partial\tilde{\psi}}{\partial x_i'}\right)=\operatorname{Im}\sum_i\left(\tilde{\psi}(x'),\ \sum_k g_{ik}T_g\Lambda_k T_g^{-1}\frac{\partial\tilde{\psi}}{\partial x_i'}\right),$$

or

$$\operatorname{Im}\sum_i\left(\tilde{\psi}(x'),\ \left(\Lambda_i-\sum g_{ik}T_g\Lambda_k T_g^{-1}\right)\frac{\partial\tilde{\psi}}{\partial x_i'}\right)=0, \tag{5"}$$

Hence *

$$\sum g_{ik}T_g\Lambda_k T_g^{-1}=\Lambda_i,$$

which coincides with (4).

Thus we have proved that if a Lagrangian function with $\varkappa\neq 0$ is invariant then the form (ψ_1, ψ_2) defining it is invariant and the matrices Λ_k satisfy condition (4).

Conversely, a Lagrangian function constructed with the aid of an invariant form (ψ_1, ψ_2) and matrices Λ_k satisfying condition (4) is invariant. This assertion is easily verified with the aid of equations (5), (5') the argument being simply reversed.

Second Case: $\varkappa=0$. In this case the Hermitian non-degenerate form (ψ_1, ψ_2) may be quite arbitrary. Here for the invariance of the Lagrangian function it is necessary and sufficient that the matrices Λ_k satisfy the conditions

$$\sum g_{ik}(T_g^+)^{-1}\Lambda_k T_g^{-1}=\Lambda_i, \tag{6}$$

where the operators T_g^+ are defined by

$$(T_g\psi_1, \psi_2)=(\psi_1, T_g^+\psi_2) \quad \text{for any } \psi_1, \psi_2. \tag{7}$$

Before coming to the proof of this assertion, we note that the operators $V_g=(T_g^+)^{-1}$ constitute a representation $g\to V_g$ of the Lorentz group: $V_e=E$ and $V_{g_1}V_{g_2}=V_{g_1g_2}$. Substituting in formula (6) the operator V_g in place of $(T_g^+)^{-1}$ we obtain the relation

$$\sum g_{ik}V_g\Lambda_k T_g^{-1}=\Lambda_i. \tag{6'}$$

This relation coincides with the necessary and sufficient condition derived in Sect.7 for the equation $\sum\limits_{k=0,1,2,3}\Lambda_k\frac{\partial\psi}{\partial x_k}=0$ be invariant under the Lorentz

---

* It may be shown that if (ψ_1, ψ_2) is a non-degenerate form (i.e. if $(\psi, \psi_0)=0$ for all ψ implies that $\psi_0=0$), then $\operatorname{Im}(\psi_1, \psi_2)$ possesses the same property $\psi_0=0$. if $\operatorname{Im}(\psi, \psi_0)=0$ for all ψ.

group (see Sect. 7, paragraph 4).

Thus as in the previous case we see that from the matrices in an invariant Lagrangian function with $\varkappa = 0$ it is possible to construct a relativistic invariant equation with $\varkappa = 0$.

We pass on to the derivation of condition (6). To this end we write the left-hand side of the equation at length

$$\mathscr{L}[\psi(x)] = \operatorname{Im}\sum_k\left(\psi,\ \Lambda_k\frac{\partial\psi}{\partial x_k}\right) = \operatorname{Im}\sum_k\left(T_g^{-1}\tilde{\psi}(x'),\ \Lambda_k T_g^{-1} g_{ik}\frac{\partial\tilde{\psi}}{\partial x'_k}\right) =$$
$$= \operatorname{Im}\sum_k\left(\tilde{\psi}(x')\,(T_g^+)^{-1}\Lambda_k T_g g_{ik}\frac{\partial\tilde{\psi}}{\partial x'_k}\right). \quad (8)$$

Since

$$\mathscr{L}[\tilde{\psi}(x')] = \operatorname{Im}\sum\left(\tilde{\psi}(x'),\ \Lambda_i\frac{\partial\tilde{\psi}}{\partial x'_i}\right)$$

and

$$\mathscr{L}[\tilde{\psi}(x')] = \mathscr{L}[\psi(x)],$$

we obtain:

$$\operatorname{Im}\sum_k\left[\tilde{\psi}(x'),\ (T_g^+)^{-1}\Lambda_k T_g \sum_i^{-1} g_{ik}\frac{\partial\psi}{\partial x'_i}\right] = \operatorname{Im}\sum_i\left(\tilde{\psi}(x')\,\Lambda_i\frac{\partial\tilde{\psi}}{\partial x'_i}\right). \quad (8')$$

Hence

$$\sum (T_g^+)^{-1}\Lambda_k T_g^{-1} g_{ik} = \Lambda_i,$$

i.e. the matrices Λ_i in the invariant Lagrangian function (1) with $\varkappa = 0$ do, in fact, satisfy condition (6).

By writing the equations (8)-(8') in the opposite order we satisfy ourselves that, conversely, from condition (6) the invariance of the Lagrangian function with $\varkappa = 0$ follows.

§ 2. Equations Arising from Invariant Lagrangian Functions

Following a formalism adopted in physics, we obtain the equations describing our particle (the equations of motion) from the condition that the variation of the action S be equal to zero.

We denote by Λ^+ the matrix possessing the property that for all ψ_1 and ψ_2

$$(\Lambda\psi_1,\ \psi_2) = (\psi_1,\ \Lambda^+\psi_2). \quad (9)$$

It is clear that the matrices Λ_k^+ $(k=0, 1, 2, 3)$ satisfy, together with the matrices Λ_k the conditions (4)*.

We have

$$\delta S = \int \delta \mathcal{L} [\psi(x)]\, d^4x,$$

$$\delta \mathcal{L} = \text{Im} \left\{ \sum_k \left(\Lambda_k \frac{\partial \psi}{\partial x_k}, \delta\psi\right) + \left(\Lambda_k \frac{\partial \delta\psi}{\partial x_k}, \psi\right) \right\} + \varkappa(\psi, \delta\psi) + \varkappa(\delta\psi, \psi) =$$

$$= \text{Im} \left\{ \sum_k \left(\Lambda_k \frac{\partial \psi}{\partial x_k}, \delta\psi\right) + \frac{\partial}{\partial x_k}(\Lambda_k \delta\psi, \psi) - \left(\Lambda_k \delta\psi, \frac{\partial \psi}{\partial x_k}\right) \right\} + 2\varkappa(\psi, \delta\psi) =$$

$$= \text{Im} \left\{ \sum_k \left(\Lambda_k \frac{\partial \psi}{\partial x_k}, \delta\psi\right) - \sum \overline{\left(\Lambda_k^+ \frac{\partial \psi}{\partial x_k}, \delta\psi\right)} \right\} + 2\varkappa(\psi, \delta\psi) +$$

$$+ \sum \frac{\partial}{\partial x_k} \text{Im}(\Lambda_k \delta\psi, \psi) =$$

$$= 2\,\text{Im} \left\{ \left(\sum_k \left(\frac{\Lambda_k + \Lambda_k^+}{2} \right) \frac{\partial \psi}{\partial x_k} + i\varkappa\psi \right), \delta\psi \right\} + \sum \frac{\partial}{\partial x_k} \text{Im}(\Lambda_k \delta\psi, \psi).$$

The last term has zero integral. Thus from the condition $\delta S = = \int \delta \mathcal{L}\, d^4x = 0$ we obtain:

$$\text{Im} \left\{ \left(\sum_k \left(\frac{\Lambda_k + \Lambda_k^+}{2} \right) \frac{\partial \psi}{\partial x_k} + i\varkappa\psi \right), \delta\psi \right\} = 0.$$

Hence

$$\sum_k \frac{\Lambda_k + \Lambda_k^+}{2} \frac{\partial \psi}{\partial x_k} + i\varkappa\psi = 0\,.^{**} \tag{10}$$

This is then the equation describing the behaviour of the particle. It is the Langrange-Euler equation for the functional $S(\psi) = \int \mathcal{L}(\psi)\, d^4x$.

If we take

---

* The fact that the matrices Λ_k^+ actually satisfy condition (4) for the matrices of a relativistic-invariant equation is easily shown thus: We write down condition (4) for the matrices Λ_k

$$\sum_k T_g \Lambda_k T_g^{-1} g_{ik} = \Lambda_i.$$

Hence taking the $+$ of both sides we obtain

$$\sum_k (T_g^{-1})^+ \Lambda_k^+ (T_g)^+ g_{ik} = \Lambda_i^+.$$

But in virtue of the invariance of the form $(T_g^{-1})^+ = T_g$ and so we obtain

$$\sum_k T_g \Lambda_k^+ T_g^{-1} g_{ik} = \Lambda_i^+, \quad \text{Q.E.D.}$$

** See footnote above.

$$L_k = \frac{\Lambda_k + \Lambda_k^+}{2}, \tag{11}$$

then equation (10) may be rewritten in the form

$$\sum_k L_k \frac{\partial \psi}{\partial x_k} + i\varkappa\psi = 0. \tag{12}$$

Since the matrices L_k satisfy the condition (4) simultaneously with Λ_k and Λ_k^+ this equation is relativistic-invariant.

It is clear that $L_k^+ = L_k$ or, what amounts to the same thing, $(L_k\psi_1, \psi_2) = (\psi_1, L_k\psi_2)$ for any ψ_1 and ψ_2.

We note that the last equation implies that the quadratic form $(L_k\psi, \psi)$ is real.

In this manner we see that not every relativistic-invariant equation

$$\sum L_k \frac{\partial \psi}{\partial x_k} + i\varkappa\psi = 0 \qquad (\varkappa \neq 0) \tag{13}$$

may be obtained from an invarient real Langrangian function; for this to be the case it is necessary that, firstly, in the space of values of the function $\psi(x)$ there should exist a non-degenerate invariant bilinear form (ψ_1, ψ_2) and secondly, that for the matrices L_k the equation

$$(L_k\psi_1, \psi_2) = (\psi_1, L_k\psi_2). \tag{14}$$

be satisfied.

Similarly it is easy to show that if a relativistic-invariant equation with $\varkappa = 0$

$$\sum L_k \frac{\partial \psi}{\partial x_k} = 0 \tag{15}$$

arises from an invariant Langrangian function, then there is a non-degenerate Hermitian form (ψ_1, ψ_2) in the space of values of the function ψ such that the matrices L_k satisfy the condition

$$(L_k\psi_1, \psi_2) = (\psi_1, L_k\psi_2). \tag{16}$$

Invariance of the Hermitian form is not required in this case.

We note that in both cases the invariant Lagrangian function itself may be constructed with the aid of the form (ψ_1, ψ_2) and the matrices L_k. In which case , the variation of the action $S = \int \mathscr{L}\, d^4x$ again leads to equation (13) for $\varkappa \neq 0$ or (15) for $\varkappa = 0$.

We note that conditions (14) and (16) will be satisfied for all matrices L_k, $k=1, 2, 3$, as soon as they are satisfied for the matrix L_0:

$$(L_0\psi_1, \psi_2)=(\psi_1, L_0\psi_2). \tag{17}$$

In fact, the matrices L_k are obtained from L_0 according to formula (6')Sect.7.

$$L_k=-[B_kL_0]=L_0B_k-B_kL_0 \qquad (k=1, 2, 3),$$

where B_k is an infinitesimal operator of the representation $g\to T_g$, corresponding to the hyperbolic screw $g_{0k}(\varphi)$ in the plane (x_0, x_k). For the operators $T_{g_{0k}(\varphi)}$ for small φ we have:

$$T_{g_{0k}(\varphi)}=E+\varphi B_k+o(\varphi).$$

Substituting the latter expression in the formula

$$(T_g\psi_1, T_g\psi_2)=(\psi_1, \psi_2)$$

and collecting first-order terms in φ, we obtain:

$$(B_k\psi_1, \psi_2)=-(\psi_1, B_k\psi_2) \quad \text{or} \quad B_k=-B_k^+. \tag{18}$$

Hence

$$L_k^+=(L_0B_k-B_kL_0)^+=B_k^+L_0^+-L_0^+B_k^+=-B_kL_0+L_0B_k=L_k,$$

i.e.

$$(L_k\psi_1, \psi_2)=(\psi_1, L_k\psi_2).$$

Summarizing the above we arrive at the final conclusion: The relativistic-invariant equation

$$L_0\frac{\partial\psi}{\partial x_0}+L_1\frac{\partial\psi}{\partial x_1}+L_2\frac{\partial\psi}{\partial x_2}+L_3\frac{\partial\psi}{\partial x_3}+i\varkappa\psi=0, \qquad \varkappa\neq 0, \tag{19}$$

may be obtained from an invariant Lagrangian function if, and only if, there exists an invariant non-degenerate Hermitian bilinear form (ψ_1, ψ_2) on the space of values of wave functions ψ, for which the relation

$$(L_0\psi_1, \psi_2)=(\psi_1, L_0\psi_2) \tag{20}$$

is satisfied for any ψ_1, ψ_2.

The relativistic-invariant equation

$$L_0\frac{\partial\psi}{\partial x_0}+L_1\frac{\partial\psi}{\partial x_1}+L_2\frac{\partial\psi}{\partial x_2}+L_3\frac{\partial\psi}{\partial x_3}=0 \tag{21}$$

may be obtained from an invariant Langrangian function if, and only if, there exists a non-degenerate Hermitian bilinear form (ψ_1, ψ_2), relative to which the matrix L_0 satisfies the condition

$$(L_0\psi_1, \psi_2) = (\psi_1, L_0\psi_2). \tag{22}$$

(Invariance of the form is not required in this case).

In both cases the Lagrangian function itself is constructed from the Hermitian form (ψ_1, ψ_2) referred to, and the matrices L_k entering into equations (19) and (22) according to the formula

$$\mathscr{L}[\psi(x)] = \operatorname{Im} \sum \left(\psi, L_k \frac{\partial \psi}{\partial x_k}\right) + \varkappa (\psi, \psi). \tag{23}$$

In concluding this paragraph we make the following observation.

We have constructed an invariant Lagrangian function $\mathscr{L}$ and with its aid the action S, and by the variation of the action S we have obtained equation (19) itself. The question arises whether it is possible to change the Lagrangian function $\mathscr{L}$ without affecting equation (19).

The following are the most important transformations of this sort.

1) Multiplication by a real constant $\mathscr{L} \to c\mathscr{L}$ (c is real),

2) the addition of an expression of the type $\operatorname{div} E(x)$, where $E(x)$ is some vector field specified in the entire space $R^{(4)}$

$$\mathscr{L} \to \mathscr{L} + \operatorname{div} E.$$

The fact that transformations 1) and 2) of the Lagrangian functions do not alter the equation obtained, is often briefly formulated thus:- it is said that the Lagrangian function is defined to within a factor (multiplier) and a divergence type term.

§ 3. Equations Arising from Invariant Lagrangian Functions (Conclusion)

In this paragraph we find the conditions imposed upon the elements of the matrix L_0 in an equation (19) $(\varkappa \neq 0)$, arising from an invariant Lagrangian function. As was shown in the previous paragraph, in order that a relativistic-invariant equation be obtainable by the variation of an invariant Lagrangian function, it is necessary and sufficient that the equation

$$(L_0\psi_1, \psi_2) = (\psi_1, L_0\psi_2). \tag{24}$$

be satisfied for some invariant form (ψ_1, ψ_2).

The general form of an invariant non-degenerate bilinear form was found in Sects.2 and 3. We proved that any form, invariant under a representation is given with respect to the canonical basis ξ^{τ}_{lm} by the equation

$$(\psi_1, \psi_2) = \sum a^{\tau\tau^*} s_l x^{\tau}_{lm} \overline{y}^{\tau^*}_{lm}, \tag{25}$$

where x^{τ}_{lm} and $y^{\tau^*}_{lm}$ are the coordinates of ψ_1 and ψ_2 in the canonical basis, $a^{\tau\tau^*}$ is different from zero only for components defined by the pairs $\tau \sim (l_0, l_1)$ and $\tau^* \sim (l_0, -l_1)$, and $a^{\tau\tau^*} = \bar{a}^{\tau^*\tau}$, $s_l = \pm 1$.

If the form (ψ_1, ψ_2) is also invariant under a representation of the complete group then the numbers $a^{\tau\tau^*}$ are subjected to the further condition:

$$a^{\tau\tau^*} = \pm a^{\dot{\tau}\dot{\tau}^*}, \tag{25'}$$

where $\dot{\tau}$ and $\dot{\tau}^*$ are the representations conjugate to the representations τ and τ^* respectively. In the case when the components τ and τ^* are equivalent to their conjugates and are consequently irreducible under the representation of the complete group, the operator S must act in the same way in these components: in the opposite case $a^{\tau\tau'} = 0$.

We proved in Sect.3 that with a suitable choice of the basis $\{\xi^{\tau}_{lm}\}$ the numbers $a^{\tau\tau^*}$ in a form invariant under a representation of the complete group, are given by:

1. $a^{\tau\tau^*} = 1$ for l_1 not real and not a pure imaginary (25'')
2. $a^{\tau\tau^*} = \pm 1$ for l_1 real or pure imaginary (25''')

After this recapitulation we return to the matrix L_0. We rewrite equation (24) for the basic vectors ξ^{τ}_{lm}

$$(L_0\xi^{\tau}_{lm}, \xi^{\tau'}_{l'm'}) = (\xi^{\tau}_{lm}, L_0\xi^{\tau'}_{l'm'}). \tag{26}$$

Substituting the expression for L_0 from (15)-(17), Sect.7 in formula (26) using (25) we find

$$c^{\tau\tau'}a^{\tau'\tau'^*} = \bar{c}^{\tau'^*\tau^*}a^{\tau\tau^*}. \tag{27}$$

If the equation with the matrix L_0 and the form (ψ, ψ) is invariant under a representation of the complete Lorentz group then, it follows from formula (25) and formulae (30) and (31) of Sect.7, that no new conditions supplementary to condition (27) arise.

Thus (27) is the final condition which the matrix L_0 (i.e. the numbers $c^{\tau\tau'}$) must satisfy in order that equation (19) be obtainable from an invariant Lagrangian function.

In particular, with a suitable choice of basis (such that $a^{\tau\tau^*}$ have the form (25'') or (25''') we find

$$c^{\tau\tau'} = \bar{c}^{\tau'^*\tau^*} \tag{27'}$$

for l_1 neither real nor pure imaginary and

$$c^{\tau\tau'} = \pm \bar{c}^{\tau'^*\tau^*} \tag{27''}$$

for l_1, real or pure imaginary.

§ 4. Quantities Formed from the Wave Function ψ and the Invariant Form*

Here we show how to use the invariant form (ψ_1, ψ_2) on the space of values of the wave function ψ, the partial derivatives $\frac{\partial\psi}{\partial x_k}$ and the matrices L_0, L_1, L_2, L_3 of the relativistic-invariant equation to construct various quantities which transform in prescribed fashion under Lorentz transformations. We define more accurately what we mean here by the word quantity. We recall that the values of the function $\psi(x)$ are transformed by a Lorentz transformation according to the representation $g \to T_g$

$$\tilde{\psi}(x') = T_g\psi(x), \qquad x' = gx,$$

where the Lorentz transformation g acts upon the argument of the function ψ, and the operator T_g upon its components.

We construct functions $\Phi(x)$, defined at each point of the four-dimensional space R^4 with values in some space $\tilde{R}$, which will be transformed in the following way by the Lorentz transformation $x' = gx$

$$\tilde{\Phi}(x') = \tau_g\Phi(x),$$

where τ_g is an operator of a representation $g \to \tau_g$, acting in a space $\tilde{R}$. We shall call the function $\Phi(x)$ a scalar, a vector, a tensor etc. according as the representation $g \to \tau_g$. is equivalent to a scalar, vector or tensor representation.

I. First of all we consider quantities not containing derivatives of the wave function ψ. We start with some examples.

a) The quantity (ψ, ψ) is a scalar in virtue of the invariance of the form.

b) Let it be possible to select an operator T, in the space of quantities ψ which commutes with all the operators T_g of a representation of the proper Lorentz group and anticommutes with the operator S corresponding to the spatial reflection s:

$$TT_g = T_gT \quad , \quad ST = -TS.$$

Then the quantity $(T\psi, \psi)$ is a pseudoscalar, i.e., it is transformed as a scalar under the action of the proper Lorentz group and changes sign under reflection.

In fact

$$(TT_g\psi, T_g\psi) = \left(T_g^{-1}TT_g\psi, \psi\right) = (\psi, \psi)$$

for proper Lorentz transformations and

---

* In the following we suppose that the form (ψ, ψ) is invariant under a representation of the complete Lorentz group.

$$(TS\psi,\ S\psi) = (S^{-1}TS\psi,\ \psi) = -(\psi,\ \psi)$$

for the spatial reflection.

c) The quantity with components $t_k = (L_k\psi,\ \psi)$ constitutes a vector i.e., is transformed in the same way as the coordinates $(x_0,\ x_1,\ x_2,\ x_3)$:

$$t'_k = \sum g_{ki} t_i. \tag{28}$$

In fact, if $x' = gx$ and $\psi' = T_g\psi$, then

$$t'_k = (L_k\psi',\ \psi') = (T_g^{-1} L_k T_g \psi,\ \psi).$$

In virtue of equation (3) Sect. 7 for the matrices $L_0,\ L_1,\ L_2,\ L_3$ occurring in a relativistic invariant equation, which may be rewritten in the form

$$T_g^{-1} L_k T_g = \sum g_{ki} L_i,$$

we obtain

$$t'_k = \left(\sum g_{ki} L_i\psi,\ \psi\right) = \sum g_{ki} (L_i\psi,\ \psi) = \sum g_{ki} t_i.$$

d) The quantity $\dot{t}_k = (TL_k\psi,\ \psi)$ is a pseudovector (the operator T commutes with the operator T_g if g belongs to the proper Lorentz group, and anticommutes with the operator S).

e) The quantity

$$t_{k_1k_2} = (L_{k_1}L_{k_2}\psi,\ \psi)$$

is transformed like a tensor of the second rank

$$t'_{k'_1k'_2} = \sum g_{k'_1k_1} g_{k'_2k_2} t_{k_1k_2}. \tag{29}$$

In fact

$$\begin{aligned} t'_{k'_1k'_2} &= \left(L_{k'_1}L_{k'_2}\psi',\ \psi'\right) = \left(T_g^{-1} L_{k'_1} L_{k'_2} T\psi,\ \psi\right) = \\ &= \left(T_g^{-1} L_{k'_1} T_g T_g^{-1} L_{k'_2} T_g\psi,\ \psi\right) = \left(\left[\sum g_{k'_1k_1} L_{k_1}\right]\left[\sum g_{k'_2k_2} L_{k_2}\right]\psi,\ \psi\right) = \\ &= \sum g_{k'_1k_1} g_{k'_2k_2} \left(L_{k_1} L_{k_2}\psi,\ \psi\right) = \sum g_{k'_1k_1} g_{k'_2k_2} t_{k_1k_2}. \end{aligned}$$

The tensor representation (29) is reducible. Therefore, the quantity $t_{k_1k_2}$ must, generally speaking, belong to some invariant subspace in the space of all tensors of the second rank.

f) A quantity with components

$$\dot{t}_{k_1k_2}=(TL_{k_1}L_{k_2}\psi,\ \psi)$$

is a pseudotensor (the operator T is the same as in examples b) and d)). The general case: The quantity $t_{k_1k_2k_3\ldots k_n}=(L_{k_1}L_{k_2}L_{k_3}\ldots L_{k_n}\psi,\ \psi)$ is transformed like a tensor of the n-th rank

$$t'_{k'_1k'_2\ldots k'_n}=\sum g_{k'_1k_1}g_{k'_2k_2}g_{k'_3k_3}\cdots g_{k'_nk_n}t_{k_1k_2\ldots k_n}.$$

Here, as in the case of a tensor of the second rank, the quantities $t_{k_1k_2\ldots k_n}$ belong to some invariant subspace in the space of all tensors of rank n. Similarly the quantity

$$\dot{t}_{k_1k_2\ldots k_n}=(TL_{k_1}L_{k_2}\ldots L_{k_n}\psi,\ \psi)$$

is a pseudotensor.

In the next section, we construct, by way of an example, the quantities considered above in the case of the Dirac equation.

II. Quantities containing partial derivatives of the wave function ψ.

Again we start with examples

a) $t^j=\left(\frac{\partial\psi}{\partial x_j},\ \psi\right)$. It is easily verified the quantity t^j is transformed according to the formula

$$t'^{j'}=\sum g^{j'j}t^j, \tag{30}$$

where the matrix $\|g^{j'j}\|=\left(\|g_{k'k}\|^{\mathrm{T}}\right)^{-1}$ (the matrix $\|g_{k'k}\|$ is the matrix of the Lorentz transformation). In virtue of the equation

$$(g^{\mathrm{T}})^{-1}=sgs^{-1},$$

where s is the spatial reflection (see paragraph 1, Sect. 1) the representation (30) is equivalent to the identity representation $g\to g$ of the Lorentz group. The quantity t^j is known as a contravariant vector to distinguish it from the ordinary vector t_k (known as a covariant vector).

b) The quantity

$$t^j_k=\left(L_k\frac{\partial\psi}{\partial x_j},\ \psi\right)$$

is transformed according to the formula

$$t'^{j'}_{k'}=\sum g^{j'j}g_{k'k}t^j_k, \tag{31}$$

where the matrix $\|g^{j'j}\|$, acts upon the upper index and the matrix $\|g_{k'k}\|$.

upon the lower. The representation specified by formula (31) is equivalent to a tensor representation of rank 2 and the quantity t_k^j is known as a tensor with one upper and one lower index.

The general case:

$$t_{k_1 \ldots k_n}^{j_1 \ldots j_s} = \left(L_{k_1} L_{k_2} \ldots L_{k_n} \frac{\partial^s \psi}{\partial x_{j_1} \partial x_{j_2} \ldots \partial x_{j_s}}, \psi \right). \tag{32}$$

This quantity is transformed according to the formula

$$t'^{j'_1 \ldots j'_s}_{k'_1 \ldots k'_n} = \sum g^{j'_1 j_1} g^{j'_2 j_2} \ldots g^{j'_s j_s} g_{k'_1 k_1} g_{k'_2 k_2} \ldots g_{k'_n k_n} t^{j_1 \ldots j_s}_{k_1 \ldots k_n} \tag{32'}$$

and is known as a tensor with s upper and n lower indices.

Representation (32') is equivalent to a tensor representation of rank $s+n$ (or, more precisely, a representation acting in some invariant subspace of the space of tensors of rank $k+n$).

We note that the tensor (32) $t_{k_1 \ldots k_n}^{j_1 \ldots j_s}$ is symmetric in the upper indices. Similarly it is possible to construct the corresponding pseudoquantities with the aid of the operator T.

§ 5. An Observation Concerning Quantities of Second Degree in the Wave Function ψ.

In the previous paragraph we constructed quantities of degree two in the wave function ψ. and its partial derivatives which transform according to some representation or other of the Lorentz group (a scalar, a vector, a tensor, etc.). Here we discuss the possible representations corresponding to the quantities of degree two in ψ, $\frac{\partial \psi}{\partial x_k}$, $\frac{\partial^2 \psi}{\partial x_i \partial x_k}$ etc.

I. We find, first of all, according to what representation a quantity Φ, of second degree in the function itself, transforms.

Let the values of the quantity Φ belong to some linear space $\tilde{R}$. We select in $\tilde{R}$ a basis η_α and denote the coordinates of the quantity Φ in this basis by $\{t_\alpha\}$. Since Φ is of second degree in the values of the wave function ψ, t_α has the form

$$t_\alpha = (\psi, \psi)_\alpha, \tag{33}$$

where $(\psi, \psi)_\alpha$ is some Hermitian quadratic form - one for each coordinate t_α

The values of the wave function ψ as ever, are transformed according to a representation $g \to T_g$, acting in the space R and consisting of the components $\tau_1, \tau_2, \ldots, \tau_k$. In this case the values of the complex conjugate of the wave function ψ are transformed according to the representation $g \to T^*_g$, consisting of the irreducible components $\tau^*_1, \tau^*_2, \ldots, \tau^*_k$ and also acting in the space R. (We recall that if the representation τ is defined by the pair (l_0, l_1), then the representation τ^* is defined by the pair $(l_0, -l_1)$).

From formula (33) it is clear that the space $\tilde{R}$, to which the values of the quantity Φ, belong, is contained in the product of the space R with itself: $\tilde{R} \subset R \times R$, and the representation τ_g, according to which the quantity Φ is transformed, is contained in the product of the representations* $T_g \times T^*_g$.

Thus, a quantity of degree two in the values of the wave function ψ, can only be transformed according to a representation which is contained in the representation $T_g \times T^*_g$.

II. We now consider the case of a quantity Φ, of second degree in the wave function ψ, and its first partial derivatives $\frac{\partial \psi}{\partial x_i}$.

We note first of all that if the wave function ψ is transformed according to the representation $g \to T_g$, i.e. if for $x' = gx$

$$\tilde{\psi}(x') = T_g \psi(x),$$

then its first-order partial derivatives are transformed according to the formula

$$\frac{\partial \tilde{\psi}(x')}{\partial x'_k} = \sum T_g \frac{\partial \psi(x)}{\partial x_i} \frac{\partial x_i}{\partial x'_k} = \sum T_g g^{ki} \frac{\partial \psi}{\partial x_i},$$

where $\|g^{ki}\| = (\|g_{ik}\|)^{-1}$

From this formula it is clear that the quantity $\left\{ \frac{\partial \psi}{\partial x_i} \right\}$ is transformed according to the representation product $\tau^1_g = T_g \times T^1_g$, where T^1_g denotes the representation $g \to (g^*)^{-1}$, equivalent to the vector representation $g \to g$ (this representation is defined by the pair $(l_0, l_1) = (0, 2)$ or in spinor notation by the pair $(k, n) = (1, 1)$).

We consider now the quantity Φ_1, which is a bilinear combination of ψ and $\frac{\partial \psi}{\partial x_i}$ (the components t_α of the quantity Φ have the form $t_\alpha = \left(\frac{\partial \psi_1}{\partial x_i}, \psi_2\right)_\alpha$, (where $(\psi_1, \psi_2)_\alpha$ is a Hermitian bilinear form).

Precisely as in the previous case we see that the quantity Φ_1 may only be transformed according to representations which are contained in the product

---

* We say that the representation $g \to T_g$ is contained in the representation $g \to \tau_g$, if the latter generates in one or other of its invariant subspaces a representation equivalent to the representation T_g.

$$\tau_g \times T_g^* = T_g \times T_g^* \times T_g^1.$$

A quantity Φ_2, which is a bilinear combination only of partial derivatives $\frac{\partial \psi}{\partial x_i}$, is clearly transformed according to the representation

$$\tau_g^1 \times (\tau_g^1)^* = T_g \times T_g^* \times T_g^1 \times T_g^1$$

(We note that the representation $g \to (T_g^1)^*$ is equivalent to the representation $g \to T_g^1$).

It is thus clear that a general quantity Φ, which is a quadratic in ψ and its partial derivatives $\frac{\partial \psi}{\partial x_i}$, is transformed according to a representation which is contained in the sum of the representations *

$$T_g \times T_g^*,\ T_g \times T_g^* \cdot T_g^{(1)} \text{and}\ T_g \times T_g^* \cdot T_g^{1)} \times T_g^{(1)}.$$

III. We consider, finally, a quantity Φ, of the second degree in the function ψ, and its partial derivatives $\frac{\partial^s \psi}{\partial x_{i_1} \partial x_{i_2} \dots \partial x_{i_s}}$ of order at most n.

We note that under Lorentz transformations the partial derivatives $\frac{\partial^s \psi}{\partial x_{i_1} \partial x_{i_2} \dots \partial x_{i_s}}$ are transformed according to the formulae

$$\frac{\partial^s \tilde{\psi}(x')}{\partial x'_{i_1} \partial x'_{i_2} \dots \partial x'_{i_s}} = \sum g^{i_1 k_1} g^{i_2 k_2} \dots g^{i_s k_s} T_g \frac{\partial^s \psi}{\partial x_{k_1} \partial x_{k_2} \dots \partial x_{k_s}}.$$

From this formulae it is clear that the representation τ_g^s, according to which the partial derivatives $\frac{\partial^s \psi}{\partial x_{i_1} \dots \partial x_{i_s}}$, are transformed is the product of the representation T_g and T_g^s, where $T_g^{(s)}$ denotes the representation acting according to the formula

$$t'_{i_1 i_2 \dots i_s} = \sum g_{i_1 k_1} g_{i_2 k_2} \dots g_{i_s k_s} t_{k_1 \dots k_s},$$

in the space of symmetric tensors of rank s,

Now, as in the previous cases, we find that a quantity Φ, which is a

* By the sum of the representations $T_g^{(1)}, \dots, T_g^{(s)}$, acting in spaces $R^{(1)}, \dots, R^{(s)}$, respectively we mean the representation τ_g, acting in the linear sum of these spaces, such that each space $R^{(i)}$ is invariant under the representation $g \to \tau_g$ and the latter generates in $R^{(i)}$ a representation coinciding with $T_g^{(i)}$.

quadratic in ψ and its partial derivatives of at most n-th order is transformed according to a representation which is contained in the representation sum of the form

$$T_g \times T_g^* \times T_g^{(l)} \times T_g^{(s)},$$

where l and s range from 0 to n.

In conclusion, we make one observation.

The representation $(T_g \times T_g^*)$, contains only one integral weight in the case of a finite representation T_g, Such a representation is clearly equivalent to a sum of tensor representations.

Any representation product $T_g \times T_g^* \times T_g^{(s)} \times T_g^{(l)}$ is also equivalent to a sum of tensor representations. Thus if the wave function ψ is transformed according to a finite representation then any quantity Φ, of second degree in ψ and its partial derivatives is transformed according to a representation whose irreducible components are equivalent to irreducible components of a tensor representation. For this reason, the quantities Φ are often called tensors of degree two in ψ and its derivatives.

We note finally that the quantity Φ is always transformed according to a unique representation of the proper or complete Lorentz group. In conclusion we consider an example. Let the representation T_g consist of the two components: $T_g^{(0,1)}$ and $T_g^{(1,0)}$:

$$T_g = T_g^{(0,1)} \dotplus T_g^{(1,0)}.$$

Here, as everywhere in this book, $T_g^{(0,1)}$ denotes an undotted spinor representation of the rank 1, and $T_g^{(1,0)}$ a dotted representation of the same rank. We recall that the representation $T_g^{(0,1)}$ is defined by the pair $\left(\frac{1}{2}, \frac{3}{2}\right)$, and the representation $T_g^{(1,0)}$ by the pair $\left(-\frac{1}{2}, \frac{3}{2}\right)$. We now find all the irreducible components τ, which are contained in the product

$$T_g \times T_g^*.$$

We note that T_g^* consists of the same components as T_g:

$$T_g^* = T_g^{(1,0)} \dotplus T_g^{(0,1)}.$$

Hence

$$T_g \times T_g^* = T_g^{(1,0)} \times T_g^{(1,0)} + T_g^{(1,0)} \times T_g^{(0,1)} + T_g^{(0,1)} \times T_g^{(1,0)} + T_g^{(0,1)} \times T_g^{(0,1)}.$$

We decompose each of the terms of this sum into irreducible components (see Sect. 6).

Decomposition	The pairs (l_0, l_1), specifying these components
1. $T_g^{(1,0)} \times T_g^{(1,0)} = T_g^{(0,0)} + T_g^{(1,0)}$	$T_g^{(0,0)} \sim \tau_0 \sim (0,1)$; $T_g^{(1,0)} \sim \tau_2 \sim (1,2)$
2. $T_g^{(1,0)} \times T_g^{(0,1)} = T_g^{(0,1)} \times T_g^{(1,0)} = T_g^{(1,1)}$	$T_g^{(1,1)} \sim \tau_1 \sim (0,2)$
3. $T_g^{(0,1)} \times T_g^{(0,1)} = T_g^{(0,0)} + T_g^{(0,2)}$	$T^{(0,0)} \sim \tau_0 \sim (0,1)$; $T_g^{(0,1)} \sim \dot{\tau}_2 \sim (-1,2)$

Thus the representation $T_g \times \overset{*}{T_g}$ of the proper Lorentz group consists of six irreducible components: two components τ_0, two components τ_1 and the components τ_2 and $\dot{\tau}_2$.

To these components there correspond the following quantities: in the first case a scalar, in the second case - a vector, finally in the third case (i.e. in the case of the reducible representation consisting of the components τ_2 and $\dot{\tau}_2$) - an antisymmetric tensor.

We note that in all three cases the representation of the proper Lorentz group may be extended to a representation of the complete group, so that in the third case, as in the first two, an irreducible representation of the complete group is obtained (for further details about this see Sect.4, paragraph 8).

Section 9. Examples of Relativistic-invariant Equations

Starting with this section we shall examine mainly the physical applications of the theory developed up to now. Here there arises a series of supplementary restrictions imposed by physical requirements which essentially determine the number of relativistic-invariant equations to be considered.

First of all we present some examples of relativistic-invarient equations. We shall later see that these equations are extremely important. For instance all the elementary particles known at the present time are apparently described by these equations only.

§ 1. The Dirac Equation

We take two representations of the proper group: τ, defined by the numbers $\left(\frac{1}{2}, \frac{3}{2}\right)$ (an undotted spinor representation of rank 1) and $\dot{\tau}$, defined by the numbers $\left(-\frac{1}{2}, \frac{3}{2}\right)$ (a dotted spinor representation of rank 1). It is easily seen that these two representations are interlocking.*

* We recall that the representations defined by the numbers (l_0, l_1) and (l_0', l_1'), are said to interlock if

$$(l_0', l_1') = (l_0 \pm 1, l_1) \quad \text{or} \quad (l_0', l_1') = (l_0, l_1 \pm 1).$$

It is therefore possible to construct a relativistic-invariant equation with respect to a quantity which is transformed according to a representation having the components τ and $\dot{\tau}$.

The matrix L_0 relative to the basis $\left\{\xi^{\tau}_{\frac{1}{2}\frac{1}{2}}, \xi^{\tau}_{\frac{1}{2}, -\frac{1}{2}}, \xi^{\dot{\tau}}_{\frac{1}{2}\frac{1}{2}}, \xi^{\dot{\tau}}_{\frac{1}{2}, -\frac{1}{2}}\right\}$ of such an equation is given by

$$L_0 = \begin{Vmatrix} 0 & 0 & c^{\tau\dot{\tau}} & 0 \\ 0 & 0 & 0 & c^{\tau\dot{\tau}} \\ c^{\dot{\tau}\tau} & 0 & 0 & 0 \\ 0 & c^{\dot{\tau}\tau} & 0 & 0 \end{Vmatrix} \tag{1}$$

(see the general formulae (15)-(19) in Sect. 7)

We note that the representation with components τ and $\dot{\tau}$ may be extended to a representation of the complete group *

Relative to the basis $\left\{\xi^{\tau}_{\frac{1}{2}m}, \xi^{\dot{\tau}}_{\frac{1}{2}m}\right\}$ the operator S corresponding to the spatial reflection s has the matrix

$$S = \begin{Vmatrix} 0 & E \\ E & 0 \end{Vmatrix}. \tag{1'}$$

If we now require that the equation with the matrix L_0 be invariant under the complete group, then we obtain

$$c^{\tau\dot{\tau}} = c^{\dot{\tau}\tau} = c,$$

i.e. the matrix L_0 has the form

$$L_0 = c\begin{Vmatrix} 0 & E \\ E & 0 \end{Vmatrix} = c\gamma_0, \quad E = \begin{Vmatrix} 0 & 1 \\ 1 & 0 \end{Vmatrix}. \tag{2}$$

We note that γ_0 coincides with the operator S (see (1')). Consequently the other matrices L_1, L_2, L_3 have the following form (see Sect.7, (18)-(19)):

$$L_3 = c\begin{Vmatrix} 0 & 0 & 1 & 0 \\ 0 & 0 & 0 & -1 \\ -1 & 0 & 0 & 0 \\ 0 & 1 & 0 & 0 \end{Vmatrix} = c\gamma_3, \quad L_2 = c\begin{Vmatrix} 0 & 0 & 0 & i \\ 0 & 0 & -i & 0 \\ 0 & -i & 0 & 0 \\ i & 0 & 0 & 0 \end{Vmatrix} = c\gamma_2;$$

$$L_1 = c\begin{Vmatrix} 0 & 0 & 0 & -1 \\ 0 & 0 & -1 & 0 \\ 0 & 1 & 0 & 0 \\ 1 & 0 & 0 & 0 \end{Vmatrix} = c\gamma_1. \tag{3}$$

* A quantity ψ which is transformed according to a representation of the complete group with components τ and $\dot{\tau}$ is a bispinor of the first rank (see Sect. 5).

It is possible to construct an invariant Hermitian bilinear form for the two components τ and $\dot{\tau}$ (just as for any finite representation of the complete group).

In the terms of the basis $\{\xi^{\tau}_{lm}, \xi^{\dot{\tau}}_{lm}\}$ the form has the matrix

$$\beta = \begin{Vmatrix} 0 & 0 & 1 & 0 \\ 0 & 0 & 0 & 1 \\ 1 & 0 & 0 & 0 \\ 0 & 1 & 0 & 0 \end{Vmatrix} = \begin{Vmatrix} 0 & E \\ E & 0 \end{Vmatrix} = \gamma_0$$

and has the following expression

$$(\psi_1, \psi_2) = \\ = x^{\tau}_{\frac{1}{2}\frac{1}{2}} \bar{y}^{\dot{\tau}}_{\frac{1}{2}\frac{1}{2}} + x^{\tau}_{\frac{1}{2}, -\frac{1}{2}} \bar{y}^{\dot{\tau}}_{\frac{1}{2}, -\frac{1}{2}} + x^{\dot{\tau}}_{\frac{1}{2}\frac{1}{2}} \bar{y}^{\tau}_{\frac{1}{2}\frac{1}{2}} + x^{\dot{\tau}}_{\frac{1}{2}, -\frac{1}{2}} \bar{y}^{\tau}_{\frac{1}{2}, -\frac{1}{2}}; \tag{4}$$

where $\{x^{\tau}_{lm}\}$ and $\{y^{\tau}_{lm}\}$ are the coordinates of ψ_1 and ψ_2 in the basis $\{\xi^{\tau}_{lm}, \xi^{\dot{\tau}}_{lm}\}$.

We note that the matrix β again coincides with the matrix of the operator S and γ_0: $\beta = S = \gamma_0$.

As is easily seen, a matrix L_0 of the form (2) for real c $(\bar{c} = c)$ satisfies the condition

$$(L_0\psi_1, \psi_2) = (\psi_1, L_0\psi_2).$$

Thus there exists an invariant Lagrangian function for the equation whose matrices L_k have the form (2), (3) for real c, namely

$$\mathcal{L}[\psi(x)] = c \operatorname{Im}\left\{\sum\left(\gamma_k \frac{\partial \psi}{\partial x_k}, \psi\right)\right\} + \varkappa(\psi, \psi).$$

Since a Lagrangian function is defined to within a real multiplier, we can arrange that $c = 1$, by changing the constant $\varkappa$ if necessary. Thus we obtain $L_k = \gamma_k$. The equation having the matrices γ_k is known as the Dirac equation.

We note that the Dirac matrices $\gamma_1, \gamma_2, \gamma_3, \gamma_0$ satisfy the following immediate relation:

$$\gamma_{k_1}\gamma_{k_2} + \gamma_{k_2}\gamma_{k_1} = \delta_{k_1k_2}E; \tag{3'}$$

where $\delta_{11} = \delta_{22} = \delta_{33} = -\delta_{00} = 1$, $\delta_{k_1k_2} = 0$ for $k_1 \neq k_2$,

We consider the matrix

$$\gamma_5 = \gamma_0\gamma_1\gamma_2\gamma_3 = -i \begin{Vmatrix} E & 0 \\ 0 & E \end{Vmatrix}. \tag{5}$$

A simple calculation shows that it commutes with all the operators T_g of the representation of the proper group; we merely recall that these operators have, in the canonical basis $\left\{\xi^{\tau}_{\frac{1}{2}}, \xi^{\tau}_{-\frac{1}{2}}, \xi^{\dot{\tau}}_{\frac{1}{2}}, \xi^{\dot{\tau}}_{-\frac{1}{2}}\right\}$ matrices of the type

$$A_g = \left\| \begin{matrix} T_g & 0 \\ 0 & T_{(g^*)^{-1}} \end{matrix} \right\|. \tag{6}$$

Moreover the matrix γ_5 anticommutes with all the matrices $\{\gamma_k\}$ ($k=0$, 1, 2, 3) and in particular with γ_0 which is at the same time the operator of the spatial reflection.

We now construct from the wave function ψ which is transformed according to the representation with the irreducible components $\tau\left(\frac{1}{2}, \frac{3}{2}\right)$ and $\dot{\tau}\left(-\frac{1}{2}, \frac{3}{2}\right)$ quantities of the second degree in ψ which yield an irreducible representation of the complete Lorentz group.

The general method for obtaining such quantities with the aid of the invariant form (ψ_1, ψ_2) and the matrices L_0, L_1, L_2, L_3 of the invariant equation, was described in the section. We now write down these quantities for the case of the Dirac equation.

1. A scalar - the invarient form (ψ, ψ) itself.

2. A pseudoscalar - $(\gamma_5\psi, \psi)$. We recall (see Sect.8, paragraph 7) that the quantity $(T\psi, \psi)$ is a pseudoscalar if the operator T commutes with the operators T'_g of a representation of the proper Lorentz group and anticommutes with the operator S. Here we take $T = \gamma_5$.

3. A vector - $t_k = (\gamma_k\psi, \psi)$. In paragraph 7, Sect.8 we showed that the quantities $t_k = (L_k\psi, \psi)$,where L_k are the matrices in the relativistic-invariant equation, are transformed like the components of a vector.

4. A pseudovector - $\dot{t}_k = (\gamma_5\gamma_k\psi, \psi)$ (see Sect.8, paragraph 7 again).

5. A tensor

$$t_{k_1k_2} = (\gamma_{k_1}\gamma_{k_2}\psi, \psi). \tag{7}$$

In paragraph 7, Sect.8 we showed that the quantity (7) is transformed like a tensor of the second rank. We recall that the space of all tensors of the second rank decomposes into the sum of three subspaces which are irreducible under the complete Lorentz group: a one-dimensional subspace (scalar), the nine-dimensional space of symmetric tensors $t_{k_1k_2} = t_{k_2k_1}$ with zero trace and the six-dimensional space of antisymmetric tensors $t_{k_1k_2} = -t_{k_2k_1}$.

Apparently, the quantity (7) may either be a scalar or an antisymmetric tensor. In fact, from equation (3') it follows that the components

$$t_{11}=t_{22}=t_{33}=-t_{00}=(\psi_1, \psi_2)$$

are unaltered by Lorentz transformations. Thus the quantities t_{kk} $(k=0, 1, 2, 3)$ are scalars.

For the components with unequal indices we have (see (3')):

$$t_{k_1k_2}=\frac{1}{2}\left(\left(\gamma_{k_1}\gamma_{k_2}-\gamma_{k_2}\gamma_{k_1}\right)\psi, \psi\right).$$

Hence it is clear that $t_{k_1k_2}=-t_{k_2k_1}$, i.e. such components constitute an antisymmetric tensor.

We note that the quantity

$$\dot{t}_{k_1k_2}=(\gamma_5\gamma_{k_1}\gamma_{k_2}\psi, \psi)$$

is either a pseudoscalar $(\dot{t}_{11}, \dot{t}_{22}, \dot{t}_{33}, \dot{t}_{00})$ or again an antisymmetric tensor $t_{k_1k_2}=-t_{k_1k_2}$ $(k_1\neq k_2)$. In the latter case $\dot{t}_{12}=-t_{34}$, $\dot{t}_{23}=-t_{01}$ etc. i.e. the antisymmetric tensor $t_{k_1k_2}$ coincides with the tensor $t_{k_1k_2}$ to within the numbering of the components and their sign.

Thus, from the wave function ψ satisfying the Dirac equation we have constructed five irreducible quantities, a scalar, a pseudoscalar, a vector, a pseudovector and an antisymmetric tensor. We saw in paragraph 7 of the previous section that from the wave function ψ which is transformed according to the representation $\tau\left(\frac{1}{2}, \frac{3}{2}\right)$ and $\dot{\tau}\left(-\frac{1}{2}, \frac{3}{2}\right)$ it is not possible to construct any other irreducible quantities of the second degree in ψ.

We now discuss whether representation (6) of the complete group may be extended to a representation of the general Lorentz group such that the Dirac equation remains invariant.

Let J be the operator corresponding to total reflection j. For the invariance of the Dirac equation under this operator it is clearly necessary that

$$J\gamma_kJ^{-1}=-\gamma_k \qquad (k=0, 1, 2, 3),$$

i.e. that j anticommutes with all the matrices γ_k and $J^2=E$. It is easily proved that there is only one such operator:

$$J=i\gamma_5.$$

It is clear from formula (5) and (6) that the operator J commutes with the operators of the representation of the proper Lorentz group.

If we now take

$$t\rightarrow T=\gamma_5\gamma_0.$$

then we may easily convince ourselves that the following equation holds

$$T\gamma_k T^{-1} = \gamma_k \qquad (k = 1, 2, 3), \qquad T\gamma_0 T^{-1} = -\gamma_0.$$

Moreover, for the operators of the representation $g \to T_g$ of the proper Lorentz group, we have

$$TT_g T^{-1} = T_{(g^*)^{-1}}.$$

Thus the operators

$$S = \gamma_0, \qquad T = \gamma_0\gamma_5, \qquad J = i\gamma_5$$

together with the operators T_g constitute a representation of the general Lorentz group which leaves the Dirac equation invariant.

However,

$$\gamma_0\gamma_5 = -\gamma_5\gamma_0, \qquad \gamma_5(\gamma_0\gamma_5) = -(\gamma_0\gamma_5)\gamma_5 \text{ and } \gamma_0(\gamma_0\gamma_5) = -(\gamma_0\gamma_5)\gamma_0,$$

i.e. all the operators T, S, J anticommute. In other words, the representation of the general group that we have obtained is two-valued (see Sects. 1, 3 and 7).

Thus a representation of the general group leaving the Dirac equation invariant is a two-valued representation. The operators, J, S, T are given by:

$$J = i\gamma_5, \qquad S = \gamma_0, \qquad T = \gamma_5\gamma_0.$$

It is possible, however, to construct an equation with respect to an eight-component function ψ which is invariant under a unique representation of the general group. In this equation the matrices L_k have the form:

$$L_k = \begin{Vmatrix} \gamma_k & 0 \\ 0 & \gamma_k \end{Vmatrix}.$$

The operators S, T, J and T_g (g is an element of the proper group) are specified by the formulae

$$J = \begin{Vmatrix} 0 & i\gamma_5 \\ i\gamma_5 & 0 \end{Vmatrix}, \qquad S = \begin{Vmatrix} \gamma_0 & 0 \\ 0 & -\gamma_0 \end{Vmatrix}, \qquad T = \begin{Vmatrix} 0 & i\gamma_5\gamma_0 \\ i\gamma_5\gamma_0 & 0 \end{Vmatrix},$$

$$T_g = \begin{Vmatrix} A_g & 0 \\ 0 & A_g \end{Vmatrix},$$

where the matrix A_g is defined by formula (6).

We note that if such an equation is considered only relative to the complete Lorentz group then it decomposes into two Dirac equations. Relative to a representation of the general group this equation is indecomposable.

§ 2. The Daffine Equation for Scalar Particles

This equation has already been quoted as an example at the beginning of

Sect.7 (see the footnote (*) on p.270). We recall that it has the form

$$\frac{\partial\psi}{\partial x_k}=\varkappa\psi_k, \qquad \sum_{k=1,2,3}\frac{\partial\psi_k}{\partial x_k}-\frac{\partial\psi_0}{\partial x_0}-\varkappa\psi=0. \tag{8}$$

As is easily seen, the quantity $\Phi=\{\psi, \psi_0, \psi_1, \psi_2, \psi_3\}$ consists of a scalar ψ and a vector $(\psi_0, \psi_1, \psi_2, \psi_3)$, i.e. the representation has the two interlocking components $\tau_0\sim(0, 1)$ and $\tau_1\sim(0, 2)$

$$\tau_0\longleftrightarrow\tau_1. \tag{8'}$$

(We use this notation to indicate the fact that the components are interlocking).

For any equation satisfying (8') the matrix L_0 has, relative to the basis

$$\{\xi_{00}^{\tau_0}, \xi_{00}^{\tau_1}, \xi_{1,-1}^{\tau_1}, \xi_{10}^{\tau_1}, \xi_{11}^{\tau_1}\}$$

the form

$$L_0=\begin{Vmatrix} 0 & c^{\tau_0\tau_1} & 0 & 0 & 0\\ c^{\tau_1\tau_0} & 0 & 0 & 0 & 0\\ 0 & 0 & 0 & 0 & 0\\ 0 & 0 & 0 & 0 & 0\\ 0 & 0 & 0 & 0 & 0 \end{Vmatrix}. \tag{8''}$$

As the invariant bilinear form we may clearly take the following *

$$(\psi_1, \psi_2)=x_{00}^{\tau_0}\bar{y}_{00}^{\tau_0}-x_{00}^{\tau_1}\bar{y}_{00}^{\tau_1}+\sum_{m=-1,0,1}x_{1m}^{\tau_1}\bar{y}_{1m}^{\tau_1}, \tag{8'''}$$

where $\{x_{lm}^{\tau}\}$ and $\{y_{lm}^{\tau}\}$ are the coordinates of ψ_1 and ψ_2 in the basis $\{\xi_{lm}^{\tau}\}$.

As we have seen, for the form (8''') to yield a Lagrangian of an equation with matrix L_0 of the form (8''), it is essential that

$$(L_0\psi_1, \psi_2)=(\psi_1, L_0\psi_2).$$

This leads to the equation

$$c^{\tau_0\tau_1}=-\bar{c}^{\tau_1\tau_0},$$

* We note that an invariant bilinear form for the representation with the components $\tau_0\sim(0, 1)$ and $\tau_1\sim(0, 2)$ $(\tau_1=\overset{*}{\tau}_1, \tau_2=\overset{*}{\tau}_2)$ may be chosen in two distinct ways

1. $a^{\tau_0\tau_0}=a^{\tau_1\tau_1}=1$,
2. $a^{\tau_0\tau_0}=-a^{\tau_1\tau_1}=1$.

We have chosen the second alternative. As we shall see in Sect.11, the first leads neither to a positive definite energy nor to a positive definite charge and is consequently devoid of physical interest.

i.e. $c^{\tau_0\tau_1}$ is purely imaginary; if we take $c^{\tau\tau_1} = i$, then we obtain a description of the system (8). We note further that any component of the quantity $\Phi = (\psi, \psi_0, \psi_1, \psi_2, \psi_3)$ satisfies the second-order equation

$$\Box\psi - \varkappa^2\psi = 0 \qquad \text{(the Klein-Gordan equation)}$$

The representations $\tau_0 \sim (0,1)$ and $\tau_1 \sim (0,2)$ may be extended in two non-equivalent ways to a representation of the complete Lorentz group (such that the system of equations (8) remains invariant as before). In one extension the scalar ψ does not change sign upon reflection (a proper scalar); in this case it is said that the equation describes scalar particles; in the other extension there is a change in the sign of ψ upon reflection; this is the case of the so-called pseudoscalar particles. Here the quantity $(\psi_0, \psi_1, \psi_2, \psi_3)$ is a vector (the first case) or a pseudovector (the second case) respectively. The operator S has, in the basis $\{\xi^{\tau}_{lm}\}$, the form

$$S\xi^{\tau_0}_{00} = \xi^{\tau_0}_{00}, \qquad S\xi^{\tau_1}_{00} = \xi^{\tau_1}_{00}, \qquad S\xi^{\tau_1}_{1m} = -\xi^{\tau_1}_{1m}$$

(for scalar particles) or

$$S\xi^{\tau_0}_{00} = -\xi^{\tau_0}_{00}, \qquad S\xi^{\tau_1}_{00} = -\xi^{\tau_1}_{00}, \qquad S\xi^{\tau_1}_{1m} = \xi^{\tau_1}_{1m}$$

(for pseudoscalar particles).

With any other method of introducing the operator S into a representation with components τ_0 and τ_1 (e.g. if we take ψ as a scalar and $(\psi_0, \psi_1, \psi_2, \psi_3)$ as a pseudovector) equation (8) does not remain invariant.

§ 3. The Daffine Equation for Vector Particles

This is the name given to the following system of equations:

$$\left.\begin{aligned} \frac{\partial\varphi_i}{\partial x_k} - \frac{\partial\varphi_k}{\partial x_i} &= \psi_{ik}, \\ \sum_i \frac{\partial\psi_{ik}}{\partial x_i} + \varkappa^2\varphi_k &= 0. \end{aligned}\right\} \tag{9}$$

Here it is clear that the quantity Ψ consists of a vector $(\varphi_0, \varphi_1, \varphi_2, \varphi_3)$ and an antisymmetric tensor ψ_{ik}, (i.e. is transformed according to a representation τ, consisting of the three components τ_1, τ_2, $\dot{\tau}_2$, defined by the pairs $(0, 2)$, $(1, 2)$, $(-1, 2)$. respectively). These components interlock according to the following scheme

$$\tau_2 \longleftrightarrow \tau_1 \longleftrightarrow \dot{\tau}_2. \tag{9'}$$

The most general form of the matrix L_0 in an equation having the scheme (9'), in terms of the basis

$$\left\{ \xi^{\dot{\tau}_2}_{11}, \ \xi^{\dot{\tau}_2}_{10}, \ \xi^{\dot{\tau}_2}_{1,-1}, \ \xi^{\tau_1}_{00}, \ \xi^{\tau_1}_{11}, \ \xi^{\tau_1}_{10}, \ \xi^{\tau_1}_{1,-1}, \ \xi^{\tau_2}_{11}, \ \xi^{\tau_2}_{10}, \ \xi^{\tau_2}_{1,-1} \right\}$$

is the following

$$L_0 = \begin{Vmatrix} 0 & 0 & 0 & 0 & c^{\dot{\tau}} & 0 & 0 & 0 & 0 & 0 \\ 0 & 0 & 0 & 0 & 0 & c^{\dot{\tau}_2\tau_1} & 0 & 0 & 0 & 0 \\ 0 & 0 & 0 & 0 & 0 & 0 & c^{\dot{\tau}_2\tau_1} & 0 & 0 & 0 \\ 0 & 0 & 0 & 0 & 0 & 0 & 0 & 0 & 0 & 0 \\ c^{\tau_1\dot{\tau}_2} & 0 & 0 & 0 & 0 & 0 & 0 & c^{\tau_1\tau_2} & 0 & 0 \\ 0 & c^{\tau_1\dot{\tau}_2} & 0 & 0 & 0 & 0 & 0 & 0 & c^{\tau_1\tau_2} & 0 \\ 0 & 0 & c^{\tau_1\dot{\tau}_2} & 0 & 0 & 0 & 0 & 0 & 0 & c^{\tau_1\tau_2} \\ 0 & 0 & 0 & 0 & c^{\tau_2\tau_1} & 0 & 0 & 0 & 0 & 0 \\ 0 & 0 & 0 & 0 & 0 & c^{\tau_2\tau_1} & 0 & 0 & 0 & 0 \\ 0 & 0 & 0 & 0 & 0 & 0 & c^{\tau_2\tau_1} & 0 & 0 & 0 \end{Vmatrix}. \tag{10}$$

The representation $g \to T_g$ with components τ_1, τ_2, $\dot{\tau}_2$ may be extended to a representation of the complete Lorentz group.

The requirement of invariance under this group leads to the equation

$$c^{\tau_1\dot{\tau}_2} = c^{\tau_1\tau_2} = c_1 \quad \text{and} \quad c^{\dot{\tau}_2\tau_1} = c^{\tau_2\tau_1} = c_2$$

(see Sect. 7 (30) and (31)).

If we now select the invariant bilinear form

$$(\psi_1, \psi_2) = \sum x_{1m}^{\tau_2} \bar{y}_{1m}^{\dot{\tau}_2} + \sum x_{1m}^{\tau_1} \bar{y}_{1m}^{\tau_1} - x_{00}^{\tau_1} \bar{y}_{00}^{\tau_1} \,{}^{*}), \tag{11}$$

then, to ensure the existence of a Lagrangian leading to an equation with a matrix of the form (10), we must take:

$$c^{\dot{\tau}_2\tau_1} = \overline{c^{\tau_1\dot{\tau}_2}},$$

or

$$c_1 = \bar{c}_2;$$

for $c_1 = i$ we arrive at a system equivalent to system (9).

We again note that the representation with components τ_1, τ_2, $\dot{\tau}_2$ may be extended to two non-equivalent representations of the complete Lorentz group:

---

* The invariant form may be selected in two ways which differ in the sign of $a^{\tau_1\tau_2}$. as in the previous example. The second way, however, leads neither to a positive definite energy nor to a postive definite charge (see Sect. 11).

1. Under reflection the components $\varphi_1, \varphi_2, \varphi_3$ of the vector $(\varphi_0, \varphi_1, \varphi_2, \varphi_3)$ change sign while the sign of φ_0 is conserved (a proper vector).

2. Reflection does not change the components $\varphi_1, \varphi_2, \varphi_3$ and changes the sign of φ_0 (a pseudovector).

§. 4. The Equation for a Two-component Neutrino

Here we consider an example of a relativistic-invariant equation with $\varkappa = 0$. We recall that such an equation can exist if together with the representation $g \to T_g$, according to which the wave function ψ, is transformed, there acts in the same space a representation $g \to V_g$, according to which the equations are transformed. In this case, the numbers $c^{\tau\tau'}_{lm, l'm'}$ which are the matrix elements of L_0, L_1, L_2, L_3, relative to the canonical bases $\{\sigma^{\tau}_{lm}\}$ of $g \to V_g$ and $\{\xi^{\tau}_{lm}\}$* of $g \to T_g$ have the same form as for equations with $\varkappa \neq 0$, they are given by formulae (13)-(19), Sect. 7.

As before the numbers $c^{\tau\tau'}$ differ from zero only for interlocking components τ and τ' of the representations $g \to V_g$ and $g \to T_g$ respectively.

We consider the two-component function ψ, which is transformed according to a two-dimensional representation τ, defined by the pair $\left(\frac{1}{2}, \frac{3}{2}\right)$ (an undotted spinor representation of rank one). In this two-dimensional space it is possible to define another irreducible representation $\dot{\tau}$, with the pair $\left(-\frac{1}{2}, \frac{3}{2}\right)$ (this is a dotted spinor representation of rank 1). The representations τ and $\dot{\tau}$ interlock. Thus we can construct a relativistic-invariant equation with $\varkappa = 0$ with respect to a quantity which is transformed according to the representation $\tau \sim \left(\frac{1}{2}, \frac{3}{2}\right)$ (in which case the equation itself is transformed according to the representation $\dot{\tau} \sim \left(-\frac{1}{2}, \frac{3}{2}\right)$).

---

* We recall that the matrix elements of L_k were defined in the following way. Let

$$\xi = L_k\psi,$$

where ψ is a vector of the space R, and ξ is its transform under the action of the operator L_k. This equation may be rewritten thus:

$$x^{\tau}_{lm} = \sum c^{\tau\tau'(k)}_{lml'm'} y^{\tau'}_{l'm'},$$

where x^{τ}_{lm} is a coordinate of the vector ξ in the canonical basis of the representation $g \to V_g$, and $y^{\tau'}_{l'm'}$ is a coordinate of ψ in the canonical basis of the representation $g \to T_g$.

The numbers $c^{\tau\tau'(k)}_{lml'm'}$ are then the matrix elements of L_k.

In the canonical bases $\left\{\xi_{\frac{1}{2}}, \xi_{-\frac{1}{2}}\right\}$ and $\left\{\eta_{\frac{1}{2}}, \eta_{-\frac{1}{2}}\right\}$ of the representations τ and $\dot{\tau}$ the numbers $c_{mm'}^{\tau\dot{\tau}(k)}$ form the matrices

$$
\begin{aligned}
L_0 &= \begin{Vmatrix} c & 0 \\ 0 & c \end{Vmatrix} = c\sigma_0 = cE, \\
L_1 &= c\begin{Vmatrix} 0 & 1 \\ 1 & 0 \end{Vmatrix} = c\sigma_1, \\
L_2 &= c\begin{Vmatrix} 0 & -i \\ i & 0 \end{Vmatrix} = c\sigma_2, \\
L_3 &= c\begin{Vmatrix} 1 & 0 \\ 0 & -1 \end{Vmatrix} = c\sigma_3.
\end{aligned} \tag{12}
$$

The matrices σ_1, σ_2, σ_3 are known as the Pauli matrices.

We note that if, in the two-dimensional space, we select the Hermitian form

$$(\psi_1, \psi_2) = x^{\dot{\tau}}_{\frac{1}{2}} \overline{y^{\tau}_{\frac{1}{2}}} + x^{\dot{\tau}}_{-\frac{1}{2}} \overline{y^{\tau}_{-\frac{1}{2}}}, \tag{13}$$

where $x^{\dot{\tau}}_{\frac{1}{2}}$, $x^{\dot{\tau}}_{-\frac{1}{2}}$ are the coordinates of ψ_1 in the basis $\left\{\eta_{\frac{1}{2}}, \eta_{-\frac{1}{2}}\right\}$, and $y^{\tau}_{\frac{1}{2}}$, $y^{\tau}_{-\frac{1}{2}}$ are the coordinates of ψ_2 in the basis $\left\{\xi_{\frac{1}{2}}, \xi_{-\frac{1}{2}}\right\}$, then

$$(V_g \psi_1, \psi_2) = (\psi_1, T_g\psi_2),$$

where V_g is an operator of the representation $\dot{\tau}$, and T_g an operator of the representation τ. As we know from the previous section, if this condition is satisfied then equation (12) with the matrices L_k may be obtained from an invariant Lagrangian constructed with the aid of the form (ψ_1, ψ_2) and the matrices $L_{\dot{k}}$. For this it is only necessary that

$$(L_0\psi, \psi) = (\psi, L_0\psi),$$

or

$$c = \bar{c},$$

i.e. that c be a real number

Thus an equation with the matrices L_k of the form (12) for real c may be obtained from an invariant Lagrangian. The latter has the form

$$\mathscr{L}[\psi(x)] = c\,\mathrm{Im}\left\{\sum_k \left(\sigma_k \frac{\partial\psi}{\partial x_k}, \psi\right)\right\}, c \text{ real} \tag{14}$$

If we take $c = 1$, then we obtain $L_k = \sigma_k$ $(k = 0, 1, 2, 3)$. An equation with the matrices σ_k is known as the equation of the "two-component neutrino".*

It has the form

$$\sigma_1 \frac{\partial\psi}{\partial x_1} + \sigma_2 \frac{\partial\psi}{\partial x_2} + \sigma_3 \frac{\partial\psi}{\partial x_3} + \sigma_0 \frac{\partial\psi}{\partial x_0} = 0. \qquad (15)$$

It is impossible to extend the representation $\tau \sim \left(\frac{1}{2}, \frac{3}{2}\right)$ of the proper group (equally for $\dot{\tau} \sim \left(-\frac{1}{2}, \frac{3}{2}\right)$) to a representation of the complete Lorentz group. Consequently it is meaningless to speak of the invariance of equation (15) with the matrices (12) under the complete Lorentz group.**

We now examine the quantities of the second degree in the wave function of a two-component neutrino which is transformed according to the representation $\tau \sim \left(\frac{1}{2}, \frac{3}{2}\right)$, i.e. according to an undotted spinor representation of rank 1.

To this end, we recall that if the wave function ψ is itself transformed according to the representation $g \to T_g$, then, as we saw in Sect.8, quantities of degree two in ψ, are transformed according to a representation entering into the product of representation $g \to T_g$ and the complex conjugate representation $g \to T^*_g$. In our case, $g \to T_g$ is an undotted spinor representation of rank 1 defined by the pair $\left(\frac{1}{2}, \frac{3}{2}\right)$. Thus the representation $g \to T^*_g$ is specified by the numbers $\left(-\frac{1}{2}, \frac{3}{2}\right)$ (see Sect.8, paragraph 8). It is a dotted

---

* Besides this the equation for a "four-component neutrino" is also considered: this is the ordinary Dirac equation with $\varkappa = 0$.

** In this connexion there used to be until recent times a principle according to which the possibility of describing a neutrino by a two-component function obeying equation (15) was rejected. It was considered that processes connected with elementary particles must be described in the same way both in left- and right-handed coordinate systems (physicists call this principle the law of conservation of parity); this means in particular that, for all elementary particles, the components of the wave function ψ must also be subjected to some linear transformation S under spatial reflection (i.e. under the change from a right- to a left-hand coordinate system) (in other words the wave function ψ must be transformed according to a representation of the complete Lorentz group), and the equations describing this particle must be invariant under reflection (or what amounts to the same thing - under the complete Lorentz group). A series of recent experiments, however, led L.D.Landau (U.S.S.R.) and Lee and Yang (U.S.A.) to the hypothesis that in some cases the invariance of processes under reflection need not occur (parity is not conserved). In particular it was permissible that such invariance be violated in those processes in which a neutrino participates. Since, in this way, invariance in relation to spatial reflection is no longer required for the neutrino (more precisely for its wave function) there emerged the possibility of describing it by a two-component function ψ, satisfying equation (15).

spinor representation of rank 1. We recall that the product of dotted and undotted spinors of rank 1 is irreducible and defined by the pair (0, 2). A quantity transforming according to such a representation is known as a vector. Thus the sole quantity that may be constituted from quadratic combinations of the wave function of a two-component neutrino, is a vector.

We note that the components of this vector may be written down in the following form:

$$t_k = (\sigma_k \psi, \psi),$$

where σ_k, for $k > 0$, are the Pauli matrices, $\sigma_0 = E$ and (ψ_1, ψ_2) is the Hermitian form (13) introduced above.

The fact that t_k is transformed like a vector under Lorentz transformations is verified in precisely the same way as was done in general form in paragraph 7, Sect. 8 for equations with

We note finally that it is impossible to construct a relativistic-invariant equation with $\varkappa \neq 0$ relative to a two-component function ψ. In the case $\varkappa = 0$, there exists another equation relative to a two-component function, besides the one already described: the function $\psi(x)$ is transformed according to the representation $\dot{\tau} \sim \left(-\frac{1}{2}, \frac{3}{2}\right)$, and the equation according to the representation $\tau \sim \left(\frac{1}{2}, \frac{3}{2}\right)$. This equation has the form

$$\sigma_1 \frac{\partial \psi}{\partial x_1} + \sigma_2 \frac{\partial \psi}{\partial x_2} + \sigma_3 \frac{\partial \psi}{\partial x_3} - \sigma_0 \frac{\partial \psi}{\partial x_0} = 0, \tag{16}$$

i.e. the matrices L_1, L_2, L_3 of this equation coincide with the corresponding matrices of equation (15) and the matrices L_0 in equations (15) and (16) differ in sign.*

The invariant form (ψ_1, ψ_2) for equation (16) is given by (13) relative to the basis $\left\{\eta_{\frac{1}{2}}, \eta_{-\frac{1}{2}}\right\}$ (for representation $\dot{\tau}$) and $\left\{\xi_{\frac{1}{2}}, \xi_{-\frac{1}{2}}\right\}$ (for representation τ).

§ 5. Maxwell's Equations for an Electromagnetic Field in Space

This is the name given to two systems of equations with $\varkappa = 0$ relative to an antisymmetric tensor F_{ik}:

I.
$$\frac{\partial F_{ik}}{\partial x_l} + \frac{\partial F_{kl}}{\partial x_i} + \frac{\partial F_{li}}{\partial x_k} = 0,$$

* Apart from equations (15) and (16) only relativistic-invariant equations with rectangular (non-square) matrices L_0, L_1, L_2, L_3 are possible for a two-component wave function (see Sect. 7).

the indices i, k, l are different; we obtain four equations in all.

$$\text{II.} \qquad \sum_{k=0}^{k=3} \frac{\partial F_{ik}}{\partial x_k} = 0.$$

This system also contains four equations. Both systems are transformed independently under Lorentz transformations.

The antisymmetric tensor F_{ik} is transformed according to the representation $g \to T_g$, consisting of the two components $\tau \sim (1,\ 2)$ and $\dot{\tau} \sim (-1,2)$. The representations $g \to V_g^{(1)}$ and $g \to V_g^{(2)}$, according to which both systems are transformed are vector representations $\tau_0 \sim (0,\ 2)$. The two systems have the same interlocking scheme:

$$\tau \longleftrightarrow \tau_0 \longleftrightarrow \dot{\tau}. \tag{17}$$

Let $\{ \xi_{lm}^{\tau},\ \xi_{lm}^{\dot{\tau}} \}$ be the canonical basis of the representation $g \to T_g$, and $\{ \sigma_{lm}^{\tau_0} \}$ be the canonical basis of the representation $g \to V_g$. We find the matrix elements of L_0 in the bases $\{ \sigma_{lm}^{\tau_0} \}$ and $\{ \xi_{lm}^{\tau},\ \xi_{lm}^{\dot{\tau}} \}$ for the equation with the interlocking scheme (17).

$$L_0 = \begin{array}{c} \\ \sigma_{00}^{\tau_0} \\ \sigma_{1,\,-1}^{\tau_0} \\ \sigma_{10}^{\tau_0} \\ \sigma_{11}^{\tau_0} \end{array} \begin{array}{c} \begin{array}{cccccc} \xi_{1,\,-1}^{\tau} & \xi_{10}^{\tau} & \xi_{11}^{\tau} & \xi_{1,\,-1}^{\dot{\tau}} & \xi_{10}^{\dot{\tau}} & \xi_{11}^{\dot{\tau}} \end{array} \\ \left\| \begin{array}{cccccc} 0 & 0 & 0 & 0 & 0 & 0 \\ c^{\tau_0\tau} & 0 & 0 & c^{\tau_0\dot{\tau}} & 0 & 0 \\ 0 & c^{\tau_0\tau} & 0 & 0 & c^{\tau_0\dot{\tau}} & 0 \\ 0 & 0 & c^{\tau_0\tau} & 0 & 0 & c^{\tau_0\dot{\tau}} \end{array} \right\| \end{array}.$$

We now stipulate that the equation with such a matrix be invariant under a representation of the complete group. Here two possibilities may arise:

1. The representation $g \to V_g$ is a pseudovector representation; in which case

$$c^{\tau_0\tau} = c^{\tau_0\dot{\tau}} = c.$$

2. The representation $g \to V_g$ is vectorial; whereupon

$$c^{\tau_0\tau} = -c^{\tau_0\dot{\tau}} = c.$$

If we require that our equation be obtained from an invariant Lagrangian, then we find that c must be real and it is therefore possible to take $c = 1$. Thus cases 1 and 2 lead to two possible matrices L_0 with the scheme of interlocking (17):

1.
$$L_0 = \begin{Vmatrix} 0 & 0 & 0 & 0 & 0 & 0 \\ 1 & 0 & 0 & 1 & 0 & 0 \\ 0 & 1 & 0 & 0 & 1 & 0 \\ 0 & 0 & 1 & 0 & 0 & 1 \end{Vmatrix} \tag{18}$$

2.
$$L_0 = \begin{Vmatrix} 0 & 0 & 0 & 0 & 0 & 0 \\ 1 & 0 & 0 & -1 & 0 & 0 \\ 0 & 1 & 0 & 0 & -1 & 0 \\ 0 & 0 & 1 & 0 & 0 & -1 \end{Vmatrix}. \tag{18'}$$

It is easily verified that the first case leads to a system of equations equivalent to system I, and the second case to a system equivalent to system II.

If we consider both systems together then the matrix L_0 has the form

$$L_0 = \begin{Vmatrix} 0 & 0 & 0 & 0 & 0 & 0 \\ 1 & 0 & 0 & 1 & 0 & 0 \\ 0 & 1 & 0 & 0 & 1 & 0 \\ 0 & 0 & 1 & 0 & 0 & 1 \\ 0 & 0 & 0 & 0 & 0 & 0 \\ 1 & 0 & 0 & -1 & 0 & 0 \\ 0 & 1 & 0 & 0 & -1 & 0 \\ 0 & 0 & 1 & 0 & 0 & -1 \end{Vmatrix}. \tag{19}$$

The matrix L_3 (to be used below) will then take the form (see Sect.7, (18) and (19)):

$$L_3 = \begin{Vmatrix} 0 & -i & 0 & 0 & -i & 0 \\ -1 & 0 & 0 & 1 & 0 & 0 \\ 0 & 0 & 0 & 0 & 0 & 0 \\ 0 & 0 & 1 & 0 & 0 & -1 \\ 0 & -i & 0 & 0 & i & 0 \\ -1 & 0 & 0 & -1 & 0 & 0 \\ 0 & 0 & 0 & 0 & 0 & 0 \\ 0 & 0 & 1 & 0 & 0 & 1 \end{Vmatrix}. \tag{19'}$$

The examples given appear to exhause all the relativistic-invariant equations that have been practically applied up to the present time. Below, in Sect.11, some unusual properties of these equations will be proved. We merely note here - and this will be important for the following - that in all the examples considered the matrix L_0 was reducible to diagonal form. We now present one more example of an invariant equation for which the matrix L_0 is no longer reducible to diagonal form. This equation is of some theoretical interest, as will appear in the sequel (see Soot.11).

§ 6. The Pauli-Fierz Equation

Let a quantity ψ be transformed according to a representation with components τ_1, $\dot{\tau}_1$, τ_2, $\dot{\tau}_2$, defined by the pairs

$$\begin{aligned} \tau_1 &\sim \left(\frac{1}{2}, \frac{3}{2}\right), \quad & \dot{\tau}_1 &\sim \left(-\frac{1}{2}, \frac{3}{2}\right), \\ \tau_2 &\sim \left(\frac{1}{2}, \frac{5}{2}\right), \quad & \dot{\tau}_2 &\sim \left(-\frac{1}{2}, \frac{5}{2}\right). \end{aligned} \tag{20}$$

These components clearly interlock according to the following scheme:

$$\begin{array}{ccc} \tau_1 & \longleftrightarrow & \dot{\tau}_1 \\ \updownarrow & & \updownarrow \\ \tau_2 & \longleftrightarrow & \dot{\tau}_2 \end{array} \tag{21}$$

(For an equation with this scheme the matrix L_0 is given on p.317). We note that the elements of the matrix L_0 with identical indices l and corresponding to one and the same pair of components τ and τ' are identical. Therefore, for the sake of simplicity, the matrix L_0 may be rewritten in the form of two compartments corresponding to the two values of l:

$$l = \frac{1}{2} \qquad\qquad\qquad\qquad l = \frac{3}{2}$$

$$\begin{array}{cccc} \tau_1 & \dot{\tau}_1 & \tau_2 & \dot{\tau}_2 \end{array} \qquad\qquad \begin{array}{cc} \tau_2 & \dot{\tau}_2 \end{array}$$

$$\left\| \begin{array}{cccc} 0 & c^{\tau_1\dot{\tau}_1} & c^{\tau_1\tau_2} & 0 \\ c^{\dot{\tau}_1\tau_1} & 0 & 0 & c^{\dot{\tau}_1\dot{\tau}_2} \\ c^{\tau_2\tau_1} & 0 & 0 & c^{\tau_2\dot{\tau}_2} \\ 0 & c^{\dot{\tau}_2\dot{\tau}_1} & c^{\dot{\tau}_2\tau_2} & 0 \end{array} \right\|; \qquad \left\| \begin{array}{cc} 0 & 2c^{\tau_2\dot{\tau}_2} \\ 2c^{\dot{\tau}_2\tau_2} & 0 \end{array} \right\|. \tag{22}$$

From the requirement of invariance under reflection we obtain:

$$c^{\tau_1\dot{\tau}_1} = c^{\dot{\tau}_1\tau_1}; \qquad c^{\dot{\tau}_2\tau_2} = c^{\tau_2\dot{\tau}_2}; \qquad c^{\tau_1\tau_2} = c^{\dot{\tau}_1\dot{\tau}_2}; \qquad c^{\tau_2\tau_1} = c^{\dot{\tau}_2\dot{\tau}_1}.$$

A bilinear Hermitian form, invariant under our representation (20) may be specified in two essentially different ways:

(a)
$$\left.\begin{aligned} (\psi_1, \psi_2) = \sum_{m=\frac{1}{2}, -\frac{1}{2}} \left(x^{\tau_1}_{\frac{1}{2}m} \bar{y}^{\dot{\tau}_1}_{\frac{1}{2}m} + x^{\dot{\tau}_1}_{\frac{1}{2}m} \bar{y}^{\tau_1}_{\frac{1}{2}m}\right) + \\ + \sum_{m=\frac{1}{2}, -\frac{1}{2}} \left(x^{\tau_2}_{\frac{1}{2}m} \bar{y}^{\dot{\tau}_2}_{\frac{1}{2}m} + x^{\dot{\tau}_2}_{\frac{1}{2}m} \bar{y}^{\tau_2}_{\frac{1}{2}m}\right) - \\ - \sum_{m=\frac{3}{2}, \frac{1}{2}, -\frac{1}{2}, -\frac{3}{2}} \left(x^{\tau_2}_{\frac{3}{2}m} \bar{y}^{\dot{\tau}_2}_{\frac{3}{2}m} + x^{\dot{\tau}_2}_{\frac{3}{2}m} \bar{y}^{\tau_2}_{\frac{3}{2}m}\right), \end{aligned}\right\}$$

$$a^{\tau_1\dot{\tau}_1} = a^{\tau_2\dot{\tau}_2} = 1$$

and

(b)

$$\left.\begin{array}{l} (\psi_1, \psi_2) = \sum\limits_{m=\frac{1}{2},\,-\frac{1}{2}} \left(x^{\tau_1}_{\frac{1}{2}m} \bar{y}^{\dot{\tau}_1}_{\frac{1}{2}m} + x^{\dot{\tau}_1}_{\frac{1}{2}m} \bar{y}^{\tau_1}_{\frac{1}{2}m} \right) - \\ \qquad - \sum\limits_{\substack{l=\frac{1}{2},\,\frac{3}{2} \\ -l \leqslant m \leqslant l}} (-1)^{[l]} \left(x^{\tau_2}_{lm} \bar{y}^{\dot{\tau}_2}_{lm} + x^{\dot{\tau}_2}_{lm} \bar{y}^{\tau_2}_{lm} \right), \\ \qquad a^{\tau_1\dot{\tau}_1} = - a^{\tau_2\dot{\tau}_2} = 1. \end{array} \right\} \qquad (23)$$

To these two ways there correspond the following conditions necessary in order that the equation with matrix L_0 may be obtained from an invariant Lagrangian:

a) $c^{\tau_1\dot{\tau}_1} = c^{\dot{\tau}_1\tau_1}$; $\quad c^{\tau_1\tau_2} = \bar{c}^{\dot{\tau}_2\dot{\tau}_1}$; $\quad c^{\dot{\tau}_1\dot{\tau}_2} = \bar{c}^{\tau_2\tau_1}$; $\quad c^{\tau_2\dot{\tau}_2} = \bar{c}^{\dot{\tau}_2\tau_2}$

and

b) $c^{\tau\dot{\tau}_1} = c^{\dot{\tau}_1\tau_1}$; $\quad c^{\tau_1\tau_2} = -\bar{c}^{\dot{\tau}_2\dot{\tau}_1}$; $\quad c^{\dot{\tau}_1\dot{\tau}_2} = -\bar{c}^{\tau_2\tau_1}$; $\quad c^{\tau_2\tau_2} = \bar{c}^{\dot{\tau}_2\tau_2}$.

Thus, in the case of an invariant equation with the interlocking scheme (21) obtainable from an invariant Lagrangian, the matrix L_0 must have one of the following forms:

a)

$$\begin{array}{cc} l = 3/2 & l = 1/2 \\ \left\| \begin{array}{cccc} 0 & c^{\tau_1\tau_1} & c^{\tau_1\tau_2} & 0 \\ c^{\tau_1\dot{\tau}_1} & 0 & 0 & c^{\tau_1\tau_2} \\ \bar{c}^{\tau_1\tau_2} & 0 & 0 & c^{\tau_2\dot{\tau}_2} \\ 0 & \bar{c}^{\tau_1\tau_2} & c^{\tau_2\dot{\tau}_2} & 0 \end{array} \right\|, & \left\| \begin{array}{cc} 0 & 2c^{\tau_2\dot{\tau}_2} \\ 2c^{\tau_2\dot{\tau}_2} & 0 \end{array} \right\|, \end{array}$$

b)

$$\begin{array}{cc} l = 3/2 & l = 1/2 \\ \left\| \begin{array}{cccc} 0 & c^{\tau_1\dot{\tau}_1} & c^{\tau_1\tau_2} & 0 \\ c^{\tau_1\dot{\tau}_1} & 0 & 0 & c^{\tau_1\tau_2} \\ -\bar{c}^{\tau_1\tau_2} & 0 & 0 & c^{\tau_2\dot{\tau}_2} \\ 0 & -\bar{c}^{\tau_1\tau_2} & c^{\tau_2\dot{\tau}_2} & 0 \end{array} \right\|, & \left\| \begin{array}{cc} 0 & 2c^{\tau_2\dot{\tau}_2} \\ 2c^{\tau_2\dot{\tau}_2} & 0 \end{array} \right\|. \end{array}$$

$\xi^{\tau_1}_{1\,\frac{1}{2}\,\frac{1}{2}}$	$\xi^{\tau_1}_{1\,\frac{1}{2},\,-\frac{1}{2}}$	$\dot{\xi}^{\tau_1}_{1\,\frac{1}{2}\,\frac{1}{2}}$	$\dot{\xi}^{\tau_1}_{1\,\frac{1}{2},\,-\frac{1}{2}}$	$\xi^{\tau_2}_{1\,\frac{1}{2}\,\frac{1}{2}}$	$\xi^{\tau_2}_{1\,\frac{1}{2},\,-\frac{1}{2}}$	$\dot{\xi}^{\tau_2}_{1\,\frac{1}{2}\,\frac{1}{2}}$	$\dot{\xi}^{\tau_2}_{1\,\frac{1}{2},\,-\frac{1}{2}}$	$\xi^{\tau_2}_{3\,\frac{3}{2}\,\frac{3}{2}}$	$\xi^{\tau_2}_{3\,\frac{3}{2}\,\frac{1}{2}}$	$\xi^{\tau_2}_{3\,\frac{3}{2},\,-\frac{1}{2}}$	$\xi^{\tau_2}_{3\,\frac{3}{2},\,-\frac{3}{2}}$	$\dot{\xi}^{\tau_2}_{3\,\frac{3}{2}\,\frac{3}{2}}$	$\dot{\xi}^{\tau_2}_{3\,\frac{3}{2}\,\frac{1}{2}}$	$\dot{\xi}^{\tau_2}_{3\,\frac{3}{2},\,-\frac{1}{2}}$	$\dot{\xi}^{\tau_2}_{3\,\frac{3}{2},\,-\frac{3}{2}}$
0	0	$c^{\tau_1\dot{\tau}_1}$	0	$c^{\tau_1\tau_2}$	0	0	0	0	0	0	0	0	0	0	0
0	0	0	$c^{\tau_1\dot{\tau}_1}$	0	$c^{\tau_1\tau_2}$	0	0	0	0	0	0	0	0	0	0
$c^{\dot{\tau}_1\tau_1}$	0	0	0	0	0	$c^{\dot{\tau}_1\dot{\tau}_2}$	0	0	0	0	0	0	0	0	0
0	$c^{\dot{\tau}_1\tau_1}$	0	0	0	0	0	$c^{\dot{\tau}_1\dot{\tau}_2}$	0	0	0	0	0	0	0	0
$c^{\tau_2\tau_1}$	0	0	0	0	0	$c^{\tau_2\dot{\tau}_2}$	0	0	0	0	0	0	0	0	0
0	$c^{\tau_2\tau_1}$	0	0	0	0	0	$c^{\tau_2\dot{\tau}_2}$	0	0	0	0	0	0	0	0
0	0	$c^{\dot{\tau}_2\dot{\tau}_1}$	0	$c^{\dot{\tau}_2\tau_2}$	0	0	0	0	0	0	0	0	0	0	0
0	0	0	$c^{\dot{\tau}_2\dot{\tau}_1}$	0	$c^{\dot{\tau}_2\tau_2}$	0	0	0	0	0	0	0	0	0	0
0	0	0	0	0	0	0	0	0	0	0	0	$2c^{\tau_2\dot{\tau}_2}$	0	0	0
0	0	0	0	0	0	0	0	0	0	0	0	0	$2c^{\tau_2\tau_2}$	0	0
0	0	0	0	0	0	0	0	0	0	0	0	0	0	$2c^{\tau_2\dot{\tau}_2}$	0
0	0	0	0	0	0	0	0	0	0	0	0	0	0	0	$2c^{\tau_2\dot{\tau}_2}$
$2c^{\dot{\tau}_2\tau_2}$	0	0	0	0	0	0	0	0	0	0	0	0	0	0	0
0	$2c^{\dot{\tau}_2\tau_2}$	0	0	0	0	0	0	0	0	0	0	0	0	0	0
0	0	$2c^{\dot{\tau}_2\tau_2}$	0	0	0	0	0	0	0	0	0	0	0	0	0
0	0	0	$2c^{\dot{\tau}_2\tau_2}$	0	0	0	0	0	0	0	0	0	0	0	0

where $c^{\tau_1\dot{\tau}_1}$, $c^{\tau_2\dot{\tau}_2}$ are arbitrary real numbers, and $c^{\tau_1\tau_2}$ is a complex number. We can further simplify this expression by passing to a new coordinate system. An admissible coordinate transformation (i.e. one conserving the form of all the infinitesimal operators H_+, H_-, H_3, F_+, F_-, F_3, the form of the transformation S, corresponding to reflection and the form of the invariant bilinear form) is in our case specified by the formulae

$$\xi'^{\tau_1}_{lm} = e^{i\theta_1}\xi^{\tau_1}_{lm}, \qquad \xi'^{\dot{\tau}_1}_{lm} = e^{i\theta_1}\xi^{\dot{\tau}_1}_{lm},$$
$$\xi'^{\tau_2}_{lm} = e^{i\theta_2}\xi^{\tau_2}_{lm}, \qquad \xi'^{\dot{\tau}_2}_{lm} = e^{i\theta_2}\xi^{\dot{\tau}_2}_{lm}. \tag{24}$$

Under such a transformation the elements $c^{\tau_1\dot{\tau}_1}$ and $c^{\tau_2\dot{\tau}_2}$ of the matrix L_0, clearly do not change; $c^{\tau_1\tau_2}$ becomes

$$c'^{\tau_1\tau_2} = c^{\tau_1\tau_2} e^{i(\theta_1-\theta_2)}. \tag{25}$$

Thus the essential parameters of the equation under discussion (having matrix L_0) will in this case be three real numbers: $c^{\tau_1\dot{\tau}_1}$, $c^{\tau_2\dot{\tau}_2}$ and $|c^{\tau_1\tau_2}|$.

By selecting θ_1 and θ_2 in equations (24) and (25) in a suitable way and dividing the entire equation by $2c^{\tau_1\dot{\tau}_2}$ (which leads to a change in the constant $\varkappa$), we can reduce our matrix L_0 to one of the following types:

a)

$$l=\frac{1}{2} \qquad\qquad l=\frac{3}{2}$$

$$\left\|\begin{matrix} 0 & \alpha & \beta & 0 \\ \alpha & 0 & 0 & \beta \\ \beta & 0 & 0 & \frac{1}{2} \\ 0 & \beta & \frac{1}{2} & 0 \end{matrix}\right\|, \qquad \left\|\begin{matrix} 0 & 1 \\ 1 & 0 \end{matrix}\right\|, \tag{26}$$

b)

$$l=\frac{1}{2} \qquad\qquad l=\frac{3}{2}$$

$$\left\|\begin{matrix} 0 & \alpha & \beta & 0 \\ \alpha & 0 & 0 & \beta \\ -\beta & 0 & 0 & \frac{1}{2} \\ 0 & -\beta & \frac{1}{2} & 0 \end{matrix}\right\|, \qquad \left\|\begin{matrix} 0 & 1 \\ 1 & 0 \end{matrix}\right\|. \tag{27}$$

where a is a real number, β is a real positive number.

Thus we have obtained the general form of the matrix L_0 for a relativistic-invariant equation with the interlocking scheme (21) obtainable from an invariant Lagrangian.

Below, in Sect.11, we shall show that one of these equations (obtained from case (b) by a definite selection of the parameters a and β) has special pre-eminence over the others. The matrix L_0 for this equation is given by:

$$\begin{array}{cc} l=\frac{1}{2} & l=\frac{3}{2} \\ \left\|\begin{array}{cccc} 0 & -\frac{1}{2} & \frac{1}{2} & 0 \\ -\frac{1}{2} & 0 & 0 & \frac{1}{2} \\ -\frac{1}{2} & 0 & 0 & \frac{1}{2} \\ 0 & -\frac{1}{2} & \frac{1}{2} & 0 \end{array}\right\|, & \left\|\begin{array}{cc} 0 & 1 \\ 1 & 0 \end{array}\right\|. \end{array} \tag{28}$$

A relativistic-invariant equation with such a matrix L_0 is known as a Pauli-Fierz equation.

§ 7. Examples of Infinite Invariant Equations

I. The simplest infinite equation (with $\varkappa \neq 0$) satisfying all the conditions of Sects.7 and 8 will be an equation in a wave function transforming according to an irreducible representation of the proper (and complete) Lorentz group, defined by the number pair $\left(0, \frac{1}{2}\right)$ or $\left(\frac{1}{2}, 0\right)$ (according to the results of Sect.7 these are the only cases of equations with $\varkappa \neq 0$, when the representation of the proper group according to which the components of the wave function are transformed is irreducible). The weights participating in the first representation $\left(0, \frac{1}{2}\right)$, assume the values $l = 0, 1, 2, \ldots$; while for the representation $\left(\frac{1}{2}, 0\right)$,

$$l = \frac{1}{2}, \frac{3}{2}, \frac{5}{2}, \ldots$$

The matrix L_0 will be diagonal in both cases; its elements have the form

$$c_{lml'm'} = \left(l + \frac{1}{2}\right)\delta_{ll'}\delta_{mm'}. \tag{29}$$

Due to the simplicity of this expression we can write the corresponding equation in full; in this case Ψ is a vector with components ψ_{lm} $(m = -l,$

$-l+1, \ldots, l-1, l)$, where $l=0, 1, 2, \ldots$ for the first and $l=\frac{1}{2}, \frac{3}{2}, \frac{5}{2}, \ldots$ for the second case respectively, and the equation has the form

$$\frac{\partial}{\partial x_0}\left(l+\frac{1}{2}\right)\psi_{lm}+\frac{1}{2}\frac{\partial}{\partial x_3}\{\sqrt{(l+m+1)(l-m+1)}\,\psi_{l+1,\,m}+$$
$$+\sqrt{(l+m)(l-m)}\,\psi_{l-1,\,m}\}+\frac{1}{4}\left(\frac{\partial}{\partial x_2}+i\frac{\partial}{\partial x_2}\right)\times$$
$$\times\{\sqrt{(l+m+1)(l+m+2)}\,\psi_{l+1,\,m+1}-\sqrt{(l-m-1)(l-m)}\,\psi_{l-1,\,m+1}\}-$$
$$-\frac{1}{4}\left(\frac{\partial}{\partial x_1}-i\frac{\partial}{\partial x_2}\right)\{\sqrt{(l-m+1)(l-m+2)}\,\psi_{l+1,\,m-1}-$$
$$-\sqrt{(l+m-1)(l+m)}\,\psi_{l-1,\,m-1}\}+i\varkappa\psi_{lm}=0. \quad (30)$$

II. As the next example we consider further equations with $\varkappa\neq 0$, for which the representation T_g, according to which the components of the function ψ are transformed, decomposes into two irreducible representations τ and $\dot{\tau}$ of the proper Lorentz group. It is obvious that such an equation satisfying all the conditions of Sects.7 and 8 can exist only if these irreducible representations τ and $\dot{\tau}$ are defined by pairs $\tau\sim\left(\frac{1}{2}, l_1\right)$ and $\dot{\tau}\sim\left(-\frac{1}{2}, l_1\right)$, where l_1 is either pure imaginary or real (in the latter case for l_1 integral or half-integral, the pairs $\left(l_1, \frac{1}{2}\right)$ and $\left(l_1, -\frac{1}{2}\right)$) are also possible. Indeed it is only in this case that the components τ and $\dot{\tau}$ interlock and the representation consisting of these components allows an invariant form:

$$\tau=\tau^* \quad \text{and} \quad \dot{\tau}=\dot{\tau}^*$$

(for l_1 real) or

$$\dot{\tau}=\tau^* \quad \text{and} \quad \tau=\dot{\tau}^*$$

(for l_1 purely imaginary).

In the case of l_1 being purely imaginary we can again define the invariant bilinear form (ψ_1, ψ_2); in two ways; we choose it positive definite. In this case the elements of the matrix L_0 for all our equations may be reduced to the form

$$c^{\tau\tau'}_{lml'm'}=c^{\tau'\tau}_{l'm'lm}=\left(l+\frac{1}{2}\right)\delta_{ll'}\delta_{mm'}. \quad (31)$$

We also note that, in the case of integral or half-integral l_1 the matrix L_0 in an infinite equation with the interlocking components $\tau\longleftrightarrow\dot{\tau}$, $\tau\sim\left(l_1, \frac{1}{2}\right)$ and $\dot{\tau}\left(l_1, -\frac{1}{2}\right)$ has the same form (31).

We note further that the infinite equation for which $\left(\frac{1}{2}, l_1\right) \longleftrightarrow \left(-\frac{1}{2}, l_1\right)$ having the matrix (31) in the case of half-integral l_1, resolves into two equations: a finite one with a wave function which transforms according to a representation with the components $\left(\frac{1}{2}, l_1\right) \longleftrightarrow \left(-\frac{1}{2}, l_1\right)$, and an infinite "tail" - the equation with the interlocking $\left(l, \frac{1}{2}\right) \longleftrightarrow \left(l_1, -\frac{1}{2}\right)$ (we recall that the representation $\left(l_1, \frac{1}{2}\right)$ and $\left(l_1, -\frac{1}{2}\right)$ are known as the "tails" of the finite representations $\left(\frac{1}{2}, l_1\right)$ and $\left(-\frac{1}{2}, l_1\right)$ (see Sect. 2).

Section 10. The Determination of the Rest Mass and Spin of a Particle

In this section we consider two physical quantities, namely the rest mass and the spin of a particle, and we shall show in what manner these quantities are connected with the matrix L_0 of the relativistic-invariant equation describing the particle.

§ 1. Plane Waves. The Energy-momentum Vector

Let the wave function of some particle satisfy the relativistic-invariant equation

$$\sum L_k \frac{\partial \psi}{\partial x_k} + i\varkappa\psi = 0. \tag{1}$$

We seek the solutions of this equation of the form

$$\psi(x_0, x_1, x_2, x_3) = \psi(p_0, p_1, p_2, p_3)\, e^{i(-p_0 x_0 + p_1 x_1 + p_2 x_2 + p_3 x_3)} \tag{2}$$

(such a solution is known as a plane wave). The quantity $\psi(p_0, p_1, p_2, p_3) = \psi(p)$ does not depend upon the coordinates x_0, x_1, x_2, x_3, and the numbers p_0, p_1, p_2, p_3 are assumed to be real.

We note that the four numbers (p_0, p_1, p_2, p_3) and the quantity $\psi(p)$ specifying the plane wave (2) are affected by passing from one orthogonal coordinate system in the four-dimensional space $R^{(4)}$ to another orthogonal coordinate system. In fact the coordinates (x_0, x_1, x_2, x_3) and (x'_0, x'_1, x'_2, x'_3) of one and the same point in the two coordinate systems are connected by the relations

$$x'_i = \sum g_{ik} x_k \qquad (i, k = 0, 1, 2, 3),$$

where $\| g_{ik} \| = g$ is the matrix of the Lorentz transformation specifying the change from one orthogonal coordinate system to the other. Upon passing from the coordinates (x_0, x_1, x_2, x_3) to the coordinates (x'_0, x'_1, x'_2, x'_3) the wave function ψ is transformed according to the formula

$$\psi(x'_0, x'_1, x'_2, x'_3) = T_g\psi(x_0, x_1, x_2\ x_3).$$

Applying this to the plane wave (2) we obtain:

$$\psi(x'_0, x'_1, x'_2, x'_3) = T_g\psi(p)\, e^{i(-p_0x_0+p_1x_1+p_2x_2+p_3x_3)}. \qquad (2')$$

On the other hand it is possible to write

$$\psi(x'_0, x'_1, x'_2, x'_3) = \psi(p')\, e^{i\left(-p'_0x'_0+p'_1x'_1+p'_2x'_2+p'_3x'_3\right)},$$

where the quantity $\psi(p') = T_g\psi(p)$, and the numbers (p'_0, p'_1, p'_2, p'_3) are connected with the numbers (p_0, p_1, p_2, p_3) by the formula

$$p'_i = \sum g_{ik}p_k \qquad (i, k = 0, 1, 2, 3). \qquad (3)$$

The expression (2') is again a plane wave, but written down in the new coordinate system (x'_0, x'_1, x'_2, x'_3). Thus if in one coordinate system the plane wave is specified by the quantity $\psi(p)$ and the four numbers (p_0, p_1, p_2, p_3), then in the other coordinate system the same plane wave is specified by the quantity $\psi(p') = T_g\psi(p)$ and the four numbers (p'_0, p'_1, p'_2, p'_3) related to the numbers (p_0, p_1, p_2, p_3) by formula (3), where $g = \| g_{ik} \|$ is the Lorentz transformation specifying the change from the first to the second coordinate system.

We see that, under a transformation of the coordinate system in the space $R^{(4)}$ the numbers (p_0, p_1, p_2, p_3) transform according to the same formulae as the coordinates of a vector (x_0, x_1, x_2, x_3) of the space $R^{(4)}$.

We now consider some other four-dimensional space $\tilde{R}^{(4)}$ and a fixed coordinary system in this space. We relate to each set of numbers (p_0, p_1, p_2, p_3) the vector p from $\tilde{R}^{(4)}$ with the coordinates (p_0, p_1, p_2, p_3) (in the coordinate system that we have selected). The vector p is known as the energy-momentum vector (p_0 is known as the energy and the three-dimensional vector (p_1, p_2, p_3) as the momentum). The space $\tilde{R}^{(4)}$ is called the momentum space.

Summarizing what has been said, we find that the plane wave (2) is specified in each orthogonal coordinate system of the space $R^{(4)}$ by some energy-momentum vector $p(p_0, p_1, p_2, p_3)$ of the momentum space $\tilde{R}^{(4)}$ and by the quantity $\psi(p)$. Upon passing from one orthogonal coordinate system to another with the aid of a Lorentz transformation g, the vector p specifying the plane wave is subjected to the transformation g and the quantity $\psi(p)$ is subjected to the transformation T_g.

Conversely, any Lorentz transformation $p' = gp$, upon the vectors p of the momentum space may be considered as a transformation from one coordinate system in the coordinate space in which the plane wave is specified by the vector p, to the coordinate system in which the same plane wave is specified by the vector p'.

We now turn to equation (1).

By substituting the plane wave (2) in equation (1) we see that the quantity $\psi(p)$ satisfies the equation

$$(-L_0p_0+L_1p_1+L_2p_2+L_3p_3)\psi(p)+\varkappa\psi(p)=0. \qquad (4)$$

If for brevity we denote $-L_0p_0+L_1p_1+L_2p_2+L_3p_3$ by $L(p)$, then equation (4) states that $\psi(p)$ is an eigenvector of the matrix $L(p)$ corresponding to the eigenvalue $-\varkappa$

$$L(p)\psi(p)=-\varkappa\psi(p). \qquad (5)$$

We shall prove that a non-zero solution $\psi(p)$ of this equation exists only for those vectors $p(p_0, p_1, p_2, p_3)$ of $\widetilde{R}^{(4)}$ for which the relation

$$p_0^2-p_1^2-p_2^2-p_3^2=\mu_i^2,$$

is satisfied, where $\mu_i=\frac{\varkappa}{\lambda_i}$, and λ_i is any real eigenvalue of the matrix L_0 different from zero.

We shall assume for the sake of simplicity that equation (1) is finite. In this case equation (5) allows a non-zero solution precisely for those vectors p for which the determinant of the matrix $L(p)+\varkappa E$ equals to zero. The determinant $\det(L(p)+\varkappa E)$ is clearly a polynomial in the variables p_0, p_1, p_2, p_3. We denote it by $D(p_0, p_1, p_2, p_3)=D(p)$.

From the relation

$$\sum_i T_g L_i T_g^{-1} g_{ki}=L_k$$

for the matrices in a relativistic-invariant equation (see Sect. 7, (3) it follows that

$$T_g L(p)\, T_g^{-1}=L(gp),$$

where gp is the image of the vector under the Lorentz transformation g. From this equation we obtain:

$$\det(L(gp)+\varkappa E)=\det(L(p)+\varkappa E)$$

or

$$D(p)=D(gp),$$

i.e. the value of the polynomial $D(p)$ does not change if the argument is subjected to a Lorentz transformation. Such a polynomial $D(p)$ is constant along a surface of transitivity of the complete or proper Lorentz group, i.e. constant on the hyperboloids

$$-S^2(p) = p_0^2 - p_1^2 - p_2^2 - p_3^2 = \text{const}$$

in the momentum space $\tilde{R}^{(4)}$. Hence it follows easily that the polynomial $D(p)$ in fact depends only upon $-S^2(p)$: $D(p) = \tilde{D}[-S^2(p)]$ where $\tilde{D}(-S^2)$ is some polynomial in one variable $-\ {}^2$.

We decompose $\tilde{D}[-S^2(p)]$ into its factors:

$$\tilde{D}[-S^2(p)] = c[-S^2(p) - \mu_1^2][-S^2(p) - \mu_2^2][-S^2(p) - \mu_3^2] \dots \\ \dots [-S^2(p) - \mu_k^2], \tag{6}$$

where $\mu_1^2, \mu_2^2, \dots, \mu_k^2$ are the roots of the polynomial $\tilde{D}$.

From this factorization it is clear that $\det(L(p) + \varkappa E)$ vanishes only if the vector p satisfies the relation

$$-S^2(p) = p_0^2 - p_1^2 - p_2^2 - p_3^2 = \mu_i^2, \tag{7}$$

where μ_i^2 is one or other of the roots of the polynomial $\tilde{D}$.

Since the numbers p_0, p_1, p_2, p_3 are real, the root μ_i^2 must also be selected real.

Thus in order that the equation $L(p)\psi(p) = -\varkappa\psi(p)$ have a non-zero solution and ipso facto that a plane wave with an energy-momentum vector $p(p_0, p_1, p_2, p_3)$, exist, the latter must satisfy relation (5) with one or other of the real roots μ_i^2 of the polynomial $\tilde{D}$.

We now find how the roots μ_i^2 are related to the eigenvalues of the matrix L_0.

We take $p_1 = p_2 = p_3 = 0$. Then $-S^2(p) = p_0^2$ and the polynomial $\tilde{D}[-S^2(p)] = \tilde{D}(p_0^2)$. The factorization (6) in this case has the form

$$D(p_0^2) = c(p_0^2 - \mu_1^2)(p_0^2 - \mu_2^2) \dots (p_0^2 - \mu_k^2) = \\ = c(p_0 - \mu_1)(p_0 + \mu_1)(p_0 - \mu_2)(p_0 + \mu_2) \dots (p_0 - \mu_k)(p_0 + \mu_k). \tag{8}$$

On the other hand, for $p_1 = p_2 = p_3 = 0$ the matrix $L(p)$ becomes $L(p) = -p_0 L_0$ and $\det(L(p) + \varkappa E) = \det(-p_0 L_0 + \varkappa E)$. The latter determinant must be of the form

$$\det(-L_0 p_0 + \varkappa E) = \tilde{c}\left(p_0 - \frac{\varkappa}{\lambda_1}\right)\left(p_0 - \frac{\varkappa}{\lambda_2}\right) \dots \left(p_0 - \frac{\varkappa}{\lambda_s}\right), \tag{9}$$

where $\lambda_1, \lambda_2, \dots, \lambda_s$ are the non-zero eigenvalues of the matrix L_0. Comparing (8) and (9) we see that, with suitable numbering, we may take:

$$\mu_1 = \frac{\varkappa}{\lambda_1}, \quad -\mu_1 = +\frac{\varkappa}{\lambda_2}, \quad \mu_2 = \frac{\varkappa}{\lambda_3} = -\frac{\varkappa}{\lambda_4} \text{etc.} \tag{10}$$

Formula (10) also gives the relation between the roots of the polynomial $\widetilde{D}$ and the eigenvalues of the matrix L_0. From this formula it is also clear that togethe with each non-zero eigenvalue λ the matrix has an eigenvalue $-\lambda$ of the same multiplicity as λ.

We now find in what case the eigenvalues of the matrix L_0 in an equation (1) obtainable from an invariant Lagrangian function, are real and in what case they are complex. As we have seen in Sect.8, the matrix L_0 of such an equation must satisfy the condition

$$(L_0\psi_1, \psi_2) = (\psi_1, L_0 \psi_2), \tag{11}$$

where (ψ_1, ψ_2) is the non-degenerate bilinear Hermitian form by means of which the Lagrangian is constructed.

Equation (11) for the eigenvector ψ_λ of the matrix L_0 with the eigenvalue λ may be rewritten thus:

$$(L_0\psi_\lambda, \psi_\lambda) = \lambda(\psi_\lambda, \psi_\lambda) = \bar{\lambda}(\psi_\lambda, \psi_\lambda);$$

whence

$$(\lambda - \bar{\lambda})(\psi_\lambda, \psi_\lambda) = 0,$$

i.e. either $\lambda = \bar{\lambda}$ and the eigenvalue λ is real, or $(\psi_\lambda, \psi_\lambda) = 0$. Thus, if the eigenvalue λ of L_0 is not real, the corresponding eigenvector ψ_λ makes the quadratic form (ψ, ψ) zero: $(\psi_\lambda, \psi_\lambda) = 0$. Such eigenvectors of the matrix L_0 are called inadmissible and we shall not consider them in the following. *

Thus we see that the eigenvalues of the matrix L_0 in an equation (1) obtainable from a Lagrangian, with the exception of the rare inadmissible cases, are real.

Let λ be an eigenvalue of the matrix L_0. Then the number $\mu = \frac{\varkappa}{\lambda}$ is also real. We may now rewrite relation (5) in the form:

$$p_0^2 - p_1^2 - p_2^2 - p_3^2 = \left(\frac{\varkappa}{\lambda}\right)^2 = \mu^2 \geqslant 0.$$

From this relation we see that the energy-momentum vector p of the plane wave (2) satisfies the inequality

$$p_0^2 - p_1^2 - p_2^2 - p_3^2 > 0 \text{ for } \varkappa \neq 0$$

---

* We shall see in the next section that a plane wave constructed with the aid of an inadmissible vector ψ_λ, has zero charge and zero energy.

and the equality

$$p_0^2 - p_1^2 - p_2^2 - p_3^2 = 0 \text{ for } \varkappa = 0 \;^{**},$$

i.e. the energy-momentum vector lies within the light cone in the momentum space $R^{(4)}$ for $\varkappa \neq 0$ (such vectors are known as timelike) and lie on the light cone for $\varkappa = 0$.

We summarize the above discussion as follows:

In order that a solution of the equation

$$\sum L_k \frac{\partial \psi}{\partial x_k} + \varkappa \psi = 0$$

exist in the form of a plane wave

$$\psi(x_0, x_1, x_2, x_3) = \psi(p)\, e^{i(-p_0x_0 + p_1x_1 + p_2x_2 + p_3x_3)},$$

it is necessary and sufficient that the energy momentum vector $p(p_0, p_1, p_2, p_3)$ satisfy a relation

$$p_0^2 - p_1^2 - p_2^2 - p_3^2 = \mu^2, \tag{12}$$

where $\mu = \frac{\varkappa}{\lambda}$, and λ is any non-zero real eigenvalue of the matrix L_0. In this case for an equation with $\varkappa \neq 0$ the energy-momentum vector is timelike and for $\varkappa = 0$ this vector lies on the light cone. The quantity $\psi(p)$ (the amplitude of the plane wave) is obtained from the equation

$$L(p)\,\psi(p) = -\varkappa \psi(p),$$

---

* In deriving relation (7) we used the fact that $D(L(p) + \varkappa E) = \det(L(gp) + \varkappa E)$. This equation was obtained from the relation

$$\sum g_{ik} T_g L_i T_g^{-1} = L_k$$

for the matrices in a relativistic-invariant equation. However, for $\varkappa = 0$ this relation becomes more general:

$$\sum V_g L_k T_g^{-1} g_{ik} = L_i.$$

It may be shown that for all the operators of any finite representation the determinant equals 1:

$$\det V_g = \det T_g = 1.$$

Whence we see that in the case $\varkappa = 0$

$$\det L(p) = \det L(gp).$$

i.e. is the eigenvector of the matrix $L(p)$ with the eigenvalue $-\varkappa$.

§ 2. The Rest System. The Rest Mass

In this and the following two systems we consider the case of an equation with $\varkappa \neq 0$. As we saw in the previous paragraph the energy momentum vector specifying the plane wave (2) is timelike in this case

$$p_0^2 - p_1^2 - p_2^2 - p_3^2 > 0.$$

We now select a Lorentz transformation g_0 which would transfer this vector into the direction of the time axis p_0 in momentum space

$$g_0 p = p' \text{ and } p_1' = p_2' = p_3' = 0.$$

In which case the equation

$$L(p)\psi(p) + \varkappa\psi(p) = 0,$$

from which the quantity $\psi(p)$, specifying the plane wave (2) is determined, is carried into the equation

$$-p_0 L_0 \psi(p_0) + \varkappa\psi(p_0) = 0, \tag{13}$$

where the quantity $\psi(p_0)$ is related to $\psi(p)$ by the relation

$$\psi(p_0) = T_{g_0}\psi(p).$$

As is seen from equation (13), the quantity $\psi(p_0)$ is the eigenvector of the matrix L_0 corresponding to the eigenvalue $\lambda = -\frac{\varkappa}{p_0}$. Obviously

$$p_0 = -\frac{\varkappa}{\lambda}.$$

Conversely the quantity $\psi(p)$ specifying the plane wave is expressed in terms of $\psi(p_0)$ according to the formula

$$\psi(p) = T_{g_0}^{-1}\psi(p_0).$$

Thus, each solution (2) in the form of a plane wave is defined by some eigenvector of the matrix L_0 with a non-zero eigenvalue.

As we observed above, any Lorentz transformation g acting upon the vectors p of the momentum space $\tilde{R}^{(4)}$ may be considered as a change of coordinates with matrix $g = \| g_{ik} \|$, from one coordinate system in the space $R^{(4)}$, in which the plane wave is specified by the vector p and by the quantity $\psi(p)$ to a coordinate system in $R^{(4)}$ in which this plane wave is defined by the vector p' and by the quantity $\psi(p') = T_g\psi(p)$. In particular, if in some coordinate system, a plane wave

$$\psi(x_0, x_1, x_2, x_3) = \psi(p)\, e^{i(-p_0x_0 + p_1x_1 + p_2x_2 + p_3x_3)} \tag{14}$$

was specified by the vector p, then the Lorentz transformation g_0 transforming the vector p into the direction of the time axis in $\tilde{R}^{(4)}$: $gp = p'$ $(p'_1 = p'_2 = p'_3 = 0)$, may be considered as a change to the coordinate system (x'_0, x'_1, x'_2, x'_3) in which the plane wave (14) takes the form

$$\psi(x'_0) = \psi(p'_0)\, e^{-ip'_0 x'_0}. \tag{15}$$

The coordinate system (x'_0, x'_1, x'_2, x'_3), in which the plane wave (14) assumes the form (15) is known as the rest system for the plane wave (14). The plane wave (15) in the rest system is often called a plane stationary wave. The quantity $\psi(p'_0)$ specifying the plane stationary wave, as we have seen, is an eigenvector of the matrix L_0 corresponding to a real non-zero eigenvalue

$$L_0\psi(p'_0) = \lambda\psi(p'_0), \text{ in which case } p'_0 = \frac{\varkappa}{\lambda} = \mu. \tag{16}$$

The numbers μ are known as the values of the rest mass of the particle *

We see that the energy p'_0 of the particle in the rest system coincides with some value of the rest mass. Relation (12) reduces then to the well-known relation in relativistic mechanics between the energy, momentum and rest mass of a particle

$$p_0^2 - p_1^2 - p_2^2 - p_3^2 = \mu^2.$$

§ 3. The Spin of a Stationary Particle

The state of a particle in the rest system is defined, as has just been shown, by an eigenvector $\tilde{\psi}^0_\lambda$ of the matrix L_0 corresponding to a real eigenvalue λ.

As a rule, the eigenvalues λ of the matrix L_0 are repeated. In fact, from the general condition which the matrix L_k of a relativistic-invariant equation satisfies (see Sect.7, (3)), it follows that the matrix L_0 is permutable with the operators $T_{\tilde{g}}$ corresponding to rotations g. But then each characteristic subspace of the matrix L_0 (i.e. a maximal subspace consisting of eigenvectors corresponding to one and the same eigenvalue) is invariant under the operators $T_{\tilde{g}}$. In fact, let $L_0\psi = \lambda\psi$, then $L_0 T_g\psi = T_g L_0\psi = \lambda T_g\psi$ i.e. $T_g\psi$ is also an eigenvector with the eigenvalue λ, i.e. belongs to the same characteristic subspace as ψ.

We denote the characteristic subspace corresponding to the eigenvalue λ

* As we saw in the previous section, with each eigenvalue λ of the matrix L_0 there is encountered the eigenvalue $-\lambda$ This means in other words that together with each value of the rest mass μ for a particle there exists a value of the rest mass $-\mu$. The state with $\mu > 0$ is known as the state of a particle with mass μ, and the state with $\mu < 0$ as the state of an antiparticle with mass $|\mu|$:. Thus in the case of finite equations, to each state of a particle there corresponds a state of its antiparticle with the same mass, or more briefly, for each particle there is an antiparticle.

by R_λ.

The space R_λ decomposes into a number of invariant subspaces R_λ^l in each of which the representation of the group of rotations is irreducible and has weight l. The numbers l are known as the spin values of the particle. It is clear that in each invariant subspace R_λ^l there exist $2l+1$ independent vectors $\tilde{\psi}^0_{\lambda l m}$ (for example the $2l+1$ eigenvectors of the operator H_3). The number m is known as the projection of the spin upon some axis or other (for example the axis x_3)* .

With the aid of the matrix L_0 we now determine the permissible values of the spin l of the particle.

Formulae (15) and (16) Sect. 7 show that the matrix L_0 decomposes into "compartments" $\|c_l^{\tau\tau'}\|$ corresponding to the different values l. Each eigenvalue $\lambda^{(l)}$ of the compartment $\|c_l^{\tau\tau'}\|$ is a $(2l+1)$-fold eigenvalue of the matrix L_0; clearly the corresponding $2l+1$ eigenvectors $\psi^0_{\lambda l m}$ $(m=-l, \ldots, l)$ of the matrix L_0 are transformed amongst themselves under spatial rotations according to an irreducible representation of weight l of the group of rotations (the vectors also form a basis in the space R_λ^l).

The eigenvectors $\psi^{(l)}_{\lambda l m}$ correspond to solutions of the form (2) describing different stationary states of the particle with spin l and mass $\frac{\varkappa}{\lambda}$ which differ in the value of the projection of the spin m onto some axis. Thus the possible spin values for a given particle are those values l for which the matrix $\|c_l^{\tau\tau'}\|$ has eigenvalues different from zero; the number of such eigenvalues multiplied by $(2l+1)$ determines the number of distinct states with spin l.

Thus with the aid of the matrix L_0 both the values of the rest masses and the spin values for a given particle are determined.

§ 4. The Spin of a Particle in an Arbitrary Coordinate System

In the previous paragraph we defined the spin of a particle in its rest system. Since any plane wave has a rest system then *ipso facto*, its spin is defined in any coordinate system: to this end it is necessary to take it equal to that value of the spin which the plane wave has with regard to the rest system.

* The spin l of a particle may be interpreted physically in the following way. The rest system for a plane wave is not uniquely defined: any rotation carries a rest system again into a rest system, such that the energy of the plane wave (the rest mass) does not change. This rotation of the rest system of the particle which does not change its mass may be interpreted as some internal degree of freedom for the particle.
This internal (rotational) degree of freedom also leads to the existence of an internal total moment - the spin of the particle l and its projection m upon some axis.

That is also related to the value of the projection of the spin m.

However, a direct definition of the spin of a particle in an arbitrary coordinate system (without transferring to the rest system) is desirable. To this end we note that the spin was determined in the previous paragraph from the following considerations: the states in the rest system are defined by the eigenvectors of the matrix L_0. The matrix L_0 commutes with the operators $T_{\tilde{g}}$ corresponding to rotations $\tilde{g}$; whence we deduced that the space R_λ of eigenvectors ψ_λ is invariant under a representation of the group of rotations. The weights of these representations l are the spin values.

We now apply these considerations to the plane wave

$$\psi(p)\, e^{i(-p_0x_0+p_1x_1+p_2x_2+p_3x_3)}$$

in an arbitrary coordinate system. The vector $\psi(p)$ is a characteristic vector for the matrix $L(p)$ with the eigenvalue $-\varkappa$

$$L(p)\psi(p) = -\varkappa\psi(p).$$

We now find these operators T_g of a Lorentz group representation which commute with the matrix $L(p)$. As we have seen, this yields the equation

$$T_g L(p) T_g^{-1} = L(gp).$$

From this equation it follows that the operator T_g is permutable with the matrix $L(p)$ if, and only if, the transformation g leaves the vector p in place: $gp = p$.

The set of such transformations $\approx\!g$ constitute a subgroup of the Lorentz group. This subgroup is known as <u>the stationary subgroup of the vector</u> p. We shall denote it by $G_0(p)$. We note that for the vector $p(p_0, 0, 0, 0)$, directed along the time axis the stationary subgroup G_0 is the group of rotations.

The representation $g \to T_g$ of the entire Lorentz group generates the representation $\overset{\approx}{g} \to T_{\overset{\approx}{g}}$ of the stationary group $G_0(p)$. Thus we have shown that the matrix $L(p)$ is only permutable with the operators $T_{\overset{\approx}{g}}$, constituting a representation of the stationary subgroup $G_0(p)$,

$$T_{\overset{\approx}{g}} L(p) T_{\overset{\approx}{g}}^{-1} = L(p).$$

Whence it follows that the characteristic subspace $R_\varkappa(p)$) of this matrix with the eigenvalue $-\varkappa$ is invariant under any operator $T_{\overset{\approx}{g}}$. In fact, let $L(p)\psi(p) = -\varkappa\psi(p)$. We apply the operator $T_{\overset{\approx}{g}}$ to both sides of this equation

$$T_{\overset{\approx}{g}} L(p)\psi(p) = -\varkappa T_{\overset{\approx}{g}}\psi(p),$$

$$T_{\overset{\approx}{g}} L(p) = L(p) T_{\overset{\approx}{g}},$$

and finally we obtain

$$L(p)\, T_{\tilde{\tilde{g}}}\psi(p) = -\varkappa T_{\tilde{\tilde{g}}}\psi(p),$$

i.e. the vector $T_{\tilde{\tilde{g}}}\psi(p)$ is again an eigenvector of the matrix $L(p)$ with the same eigenvalue $-\varkappa$. Thus the space $R_\varkappa(p)$ is invariant under the representation $\tilde{\tilde{g}} \to T_{\tilde{\tilde{g}}}$ of the stationary group $G_0(p)$.

We now prove that for the timelike vector p the group $G_0(p)$ is isomorphic to the group of rotations. In fact let g_0 be a transformation carrying the vector p into the direction of the time axis.

$$g_0 p = p', \qquad p'_1 = p'_2 = p'_3 = 0.$$

It is clear that any transformation

$$\tilde{\tilde{g}} = g_0^{-1} \tilde{g} g_0,$$

where $\tilde{g}$ is a rotation, leaves the vector p in place. Thus the group $G_0(p)$ is obtained from the group of rotations G_0 as follows

$$G_0(p) = g_0^{-1} G_0 g_0.$$

Whence it follows that the groups $G_0(p)$ and G_0 are isomorphic. The latter means that any representation of the group $G_0(p)$, in particular the representation in the space $R_\varkappa$, may be considered as a representation of the group of rotations. Hence it follows that a representation of the stationary group in the space $R^{(p)}_{-\varkappa}$ breaks down into irreducible representations with integral or half-integral weight l, acting in the subspaces $R^l_\varkappa(p)$. The values l are then the spin values in the coordinate system (x_0, x_1, x_2, x_3).

It is easily shown that these spin values l coincide with the spin values in the rest system. Each spin l corresponds to $2l+1$ linearly independent eigenvectors of the matrix $L(p)$ (with eigenvalue $-\varkappa$).

It is convenient to select them all in the following fashion.

We consider some direction γ in three-dimensional space. It can be shown that in the stationary group $G_0(p)$ of the vector p there exists a one-paramete subgroup $G_{0\gamma}(p)$ of transformations leaving the axis γ in place, this subgroup $G_{0\gamma}(p)$ being isomorphic to the group of rotations about a fixed axis (for example, the axis p_3)*

* In fact the transformation g_0 carrying the direction of the time axis into the vector p may be selected such that the plane (p_0, p_3) is carried under the action of the transformation g_0, into the plane containing the vector p and the axis γ. In this case the rotations $\tilde{g}_{12}$ about the axis p_3. leaving p_3, any vector of the plane (p_0, p_3) in place, correspond, according to the automorphism

In the space $R_{\varkappa}^{l}(p)$, where an irreducible representation of the stationary group acts, there is specified ipso facto a representation of the one-parameter subgroup $G_{0\gamma}(p)$. We denote an infinitesimal operator corresponding to this group by $H_{\gamma}(p)$. Its eigenvectors constitute the basis in $R_{\varkappa}^{l}(p)$, and the eigenvalues m have the form $m=-l, -l+1, \ldots, +l$. As the eigenvectors of the matrix $L(p)$ corresponding to the spin l, it is possible to select the eigenvalues $\psi_{\mu l m}^{\gamma}(p)$ of the operator $H_{\gamma}(p)$. Plane waves constructed with the aid of the vectors $\psi_{\mu l m}^{\gamma}(p)$ are known as plane waves of spin l "polarized in the direction γ". In which case the numbers "m" are known as the projections of the spin l on the axis γ or otherwise as the polarization values.

In the case when the direction γ coincides with the direction of the momentum p_1, p_2, p_3, it is said that the plane wave is polarized in the direction of motion.

Thus a plane wave is specified in an arbitrary coordinate system by the energy-momentum vector p satisfying relation (12), by the spin l and by the projection m of the spin onto some direction γ (the polarization value).

§ 5. Particles with Zero Rest Mass

Up until now we have considered the equation $\sum_i L_i \frac{d\psi}{dx_i} + i\varkappa\psi = 0$ with $\varkappa \neq 0$. From formula (16), defining the values of the rest mass of such a particle, it follows that $\mu \neq 0$, i.e. for $\varkappa \neq 0$ all rest masses are different from zero. We now pass on to equations with $\varkappa = 0$. From formula (16) it follows that in this case $\mu = 0$, i.e. for such equations all rest masses equal zero.

We note that for equations with $\varkappa = 0$ no rest system exists.

In fact, we saw in paragraph 1 that the energy-momentum vector p of a plane wave satisfies the relation

$$p_0^2 - p_1^2 - p_2^2 - p_3^2 = 0,$$

i.e. the vector p lies on the light cone. A vector on the light cone cannot be carried into the time axis by any Lorentz transformation; in other words

---

• (Cont'd).

$$\tilde{\tilde{g}}_{\gamma} = g_0 \tilde{g}_{12} g_0^{-1} \tag{17}$$

to transformations $\tilde{\tilde{g}}_{\gamma}$ leaving any vector of the plane (p, γ) in place, in particular also the direction γ itself. These transformations $\tilde{\tilde{g}}_{\gamma}$ also constitute a one-parameter subgroup $G_{0\gamma}(p)$ of the stationary group $G_0(p)$, which leaves the axis γ in place. From equation (13) it is clear that the group $G_{0\gamma}(p)$ is isomorphic to the group of rotations about the axis p_0

there exists no orthogonal coordinate system in which the last three coordinates of the vector p vanish (obviously except for the trivial case when $p_0 = p_1 = p_2 = p_3 = 0$, i.e. when any coordinate system is a rest system. This case is devoid of interest). Thus we see that for particles described by an equation with $\varkappa = 0$ the rest mass equals zero and rest systems do not exist. Thus the term "rest mass" has only a conventional meaning here.

§ 6. The Polarization of Particles with Zero Rest Mass

Inasmuch as a rest system does not exist for particles with $\varkappa = 0$, the definition of the spin of such particles as the weight of the representation of the group of rotations which acts in the space of the eigenvectors of the matrix L_0 describing the stationary plane wave, loses its meaning.

It is nonetheless possible to define the polarization for particles with $\varkappa = 0$ as for particles with $\varkappa \neq 0$.

We introduce it precisely in the same way as in paragraph 3 where we defined the spin and the polarization of a particle with $\varkappa \neq 0$ in an arbitrary coordinate system with the aid of a representation of the stationary subgroup $G_0(p)$ for the energy-momentum vector p, acting in a characteristic subspace of the matrix $L(p)$. Thus, we seek, in some coordinate system, the solutions of the equation $\sum L_i \frac{d\psi}{dx_i} = 0$ in the form of a plane wave

$$\psi(x_0, x_1, x_2, x_3) = \psi(p_0, p_1, p_2, p_3)\, e^{i(-p_0x_0 + p_1x_1 + p_2x_2 + p_3x_3)}.$$

The quantity $\psi(p)$ is determined as usual from the equation

$$L(p)\psi(p) = 0 \tag{18}$$

and the energy-momentum vector p lies on the light cone

$$p_0^2 - p_1^2 - p_2^2 - p_3^2 = 0.$$

The solutions of equation (18) constitute a subspace $R(p)$ of space R of the values of the wave functions. Precisely as in paragraph 3 it can be shown that the subspace $R(p)$ is invariant under precisely those operators T_g of the Lorentz group representation $g \to T_g$ which correspond to transformations g leaving the vector p in place. In other words, the space $R(p)$ is invariant under a representation of the stationary group $G_0(p)$ of the vector p.

In paragraph 3 we saw that the stationary subgroup of any timelike vector is isomorphic to the group of rotations. For a vector p lying on the light cone, the stationary subgroup $G_0(p)$ is organized entirely differently from the group of rotations.

We now describe this group briefly. We note that the stationary groups of any two vectors lying on the light cone are isomorphic. In this connexion we construct the stationary subgroup of the vector p with the coordinates

$(p_0, 0, 0, p_3), p_0 = p_3$. Any subgroup of the proper Lorentz group corresponds to some subgroup of the group $\mathfrak{A}$ of complex second-order matrices a with determinant equal to 1. We now write out the subgroup of the group $\mathfrak{A}$ corresponding to the stationary subgroup $G_0(p_0, 0, 0, p_3)$. We denote it by $\mathfrak{A}_0(p)$ (in which case we have obviously used the correspondence $g_a \to a$ between proper Lorentz transformations and the complex second-order matrices a (det $a = 1$), which we established in Sect.1).
The group $\mathfrak{A}_0(p)$ consists of three one-parameter subgroups:

$$a_1(\varphi) = \begin{pmatrix} e^{\frac{i\varphi}{2}} & 0 \\ 0 & e^{-\frac{i\varphi}{2}} \end{pmatrix}, \qquad a_2(t) = \begin{pmatrix} 1 & t \\ 0 & 1 \end{pmatrix}, \qquad a_3(s) = \begin{pmatrix} 1 & is \\ 0 & 1 \end{pmatrix}.$$

We note that the subgroup $a_1(\varphi)$ corresponds to rotations about the x_3 axis. (These rotations, as is easily seen, belong to the stationary subgroup $G_0(p_0, 0, 0, p_3)$.) It is easily seen that the group $\mathfrak{A}_0(p)$ is isomorphic to the group of all movements of the plane (x, y). In fact, if to the subgroup $a_1(\varphi)$ we relate rotations in the plane (x, y) about the origin through the angle φ, and to subgroups $a_2(t)$ and $a_3(s)$ displacements along the axes x and y through distances t and s respectively, then in this way the isomorphism between the group of motions in the plane and the group $\mathfrak{A}_0(p)$ will be established.

We note that the stationary group $G_0(p)$ contains a rotation about the direction of the momentum (p_1, p_2, p_3). For the sake of convenience we now select a coordinate system in four-dimensional space such that the positive direction of the x_3 axis coincides with the vector (p_1, p_2, p_3). In this system the energy-momentum vector has the coordinates $(p_0, 0, 0, p_3)$. With such a choise of the x_3 axis a rotation about the direction of the momentum coincides with a rotation about the x_3 axis and, consequently, the space $R(p)$ is invariant under the operators $T_{g_{12}}$ corresponding to these rotations.

Ipso facto the space $R(p)$ is also invariant under the infinitesimal operator H_3; and in the space $R(p)$ it is possible to select a basis from the eigenvectors of this operator. We denote these vectors by $\psi_m(p)$ where the m are the eigenvalues of the operator H_3. As we know, the numbers m may be all integral or all half-integral in value.

The values of the numbers $m = m_1, m_2, \ldots, m_k$ are known as the polarization values of a particle with $\varkappa = 0$. The eigenvectors $\psi_m(p)$ correspond to plane waves with the defined polarization values m. In this case it is said that the plane wave $\psi_m(p)$ is polarized in the direction of motion. *

---

* We shall not consider here the case of the polarization of plane waves with $\varkappa = 0$ along a direction other than the direction of the momentum. We merely note that, in contrast with the case $\varkappa \neq 0$, plane waves with $\varkappa = 0$ not polarized along the motion do not always exist. We shall convince ourselves of this below by the example of the two component neutrino.

We note that in the case when the direction of the x_3 axis does not coincide with the direction of the momentum (p_1, p_2, p_3), as we have just assumed, instead of the operator H_3 it is necessary to select the infinitesimal operator H_p corresponding to the one-parameter group of rotations about the direction of the momentum (p_1, p_2, p_3). In this case it is necessary to take the eigenvectors of this operator as the vectors $\psi_m(p)$.

Thus, every plane wave for a particle with $\varkappa = 0$ is specified by the selection of an energy-momentum vector laying in the light cone

$$p_0^2 - p_1^2 - p_2^2 - p_3^2 = 0,$$

and by some vector $\psi_m(p)$ of the subspace annihilating the matrix $L(p)$:

$$L(p)\psi_m(p) = 0$$

with a definite polarization value m.

§ 7. The Rest Mass and the Spin of Particles Described by the Equations of the Previous Section

I. The Dirac equation. The matrix

$$L_0 = \begin{Vmatrix} 0 & E \\ E & 0 \end{Vmatrix}$$

consists of one compartment

$$\left\| c_{\frac{1}{2}}^{\tau\dot{\tau}} \right\| = \begin{Vmatrix} 0 & 1 \\ 1 & 0 \end{Vmatrix},$$

where $l = \frac{1}{2}$. Thus the Dirac equation describes particles with spin $l = \frac{1}{2}$. The mass assumes the two values $\mu = +\varkappa$ and $\mu = -\varkappa$. Since for $l = \frac{1}{2}$ the projection of the spin m assumes the two values $\frac{1}{2}$ and $-\frac{1}{2}$ there exist four linearly independent vectors of the matrix L_0 namely:

$$\psi_{\varkappa, \frac{1}{2}, \frac{1}{2}}, \quad \psi_{\varkappa, \frac{1}{2}, -\frac{1}{2}}, \quad \psi_{-\varkappa, \frac{1}{2}, \frac{1}{2}}, \quad \psi_{-\varkappa, -\frac{1}{2}, -\frac{1}{2}},$$

i.e. the state with mass $\varkappa$ and spin projection $\frac{1}{2}$, the state with mass $\varkappa$ and spin projection $-\frac{1}{2}$, and so on.

II. The Daffine equation for scalar particles.

Here there is only one compartment different from zero, (see Sect.9, (8")), $\left\| c_0^{\tau_1\tau_2} \right\| = \begin{Vmatrix} 0 & i \\ -i & 0 \end{Vmatrix}$, i.e. the spin of the particle assumes only the value $l = 0$. The mass assumes the values $\mu_1 = \varkappa$ and $\mu_2 = -\varkappa$. There are thus only two independent states in the rest system.

III. The Daffine equation for vector particles.

Here again there is just one compartment (see Sect.9, (10)).

$$\| c_1^{\tau\tau} \| = \begin{Vmatrix} 0 & i & i \\ -i & 0 & 0 \\ -i & 0 & 0 \end{Vmatrix}.$$

Consequently the spin $l = 1$.

The non-zero eigenvalues of this matrix are $\lambda_1 = \sqrt{2}, \lambda_2 = -\sqrt{2}$. The rest mass equals $\mu_{1,2} = \pm \frac{\varkappa}{\sqrt{2}}$. In all there are six linearly independent states in the rest system

$$\psi_{\frac{\varkappa}{\sqrt{2}}, 1, m} \text{ and } \psi_{-\frac{\varkappa}{\sqrt{2}}, 1, m} \qquad (m = -1, 0, 1).$$

IV. The Pauli-Fierz equation.

Here there are two compartments:

$$l = \frac{1}{2} \qquad\qquad l = \frac{3}{2}$$

$$\begin{Vmatrix} 0 & -\frac{1}{2} & \frac{1}{2} & 0 \\ -\frac{1}{2} & 0 & 0 & \frac{1}{2} \\ -\frac{1}{2} & 0 & 0 & \frac{1}{2} \\ 0 & -\frac{1}{2} & \frac{1}{2} & 0 \end{Vmatrix}, \qquad \begin{Vmatrix} 0 & 1 \\ 1 & 0 \end{Vmatrix}.$$

All the eigenvalues of the first of them equal zero, and the eigenvalues of the second equal ± 1.

Thus the spin for the Pauli-Fierz equation equals $l = \frac{3}{2}$, and the mass $\mu_{1,2} = \pm \varkappa$.

In the rest system there are eight linearly independent states:

$$\psi_{\varkappa, \frac{3}{2}, m} \text{ and } \psi_{-\varkappa, \frac{3}{2}, m} \qquad \left(m = \frac{3}{2}, \frac{1}{2}, -\frac{1}{2}, -\frac{3}{2}\right).$$

Thus the equations with $\varkappa \neq 0$ presented in Sect.9 are the equations for particles with spin $l = \frac{1}{2}, 0, 1, \frac{3}{2}$.

V. The two-component neutrino (see Sect.9, paragraph 4). The equation for this has the form:

$$\sigma_1 \frac{\partial \psi}{\partial x_1} + \sigma_2 \frac{\partial \psi}{\partial x_2} + \sigma_3 \frac{\partial \psi}{\partial x_3} + \sigma_0 \frac{\partial \psi}{\partial x_0} = 0, \tag{19}$$

or

$$\sigma_1 \frac{\partial \psi}{\partial x_1} + \sigma_2 \frac{\partial \psi}{\partial x_2} + \sigma_3 \frac{\partial \psi}{\partial x_3} - \sigma_0 \frac{\partial \psi}{\partial x_0} = 0. \tag{20}$$

In the first case the quantity ψ is transformed according to the representation $\tau\left(\frac{1}{2}, \frac{3}{2}\right)$ (an undotted spinor representation of rank 1) and in the second case according to the representation $\dot{\tau}\left(-\frac{1}{2}, \frac{3}{2}\right)$ (a dotted spinor representation of rank 1). The mass of the two component neutrino equals zero. We now find the polarization.

We consider, first of all, the case of equation (19). We select, as in the general case, a coordinate system such that $p_1 = p_2 = 0$, $p_3 > 0, p_0 = \pm p_3$. We consider, first of all, the case $p_0 = p_3$. Equation (18) for determining the quantities $\psi^0(p_3, p_0)$ assumes the form

$$(p_3\sigma_3 + p_0\sigma_0)\psi^0(p_3, p_0) = 0.$$

Whence $\psi^0(p_3, p_0) = \alpha \begin{pmatrix} 1 \\ 0 \end{pmatrix}$, i.e. the neutrino can exist only in one state with a given momentum and energy (the space $R_0(p_3, p_0)$ is one-dimensional). The vector $\psi(p_3, p_0)$ being an eigenvector of the operator H_3 in this case, must coincide with one or other of the vectors of the canonical basis of the representation $\tau\left(\frac{1}{2}, \frac{3}{2}\right)$. From the results of 4, paragraph 1, it follows in fact that the vector $\psi^0 = \alpha \begin{pmatrix} 1 \\ 0 \end{pmatrix}$ coincides with the vector $\xi_{\frac{1}{2}}$ (corresponding to the eigenvalue $m = \frac{1}{2}$) of the canonical basis of the representation $\tau\left(\frac{1}{2}, \frac{3}{2}\right)$.

Thus, in the case of positive energy $p_0 > 0$. the polarization value for a neutrino obeying equation (19) equals $\frac{1}{2}$. In the case $p_0 < 0$ the polarizations equals $-\frac{1}{2}$.

If the wave function of a neutrino satisfies equation (20) we obtain opposite signs of the polarization: for $p_0 > 0, m = -\frac{1}{2}$ and for $p_0 < 0$, $m = \frac{1}{2}$. Physicists call the state of a neutrino with $p_0 < 0$ an antineutrino (see the footnote on p.328). We now construct a table:

Particle	Representation	
	$\tau\left(\frac{1}{2}, \frac{3}{2}\right)$ equation(19)	$\dot{\tau}\left(-\frac{1}{2}, \frac{3}{2}\right)$ equation (20)
Neutrino $p_0 > 0$	$m = \frac{1}{2}$	$m = -\frac{1}{2}$
Antineutrino $p_0 < 0$	$m = -\frac{1}{2}$	$m = \frac{1}{2}$

The question of which of the two variants of the neutrino occurs in reality (if it is a two component neutrino in general) is answered by experiment.

VI. The polarization of photons. We consider a plane wave for Maxwell's equations describing photons. The rest mass of a photon equals zero. We now determine the polarization. We select the x_3 axis along the positive direction of the photon momentum. In such a coordinate system the energy-momentum p has the coordinates $p(p_0, 0, 0, p_3)$, $p_3 > 0$ and the plane wave has the following form:

$$F_{ik}(x) = F_{ik}(p)\, e^{ip_3x_3 - ip_0x_0}.$$

The quantity $F_{ik}(p)$ satisfies the equation

$$(p_3L_3 - p_0L_0)\, F_{ik}(p) = 0,$$

i.e. belongs to the subspace annihilating the matrix $p_3L_3 - p_0L_0$. We write this matrix out (See Sect.9, (19) and (19'))

$$
\begin{matrix} & \xi^{\tau}_{1,-1} & \xi^{\tau}_{10} & \xi^{\tau}_{11} & \xi^{\dot{\tau}}_{1,-1} & \xi^{\dot{\tau}}_{10} & \xi^{\dot{\tau}}_{11} \end{matrix}
$$

$$
p_3L_3 - p_0L_0 = \left\| \begin{matrix}
0 & -ip_3 & 0 & 0 & -ip_3 & 0 \\
-p_3 - p_0 & 0 & 0 & p_3 - p_0 & 0 & 0 \\
0 & -p_0 & 0 & 0 & -p_0 & 0 \\
0 & 0 & p_3 - p_0 & 0 & 0 & -p_3 - p_0 \\
0 & -ip_3 & 0 & 0 & ip_3 & 0 \\
-p_3 - p_0 & 0 & 0 & -p_3 + p_0 & 0 & 0 \\
0 & -p_0 & 0 & 0 & p_0 & 0 \\
0 & 0 & p_3 - p_0 & 0 & 0 & p_3 + p_0
\end{matrix} \right\|
$$

The first row of this matrix is proportional to the third, and the fifth to the seventh. If we strike out two of these (for example the 1st and the 5th), then we obtain a square 6 x 6 matrix

$$
\begin{array}{cccccc}
\xi^{\tau}_{1,-1} & \xi^{\tau}_{10} & \xi^{\tau}_{11} & \dot{\xi}^{\tau}_{1,-1} & \dot{\xi}^{\tau}_{10} & \dot{\xi}^{\tau}_{11}
\end{array}
$$

$$
\left\|\begin{array}{cccccc}
-p_3-p_0 & 0 & 0 & p_3-p_0 & 0 & 0 \\
0 & -p_0 & 0 & 0 & -p_0 & 0 \\
0 & 0 & p_3-p_0 & 0 & 0 & -p_3-p_0 \\
-p_3-p_0 & 0 & 0 & -p_3+p_0 & 0 & 0 \\
0 & -p_0 & 0 & 0 & p_0 & 0 \\
0 & 0 & p_3-p_0 & 0 & 0 & p_3+p_0
\end{array}\right\| .
$$

By interchanging rows and columns it is possible to reduce this matrix to compartment form

$$
\begin{array}{cccccc}
\xi^{\tau}_{1,-1} & \dot{\xi}^{\tau}_{1,-1} & \xi^{\tau}_{10} & \dot{\xi}^{\tau}_{10} & \xi^{\tau}_{11} & \dot{\xi}^{\tau}_{11}
\end{array}
$$

$$
\left\|\begin{array}{cccccc}
-p_3-p_0 & p_3-p_0 & 0 & 0 & 0 & 0 \\
-p_3-p_0 & -p_3+p_0 & 0 & 0 & 0 & 0 \\
0 & 0 & -p_0 & -p_0 & 0 & 0 \\
0 & 0 & -p_0 & p_0 & 0 & 0 \\
0 & 0 & 0 & 0 & p_3-p_0 & -p_3-p_0 \\
0 & 0 & 0 & 0 & p_3-p_0 & p_3+p_0
\end{array}\right\| . \tag{21}
$$

As the matrix is reducible to such a form, it is clear that for $p_0 = p_3 \neq 0$ its annihilating subspace $R(p_3, p_3)$ is two-dimensional and contains the vectors $\dot{\xi}^{\tau}_{1,-1}$ and ξ^{τ}_{11}. In this case $p_0 = -p_3 \neq 0$ the annihilating subspace $R(p_3, -p_3)$ is also two-dimensional and contains the vectors $\xi^{\tau}_{1,-1}$ and $\dot{\xi}^{\tau}_{11}$. In both cases the annihilating subspaces are invariant under the operator H_3 and the eigenvalues of this operator equal $m = 1, -1$. The polarization value $m = 0$ is excluded since a linear combination of the vectors $\xi^{\tau}_{10}, \dot{\xi}^{\tau}_{10}$ can belong to the annihilating subspace of matrix (21) only if $p_3 = p_0 = 0$. Thus the polarization of a photon may only assume the values $m = 1, -1$ or as physicists say, a photon is always polarized transversely.

§ 8. Infinite Equations

Formulae (13)-(16), Sect.2 for infinite irreducible representations of the Lorentz group (more precisely the set of values l in these formulae) show that equations in the wave function ψ which are transformed according to an infinite representation, will, generally speaking, describe particles capable of existing in states with any integral or half-integral spin which is larger than some minimum value l_0. In this case if l_0 is integral then the spin also assumes integral values; if l_0 is half-integral then the spin is half-integral. We shall now assume that the representation T_g, according to which the quantities ψ are transformed, decomposes into a finite number of irreducible representations, amongst which there are some infinite ones. In this case all the matrices $\| c_l^{\tau\tau'} \|$ will be matrices of finite order and their orders clearly coincide for all l with the possible exception of some very low values of l.

From the basic formulae (15) and (16), Sect.7, it follows that the coefficients of the characteristic equations for the matrices $\| c_l^{\tau\tau'} \|$ will be equal to the square roots of polynomials in l*.

Hence it follows that this equation will in the general case have roots increasing without limit in absolute value as $l \to \infty$**; only in an exceptional case can all the coefficients of the characteristic equation not depend upon l at all, i.e. can all the eigenvalues of the matrix $\| c_l^{\tau\tau'} \|$ coincide for all l, with the possible exception of some very low values.

Thus, infinite equations of such a type will correspond to a spectrum of masses either tending to zero as $l \to \infty$ (general case), or coinciding for all sufficiently high values of l (exceptional case).

This result explains the failure of all attempts to construct relativistic-invariant equations with an ascending mass spectrum. It is true that it is possible to construct equations with an ascending mass spectrum using an infinite number of irreducible representations: by adding for each spin l a sufficient number of new irreducible representations with $l_0 = l$, we can, without altering the states of the particle with spin less than l, attain any values of the mass for this spin. Such a construction however is rather complicated; as a rule, equations for which T_g decomposes into an infinite number of irreducible representations will also have a descending mass spectrum. In the examples of infinite equations considered in the previous paragraph the mass spectrum in both cases is of this kind: the spin l corresponds to the mass $\mu^{(l)}$, where

$$\mu^{(l)} = \frac{\varkappa}{l + \frac{1}{2}}.$$

The spin in each of these examples assumes either all integral or all half-integral values.

Section 11. The Charge and Energy of Relativistic Particles

In this section we shall everywhere assume that the relativistic invariant equation, describing the field ψ of the particle, is obtained from some invariant Lagrangian. As we saw in Sect.8 this means that it is possible to construct an invariant non-degenerate Hermitian form (ψ_1, ψ_2) from the components of the wave function ψ and that the matrices L_1, L_2, L_3, L_0 of the relativistic invariant equation satisfy the relation

---

* It may be shown, as a matter of fact, that they will always be polynomials in l.

** Generally speaking the eigenvalues of the matrix $\| c_l^{\tau\tau'} \|$ increase as l when $l \to \infty$ (in the sense of order of growth).

$$(L_k\psi_1, \psi_2) = (\psi_1, L_k\psi_2)$$

for all ψ_1 and ψ_2 or, what amounts to the same thing, the Hermitian quadratic form $(L_k\psi, \psi)$ is real.

Moreover, in this section we consider only the case of equations with $\varkappa \neq 0$.

§ 1. The Definition of Charge and Energy

With each equation arising from an invariant Lagrangian function there corresponds a vector with the components s_k:

$$s_k = (L_k\psi, \psi)\ ^* \qquad (k = 0,\ 1,\ 2,\ 3). \tag{1}$$

The vector s_k is known as the current vector**

Its last component

$$s_0 = (L_0\psi, \psi) \tag{2}$$

is known as the charge density. The total charge s is given by

$$s = \int s_0\, d^4x. \tag{3}$$

Besides the current vector there is associated with the relativistic-invariant equation the energy-momentum tensor

$$T_k^j = \mathrm{Im}\left(L_k \frac{\partial\psi}{\partial x_j}, \psi\right)^{***}, \tag{4}$$

The component T_0^0, namely

$$T_0^0 = \mathrm{Im}\left(L_0 \frac{\partial\psi}{\partial x_0}, \psi\right) \tag{5}$$

specifies the energy density $W(x) = -T_0^0$.

The total energy of the field $\psi(x)$ equals

$$E = \int W(x)\, d^4x. \tag{6}$$

---

* The result that the four numbers s_k constitute a vector, i.e. that a Lorentz transformation changes them in the same way as the coordinates x_0, x_1, x_2, x_3, was proved in Sect.8, paragraph 7.

** See for example W.Pauli, The relativistic theory of elementary particles, or A.I.Ahiezer and V.B.Berestetskii, Quantum electrodynamics, Gostekhizdat, p.393, 1953.

*** The sixteen components of the tensor T_k^j are transformed as the product of two vectors, i.e. according to a completely reducible tensor representation of the first rank (see Sect.8, paragraph 7).

We recall that in the rest system of the particle its wave function may be represented in the form of a plane wave

$$\psi(x_0, x_1, x_2, x_3) = \psi(p_0)\, e^{-ip_0x_0}, \tag{7}$$

where $\psi(p_0)$ is an eigenvector of the matrix L_0 with a non-zero eigenvalue λ

$$p_0 L_0 \psi(p_0) = \varkappa\, \psi(p_0), \qquad \lambda = \frac{\varkappa}{p_0}.$$

The charge density for relation (7) is clearly equal to

$$s_0 = (L_0 \psi(p_0)\, e^{-ip_0x_0}, \quad \psi(p_0)\, e^{-ip_0x_0}) = \lambda(\psi(p_0), \psi(p_0)). \tag{8}$$

We see that the charge density for relation (7) does not depend upon (x_0, x_1, x_2, x_3). The energy density for the plane wave (7) also does not depend upon (x_0, x_1, x_2, x_3) and is in fact equal to

$$W = (p_0 L_0 \psi(p_0), \psi(p_0)) = \varkappa(\psi(p_0), \psi(p_0)). \tag{9}$$

We now seek out finite equations with $\varkappa \neq 0$ for which either the charge density s_0 or the energy density is positive for all plane waves (7); this means that, for eigenvectors $\psi(p_0)$ of the matrix L_0 with non-zero real eigenvalues λ, one of the forms $(L_0 \psi(p_0), \psi(p_0))$, and $(\psi(p_0), \psi(p_0))$ is positive.

We shall assume from the very beginning that all the eigenvalues of the matrix L_0 are real. This restriction is reasonable in view of the fact that an eigenvector ψ_λ of the matrix L_0 with a non-real eigenvalue λ (in the previous section we called such vectors ψ_λ inadmissible) corresponds to zero energy and charge densities.

In fact, if

$$L_0 \psi_\lambda = \lambda \psi \quad \text{and } \lambda \text{ is non-real}$$

then

$$(L_0 \psi_\lambda, \psi_\lambda) = \lambda(\psi_\lambda, \psi_\lambda) = \lambda(\psi_\lambda, \psi_\lambda) = (\psi_\lambda, L_0\psi_\lambda).$$

Whence we obtain, since $\lambda \neq \overline{\lambda}$,

$$(\psi_\lambda, \psi_\lambda) = 0,$$

and in this case $s_0 = 0$ and $W = 0$.

It so happens that the finite equations ($\varkappa \neq 0$) with positive charge density or positive energy density are easily determined if we assume that the matrix L_0 is diagonable. We see that these are equations with which we are already familiar from Sect.9 : the Dirac equation for particles with spin $l = \frac{1}{2}$ (positive charge) and the Daffine equations for particles with spin 0 or 1 (positive energy).

§ 2. Finite Equations with Positive Charge and Diagonable Matrix L_0

The condition for positive charge density is clearly that for all eigenvectors $\psi_\lambda(p)$ of the matrix L_0 with a non-zero eigenvalue λ,

$$(L_0\psi_\lambda, \psi_\lambda) > 0. \tag{10}$$

We shall show that if the matrix L_0 reduces to diagonal form then condition (10) is equivalent to the stipulation that

$$(L_0\psi, \psi) \geqslant 0$$

for all the vectors ψ of the space R, i.e. that the form $(L_0\psi, \psi)$ be definite non-negative.

In fact the set of all linearly independent eigenvectors of a diagonable matrix L_0 constitute a basis of the space R. In this case, any vector ψ of R can be resolved into a sum

$$\psi = c_1\psi_1 + c_2\psi_2 + \ldots + c_k\psi_k + \ldots,$$

where $\psi_1, \ldots, \psi_k$, are the eigenvectors of the matrix L_0 with the various eigenvalues $\lambda_1, \lambda_2, \ldots, \lambda_k$. We note that for two eigenvectors ψ_i and ψ_k with different eigenvalues, the equation

$$(L_0\psi_i, \psi_k) = 0.$$

is satisfied. This equation is derived as follows:

$$(L_0\psi_i, \psi_k) = \lambda_i(\psi_i, \psi_k)$$

and

$$(L_0\psi_i, \psi_k) = (\psi_i, L_0\psi_k) = \lambda_k(\psi_i, \psi_k)$$

(we have assumed that λ_k is real).

Since $\lambda_i \neq \lambda_k$, $(\psi_i, \psi_k) = 0$ and $(L_0\psi_i, \psi_k) = 0$. We now calculate the value of the form $(L\psi, \psi)$ explicitly:

$$(L_0\psi, \psi) = \sum_{i,k} c_i\bar{c}_k(L_0\psi_i, \psi_k) = \sum |c_i|^2 (L_0\psi_i, \psi_i).$$

In virtue of $(L_0\psi_i, \psi_i) \geqslant 0$ we have

$$(L_0\psi, \psi) \geqslant 0,$$

i.e. the form $(L_0\psi, \psi)$ is definite non-negative.

Thus we must find finite equations for which the form $(L_0\psi, \psi)$ is definite non-negative.

With the aid of formulae (15)-(16), Sect.7, and (20), Sect.8, the form $(L_0\psi, \psi)$ has the following expression in terms of the basis $\{\xi^{\tau}_{lm}\}$ which is canonical for the representation T_g which transforms the components of the wave function,

$$(L_0\psi, \psi) = \sum a_l^{\tau\tau^*} c_l^{\tau\tau'} y^{\tau'}_{lm} \bar{y}^{\tau^*}_{lm}, \tag{11}$$

where the numbers $a_l^{\tau\tau^*}$ and $c_l^{\tau\tau'}$ are defined by formulae (15)-(16) of 7 and (16)-(17) of Sect.8. We recall that $c_l^{\tau\tau'} \neq 0$ only for interlocking components and $a_l^{\tau\tau^*} \neq 0$ for the components $\tau \sim (l_0, l_1)$ and $\tau^* \sim (l_0, -\bar{l}_1)$, l_1 being real in a finite case, and therefore $\tau^* = \dot{\tau} = (-l_0, l_1)$.

From expression (11) it is clear that if the form $(L_0\psi, \psi)$ is non-negative the components τ and τ^* must interlock, for otherwise the form would not contain the terms with $|y^{\tau^*}_{lm}|^2$ and $|y^{\tau'}_{lm}|^2$ but would only contain the products of different coordinates $y^{\tau^*}_{lm}\bar{y}^{\tau'}_{lm}$. It is easy to see that such a form can assume values of different signs. The finite components τ and τ^* interlock only if $l_0 = -l_0 \pm 1$, i.e. $l_0 = \pm\frac{1}{2}$. Thus the representation T_g contains only irreducible components of the form $\left(\pm\frac{1}{2}, l_1\right)$ which interlock according to the scheme

$$\left(\frac{1}{2}, l_1\right) \longleftrightarrow \left(-\frac{1}{2}, l_1\right). \tag{12}$$

The form $(L_0\psi, \psi)$ in this case becomes

$$(L_0\psi, \psi) = \sum a_l^{\tau\tau'} c_l^{\tau\tau'} |y^{\tau'}_{lm}|^2. \tag{13}$$

If now any pair of interlocking components contain a weight l greater than one, $\left(\text{i.e. if } l_1 > \frac{3}{2}\right)$, then the terms $|y^{\tau'}_{lm}|^2$ and $|y^{\tau'}_{l+1,m}|^2$ enter into equations (13) with different signs (in fact, for finite representations $a_l^{\tau\tau'} = -a_{l+1}^{\tau\tau'}$, (see formula (38) Sect.2), and $c_l^{\tau\tau'}$ and $c_{l+1}^{\tau\tau'}$ have the same sign). Consequently the form is non-negative only when the representation T_g contains the components $\tau \sim \left(\pm\frac{1}{2}, \frac{3}{2}\right)$.

It is easily verified that in the case when each of these components appears in the representation T_g more than once, the equation and the invariant form also will be degenerate. Thus for a non-degenerate equation the representation $g \to T_g$ contains only two interlocking components,

$\tau\left(\frac{1}{2},\frac{3}{2}\right)$ and $\tau^*=\dot{\tau}\left(-\frac{1}{2},\frac{3}{2}\right)$. As we have seen in Sect.9, this leads to the Dirac equation. For the Dirac equation the form $(L_0\psi,\psi)$ is given by (see (2) and (4), Sect.9).

$$(L_0\psi,\psi)=\sum_{m=\frac{1}{2},-\frac{1}{2}}\left\{\left|y^{\tau}_{\frac{1}{2}m}\right|^2+\left|y^{\dot{\tau}}_{\frac{1}{2}m}\right|^2\right\}>0. \tag{14}$$

Thus, of the finite equations ($\varkappa\neq 0$) with a diagonable matrix L_0, only the Dirac equation has a positive charge density.

§ 3. Finite Equations with Positive Energy and a Diagonable Matrix L_0

As we have seen, for positive energy, the following form must be positive

$$W=\varkappa(\psi_\lambda,\psi_\lambda)>0 \tag{15}$$

whenever the eigenvector ψ_λ of the matrix L_0 corresponds to a non-zero eigenvalue. In this case when the matrix L_0 can be reduced to diagonal form, this condition is equivalent to the requirement that

$$(L_0\psi,L_0\psi)\geqslant 0 \tag{16}$$

for all vectors ψ. This circumstance is tested in precisely the same way as in the previous section for the case of the form $(L_0\psi,\psi)$.

In the canonical basis of the representation $g\to T_g$ this form is given by:

$$(L_0\psi,L_0\psi)=\sum_{l,\tau,m} a_l^{\tau\tau^*}c_l^{\tau\tau'}\bar{c}_l^{\tau^*\tau^{*\prime}}y^{\tau'}_{lm}\bar{y}^{\tau^{*\prime}}_{lm}; \tag{17}$$

Here, $\tau^{*\prime}$ denotes components interlocking with τ^*. In order that this expression be non-negative the components τ and $\tau^{*\prime}$ must interlock, for otherwise the form (17) would not contain the terms $\left|y^{\tau'}_{lm}\right|^2$ and $\left|y^{\tau^*}_{lm}\right|^2$ but would contain only the products $y^{\tau'}_{lm}\bar{y}^{\tau^*}_{lm}$ and, consequently, could assume values of either sign.

In the finite case, the components τ and τ^* interlock only if

a) $l_0\pm 1=-l_0\pm 1$, i.e. $l_0=0,\ 1,\ -1$,

and

b) $l_0=0$, i.e. $\tau\sim(0,l)$ and $\tau^{*\prime}\sim(0,l\pm 1)$.

Thus, the representation T_g contains components of the form $(1,l_1)$, $(-1,l_1)$, $(0,l_1)$, which interlock according to the scheme

a) $$(-1,l_1)\longleftrightarrow(0,l_1)\longleftrightarrow(1,l_1)$$

and

b) $$(0,\, l_1-1) \longleftrightarrow (0,\, l_1).$$

We shall show that in both cases $l_1 = 2$. In fact for $l_1 > 2$ each of the components $(1,\, l_1)$, $(-1,\, l_1)$, $(0,\, l_1)$, $(0,\, l_1-1)$ would contain more than unit weight l and the terms $|y_{lm}^{\tau_1}|^2$ and $|y_{l-1,m}^{\tau_1}|^2$ would enter into expression (17) with different signs $\left(a_{l-1}^{\tau\tau^*} = -a_l^{\tau\tau^*}\right)$.

Thus the representation T_g contains only the components of the form $\tau_0 \sim (0,\, 1), \tau_1 \sim (0,\, 2),\ \tau_2 \sim (-1,\, 2),\ \tau_2 \sim (1,\, 2)$. They cannot all enter simultaneously into the interlocking scheme

$$\begin{array}{ccccc} \dot{\tau}_2 & \longleftrightarrow & \tau_1 & \longleftrightarrow & \tau_2 \\ & & \updownarrow & & \\ & & \tau_0 & & \end{array},$$

since, in, this case, the coefficient in form (17) for $|y_{00}^{\tau_0}|^2$ (equal to $a_0^{\tau_1\tau_1}|c_0^{\tau_1\tau_0}|^2$) and the coefficient for $|y_{1m}^{\tau_2}|^2$ (equal to $a_1^{\tau_1\tau_1}|c_1^{\tau_1\tau_2}|^2$) would be of different sign. Moreover in the case when any of the components enters into T_g more than once the equation is degenerate. Thus for a non-degenerate equation only two schemes of interlocking are possible:

a) $$(0,\, 2) \longleftrightarrow (0,\, 1)$$

and

b) $$(-1,\, 2) \longleftrightarrow (0,\, 2) \longleftrightarrow (1,\, 2).$$

The first of them leads to the Daffine equation (see Sect. 9) for particles with spin 0, and the second to the Daffine equation for particles with spin 1. The form $(L_0\psi,\, L_0\psi)$ in each of these cases equals (see Sect. 9).

a) $$(L_0\psi,\, L_0\psi) = |y_{00}^{\tau_0}|^2 + |y_{00}^{\tau_1}|^2 \geqslant 0$$

and

b) $$(L_0\psi,\, L_0\psi) = 2\sum_m |y_{1m}^{\tau_1}|^2 + \sum_m \{|y_{1m}^{\tau_2}|^2 + |y_{1m}^{\dot{\tau}_2}|^2 + y_{1m}^{\tau_2}\overline{y_{1m}^{\dot{\tau}_2}} + y_{1m}^{\dot{\tau}_2}\overline{y_{1m}^{\tau_2}}\} \geqslant 0,$$

so that the energy is in fact positive.

Thus, of the finite equations with a diagonable matrix L_0 only the Daffine equations (for particles with spin $l = 0$ or for particles with spin $l = 1$) have a positive energy.

However, finite equations with positive energy or charge are possible for which the matrix L_0 does not reduce to diagonal form.

We now consider one such equation with positive charge.

§ 4. Equations with Positive Charge and a Matrix L_0 not Reducing to Diagonal Form

Let the representation T, which transforms the components of the wave function consist of the components

$$\tau_1 \sim \left(\frac{1}{2}, \frac{3}{2}\right), \quad \dot{\tau}_1 \sim \left(-\frac{1}{2}, \frac{3}{2}\right), \quad \tau_2 \sim \left(\frac{1}{2}, \frac{5}{2}\right), \quad \dot{\tau}_2 \sim \left(-\frac{1}{2}, \frac{5}{2}\right),$$

which interlock according to the scheme

$$\begin{array}{ccc} \tau_1 & \longleftrightarrow & \dot{\tau}_1 \\ \updownarrow & & \updownarrow \\ \tau_2 & \longleftrightarrow & \dot{\tau}_2 \end{array}. \tag{18}$$

We recall that we have already encountered such a scheme in Sect. 9, paragraph 5, where we also found the general form of the invariant equations corresponding to this scheme arising from an invariant Lagrangian function. The matrix L_0 of such equations consists of the following compartments

$$\left.\begin{array}{lll}
\text{a)} & l=\dfrac{1}{2} & l=\dfrac{3}{2} \\
& \left\|\begin{array}{cccc} 0 & \alpha & \beta & 0 \\ \alpha & 0 & 0 & \beta \\ \beta & 0 & 0 & \frac{1}{2} \\ 0 & \beta & \frac{1}{2} & 0 \end{array}\right\|; & \left\|\begin{array}{cc} 0 & 1 \\ 1 & 0 \end{array}\right\|, \\
\text{or} & & \\
\text{b)} & l=\dfrac{1}{2} & l=\dfrac{3}{2} \\
& \left\|\begin{array}{cccc} 0 & \alpha & \beta & 0 \\ \alpha & 0 & 0 & \beta \\ -\beta & 0 & 0 & \frac{1}{2} \\ 0 & -\beta & \frac{1}{2} & 0 \end{array}\right\|; & \left\|\begin{array}{cc} 0 & 1 \\ 1 & 0 \end{array}\right\|.
\end{array}\right\} \tag{19}$$

Here α and β are real numbers, β being greater than zero.

To each of the cases a) and b) there correspond different invariant forms. We now write them out:

$$
\left.
\begin{aligned}
&\text{a)}\quad (\psi_1, \psi_2) = \sum_{m=\frac{1}{2}, -\frac{1}{2}} \left\{ x^{\tau_1}_{\frac{1}{2} m} \bar{y}^{\dot{\tau}_1}_{\frac{1}{2} m} + x^{\dot{\tau}}_{\frac{1}{2} m} \bar{y}^{\tau_1}_{\frac{1}{2} m} \right\} + \\
&\qquad + \sum_{l, m} (-1)^{[l]} \left\{ x^{\tau_2}_{lm} \bar{y}^{\dot{\tau}_2}_{lm} + x^{\dot{\tau}_2}_{lm} \bar{y}^{\tau_2}_{lm} \right\}; \\
&\text{b)}\quad (\psi_1, \psi_2) = - \sum_{m=\frac{1}{2}, -\frac{1}{2}} \left\{ x^{\tau_1}_{\frac{1}{2} m} \bar{y}^{\dot{\tau}_1}_{\frac{1}{2} m} + x^{\dot{\tau}_1}_{\frac{1}{2} m} \bar{y}^{\tau_1}_{\frac{1}{2} m} \right\} + \\
&\qquad + \sum_{l, m} (-1)^{[l]} \left\{ x^{\tau_2}_{lm} \bar{y}^{\dot{\tau}_2}_{lm} + x^{\dot{\tau}_2}_{lm} \bar{y}^{\tau_2}_{lm} \right\}.
\end{aligned}
\right\}
\tag{20}
$$

We note that the energy W in both cases is not positive definite. In fact in the states $\psi_{-\varkappa \frac{3}{2} m}$ and $\psi_{\varkappa \frac{3}{2} m}$, corresponding to the spin $l = \frac{3}{2}$ and to the eigenvalues of the compartment

$$
\begin{Vmatrix} 0 & 1 \\ 1 & 0 \end{Vmatrix},
$$

equal to ± 1, the energy has different signs:

$$
(\psi_{-\varkappa \frac{3}{2} m}, \psi_{-\varkappa \frac{3}{2} m}) = -(\psi_{\varkappa \frac{3}{2} m}, \psi_{\varkappa \frac{3}{2} m}).
$$

We now consider the charge. In case a) the matrix L_0 reduces to diagonal form for any values α and β and, consequently, the charge for the corresponding equations is not positive definite.

We now turn to case b). Since we are looking for a matrix L_0 which does not reduce to diagonal form, it must have multiple eigenvalues. This means that the characteristic equation of the matrix $\| c^{\tau\tau'}_{\frac{1}{2}} \|$

$$
\lambda^4 - \left(\alpha^2 + \frac{1}{4} - 2\beta^2 \right) \lambda^2 + \left(\frac{\alpha}{2} + \beta^2 \right)^2 + 0 \tag{21}
$$

has multiple roots

Three cases are possible here:

1) $\quad \alpha = -\frac{1}{2}; \quad$ the roots: $\lambda_{1,2} = \sqrt{\frac{1}{4} - \beta^2} = -\lambda_{3,4}$,

2) $\quad \alpha = \frac{1}{2} \pm 2\beta; \quad$ the roots: $\lambda_{1,2} = \frac{1}{2} \pm \beta = -\lambda_{3,4}$,

3) $\alpha = -2\beta^2$; the roots: $\lambda_{1,2} = 0$, $\lambda_3 = -\lambda_4 = \frac{1}{2} - 2\beta^2$. (22)

Cases 1) and 2) for $\beta \neq \frac{1}{2}$ give a matrix L_0 reducible to diagonal form, and consequently the charge in these cases is not positive definite. In case 3) the matrix L_0 no longer reduces to diagonal form. In this case for all the eigen vectors $\psi^0_{\lambda \frac{1}{2} m}$ of the compartment $\|c^{\tau\tau'}_{\frac{1}{2}}\|$ the form $(\psi^0, \psi^0) = 0$. Thus in case 3) the charge is non-negative. However for $\beta \neq \frac{1}{2}$ there exist non-zero eigenvalues λ of the compartment $\|c^{\tau\tau'}_{\frac{1}{2}}\|$ i.e. states (plane waves) are possible with spin $l = \frac{1}{2}$, whose energy and charge are equal to zero $W = \varkappa(\psi^0, \psi^0) = s_0 = \lambda_i(\psi^0, \psi^0) = 0$. Only for $\beta = \frac{1}{2}$ is this not so for such states (all the eigenvalues of the compartment $\|c^{\tau\tau'}_{\frac{1}{2}}\|$ corresponding to the spin $l = \frac{1}{2}$ for $\beta = \frac{1}{2}$ are zero). For $\beta = \frac{1}{2}$ of formulae (22) 1) and 2) we find that $\alpha = -\frac{1}{2}$ and we arrive at the Pauli-Fierz equation (see Sect.9). The charge for this equation (in the states $l = \frac{3}{2}$) has the form

$$s_0 = \sum_{m=-\frac{3}{2}}^{m=+\frac{3}{2}} \left\{ |y^{\tau_2}_{\frac{3}{2} m}|^2 + |\dot{y}^{\tau_2}_{\frac{3}{2} m}|^2 \right\} > 0.$$

Thus, of all equations with the interlocking scheme (18) the Pauli-Fierz equation is the only one for which all plane waves in the rest system have positive charge (and spin $l = \frac{3}{2}$).

§ 5. Pauli's Theorem

In the previous section we saw that in the case of finite equations there do not exist equations with integral spin and positive definite charge density, nor particles with half-integral spin having positive definite energy density. This fact is a consequence of a more general theorem due to Pauli. Before formulating this theorem we recapitulate a little.

In Sect.8, § 8, we considered the functions $\Phi(x)$ whose values at each point x depend upon the values of the wave function $\psi(x)$ and its partial derivatives at this point. Under the Lorentz transformation $x' = gx$ and $\psi(x') = T_g\psi(x)$ the quantities $\Phi(x)$ are also transformed according to some representation $g \to \tau_g$

$$\tilde{\Phi}(x') = \tau_g \Phi(x).$$

We showed in Sect.8 that the representation $g \to \tau_g$ according to which the quantities $\Phi(x)$ are transformed is contained in the sum of the representations

We note that both the representations τ_1 and τ_2 are unitary (see Sect.2). Thus the scalar product (ψ_1, ψ_2) is an invariant form (ψ, ψ) being positive. Hence the charge (for plane waves in the rest system)

$$s_0 = (L_0\psi^0, \psi^0) = \left(l + \frac{1}{2}\right)(\psi^0, \psi^0)$$

and the energy

$$W = \varkappa(\psi^0, \psi^0)$$

are positive.

Thus we have obtained two representations: one for particles with integral spin (representation τ_1), the other for particles with half-integral spin (the pair τ_2), for which the energy and the charge are positive.

II. The equation with the interlocking scheme

$$\left(\frac{1}{2}, l_1\right) \longleftrightarrow \left(-\frac{1}{2}, l_1\right), \tag{25}$$

where l_1 is purely imaginary or real. The matrix L_0 consists of the compartments $\left\|c_l^{\tau\tau'}\right\|$

$$\left\|c_l^{\tau\tau'}\right\| + \begin{Vmatrix} 0 & l+\frac{1}{2} \\ l+\frac{1}{2} & 0 \end{Vmatrix}. \tag{26}$$

For l_1 purely imaginary it is possible to select the invariant form (ψ, ψ) positive definite: $(\psi, \psi) > 0$.

In this case the energy is positive and the charge may assume values of either sign. The spin of the particle is half-integral.

For real l_1 we obtain an equation with an energy of both signs: for $0 \leqslant |l_1| \leqslant \frac{3}{2}$ the charge will be positive definite and for $|l_1| > \frac{3}{2}$ the sign of the charge will be indefinite; the spin, as before, is half-integral.

Finally, for the equations with the interlocking scheme

$$\left(l_1, \frac{1}{2}\right) \longleftrightarrow \left(l_1, -\frac{1}{2}\right)$$

(where l_1 is integral or half-integral; the matrix L_0 for such equations has the previous form (26)) we will have positive definite charge and energy of both signs: the spin corresponding to these equations will be integral or half-integral simultaneously with l_1.

Thus, from the example cited it is clear that for infinite equations it is possible to have, both for integral and for half-integral spin, either the energy or the charge positive, or both quantities simultaneously.

SUPPLEMENTS

I. Irreducible Representations of the Group of Orthogonal Matrices

In the first part of the book formulae were given for the infinitesimal operators of any irreducible representation of the three-dimensional group of rotations, or what comes to the same thing, of the group of orthogonal matrices of the third order. In this supplement we shall give the explicit formulae specifying the infinitesimal operators of irreducible representations of the group of orthogonal matrices of any order.

We recall that a matrix $\|g_{ik}\|$ is said to be orthogonal if the linear transformation specified by it, namely

$$x'_i = \sum_{k=1}^{n} g_{ik} x_k \qquad (i = 1, 2, \ldots, n)$$

conserves the form $\sum_{i=1}^{n} x_i^2$, i.e. if $\sum_{i=1}^{n} (x'_i)^2 = \sum_{i=1}^{n} x_i^2$.

It is clear that the n-th order orthogonal matrices constitute a group, which we shall denote by K_n. In the group K_n it is possible to select $n(n-1)/2$ distinct orthogonal subgroups just as we were able to do for the group of rotations and the proper Lorentz group. We consider the subgroup of orthogonal matrices which transform only the variables x_i and x_k $(i \neq k)$ and does not alter the remaining variables. Inasmuch as each such matrix conserves the sum $x_i^2 + x_k^2$, there corresponds to it a rotation in the plane $(x_i x_k)$. The matrix itself has the form

$$g_{ik}(t) = \left\| \begin{array}{cccccccc} & & & (i) & & (k) & & \\ 1 & 0 \ldots 0 & & 0 & \ldots & 0 & \ldots & \\ 0 & 1 & \ldots & \ldots & \ldots & \ldots & \ldots & \\ \ldots & \ldots & \ldots & \ldots & \ldots & \ldots & \ldots & \\ 0 & 0 \ldots 1 & & 0 & \ldots & 0 & \ldots & \\ \ldots & \ldots 0 & & \cos t & \ldots & \sin t & \ldots & (i) \\ \ldots & \ldots 0 & & 0 & \ldots & 0 & \ldots & \\ \ldots & \ldots & \ldots & \ldots & \ldots & \ldots & \ldots & \\ \ldots & \ldots 0 & & -\sin t & \ldots & \cos t & 0 \ldots & (k) \\ & & & & & 0 & 1 \ldots & \\ \ldots & \ldots & \ldots & \ldots & \ldots & \ldots & \ldots 1 & \end{array} \right\| \qquad (1)$$

For a given pair of indices i, k, the set of all matrices of type (1) clearly constitutes a one-parameter subgroup of the group of orthogonal matrices. In all it is possible to construct $n(n-1)/2$ different one-parameter subgroups in this manner. It can be shown that every orthogonal matrix can be represented as a product of matrices $g_{ik}(t)$ from these subgroups.

We now consider a representation $g \to T_g$ of the group K_n. To each one-parameter subgroup $g_{ik}(t)$ there corresponds an infinitesimal operator I_{ik} $(i < k = 1, \ldots, n)$ in this representation. We shall write down the commutation relations between these operators. We note that for the identity representation $g \to g$ of the group K_n the infinitesimal operator corresponding to the subgroup $g_{ik}(t)$ is the matrix e_{ik} for which the elements $g_{ik} = -g_{ki} = 1$, and the remaining elements equal zero. It is easily verified that the commutators of these matrices are given by

$$[e_{i_1k_1}, e_{i_2k_2}] = \delta_{k_1i_2}e_{i_1k_2} + \delta_{i_1k_2}e_{k_1i_2} - \delta_{k_1k_2}e_{i_1i_2} - \delta_{i_1i_2}e_{k_1k_2}. \tag{2}$$

Analogous commutation relations hold also between the infinitesimal operators I_{ik} of any representation of the orthogonal group.

We note that any representation of an orthogonal group is unitary. In this case the operators I_{ik} and I_{ki} are connected by the relation $I_{ik} = -(I_{ki})^*$. From relations (2) it is clear that for the description of the representation it is sufficient to know how the operators $I_{21}, I_{32}, I_{43}, \ldots, I_{n,\,n-1}$ act since the others can be expressed in terms of them with the aid of relations (2).

Thus we construct the infinitesimal operators $I_{k+1,\,k}$ for all the irreducible representations of the group of n-th order orthogonal matrices.

We start with some particular cases:

1. We recall first of all the case $n = 3$. Each irreducible representation of this group is specified by an integral or half-integral number l. In the space R where such a representation acts, it is possible to select a canonical basis $\{\xi_m\}$ where the index m assumes all values, which are integral or half-integral simultaneously with l, and are included in the range $l \geqslant m \geqslant -l$. In the basis $\{\xi_m\}$ the operators I_{21} and I_{32} are specified by the formulae

$$\left.\begin{aligned} I_{21}\xi_m &= im\xi_m, \\ I_{32}\xi_m &= \sqrt{(l-m)(l+m+1)}\,\xi_{m+1} - \sqrt{(l-m+1)(l+m)}\,\xi_{m-1}. \end{aligned}\right\} \tag{3}$$

2. We consider the case $n = 4$. Each irreducible representation of the group of 4-th order orthogonal matrices generates a representation $\tilde{g} \to T_{\tilde{g}}$ of the group of 3rd-order matrices (generally speaking, reducible). The space R, where the representation $g \to T_g$ acts, in this case breaks up into a sum of subspaces R_l in each of which the representation $\tilde{g} \to T_{\tilde{g}}$ is irreducible and is specified by the weight l.

It can be shown that the weight l assumes all integral or half-integral values between two numbers m_1 and $|m_2|$ once only: $m_1 \geqslant l \geqslant |m_2|$, the number m_1 and m_2 specifying the representation are simultaneously integral or half-integral. In each of the subspaces R_l it is possible to select a canonical basis $\{\xi^l_m\}$, and all these bases together form a basis of the entire space R. The operators I_{12} and I_{23} act on the basis $\{\xi^l_m\}$ according to formula (3). The operator I_{43} is specified by the formula

$$I_{43}\xi^l_m = \sqrt{\frac{(l+m+1)(l-m+1)(m_1-l)(m_1+l+2)(l-m_2+1)(l+m_2+1)}{(2l+1)(2l+3)(l+1)^2}} \times$$

$$\times \xi^{l+1}_m + im\frac{(m_1+1)m_2}{(l+1)l}\xi^l_m -$$

$$-\sqrt{\frac{(l+m)(l-m)(m_1-l+1)(m_1+l+1)(l-m_2)(l+m_2)}{(2l+1)(2l-1)l^2}}\xi^{l-1}_m. \quad (4)$$

3. Finally, we consider the case $n = 5$. Each representation of the fifth-order group generates a representation of the fourth-order group. The latter may be decomposed into irreducible components defined by pairs m_1, m_2. The numbers m_1, m_2 independently assume once only all the values included within the limits $n_1 \geqslant m_1 \geqslant n_2 \geqslant m_2 \geqslant -n_2$, where the numbers n_1 and n_2 (simultaneously integral or half-integral) specify the representation. In each of the irreducible subspaces $R(m_1 m_2)$ it is possible to select a canonical basis $\xi\begin{pmatrix} m_1 & m_2 \\ & l & \\ & m & \end{pmatrix}$. The operators I_{12}, I_{23}, I_{34} act in this basis according to formula (3) and (4) (the indices m_1 and m_2 do not change).

The operator I_{54} is specified by the formula

$$I_{54}\xi\begin{pmatrix} m_1 & m_2 \\ & l & \\ & m & \end{pmatrix} =$$

$$= \sqrt{\frac{(m_1-l+1)(m_1+l+2)(n_1-m_1)(n_1+m_1+2)(m_1-n_2+1)(m_1+n_2+2)}{(m_1+m_2+1)(m_1+m_2+2)(m_1-m_2+1)(m_1-m_2+2)}} \times$$

$$\times \xi\begin{pmatrix} m_1+1, & m_2 \\ & l & \\ & m & \end{pmatrix} +$$

$$+\sqrt{\frac{(l-m_2)(m_2+l+1)(n_2-m_2)(n_2+m_2+1)(n_1-m_2+1)(n_1+m_2+2)}{(m_1+m_2+1)(m_1+m_2+2)(m_1-m_2)(m_1-m_2+1)}} \times$$

$$\times \xi\begin{pmatrix} m_1, & m_2+1 \\ & l & \\ & m & \end{pmatrix} -$$

$$-\sqrt{\frac{(m_1+l+1)(m_1-l)(n_1-m_1+1)(n_1+m_1+2)(m_1-n_2)(m_1+n_2+1)}{(m_1+m_2)(m_1+m_2+1)(m_1-m_2)(m_1-m_2+1)}}\times$$

$$\times\,\xi\begin{pmatrix} m_1-1, & m_2 \\ l & \\ m & \end{pmatrix}-$$

$$-\sqrt{\frac{(l-m_2+1)(m_2+l)(n_2-m_2+1)(n_2+m_2)(n_1-m_2+2)(m_2+n_1+1)}{(m_1+m_2)(m_1+m_2+1)(m_1-m_2+2)(m_1-m_2+1)}}\times$$

$$\times\,\xi\begin{pmatrix} m_1, & m_2-1 \\ l & \\ m & \end{pmatrix}.$$

4. We now consider the general case. Let n be even: $n=2k+2$. We form the array

$$\alpha=\left\{\begin{array}{ccccccc}
m_{2k,1} & & m_{2k,2} & \dots & \dots & \dots & m_{2k,k} \\
m_{2k-1,1} & & m_{2k-1,2} & \dots & \dots & \dots & m_{2k-1,k} \\
& m_{2k-2,1} & & m_{2k-2,2} & \dots & m_{2k-2,k-1} & \\
& m_{2k-3,1} & & m_{2k-3,2} & \dots & m_{2k-3,k-1} & \\
\dots & \dots & \dots & \dots & \dots & \dots & \dots \\
\dots & \dots & \dots & \dots & \dots & \dots & \dots \\
& & m_{41} & & m_{42} & & \\
& & m_{31} & & m_{32} & & \\
& & & m_{21} & & & \\
& & & m_{11} & & &
\end{array}\right\},$$

where the numbers m_{ij} are simultaneously integral or half-integral. In this case the numbers m_{ij} assume all the values

$$n_1 \geqslant m_{2k,1} \geqslant n_2 \geqslant m_{2k,2} \geqslant \dots \geqslant n_k \geqslant m_{2k,k} \geqslant |n_{k+1}|,$$

$$m_{2k,1} \geqslant m_{2k-1,1} \geqslant m_{2k,2} \geqslant \dots \geqslant m_{2k-1,k} \geqslant -m_{2k-1,k},$$

etc. $m_{2k-1,1} \geqslant m_{2k-2,1} \geqslant m_{2k-1,2} \geqslant \dots \geqslant m_{2k-2,k-1} \geqslant |m_{2k-1,k}|$

$$\left.\begin{array}{llll}
m_{2p+1,i} \geqslant m_{2p,i} & \geqslant m_{2p+1,i+1}, & & i=1,\dots,p-1, \\
m_{2p+1,p} \geqslant m_{2p,p} & \geqslant |m_{2p+1,p+1}|, & & \\
m_{2p,i} \geqslant m_{2p-1,i} & \geqslant m_{2p,i+1}, & & i=1,\dots,p-1, \\
m_{2p,p} \geqslant m_{2p-1,p} & \geqslant -m_{2p,p}. & &
\end{array}\right\} \quad (5)$$

Here $n_1, n_2, \dots, n_{k+1}$ are fixed numbers (simultaneously integral or half-integral), which define the representation.

In the case $n=2k+1$ the array α has the form

$$\alpha = \begin{pmatrix} m_{2k-1,1} & & m_{2k-1,2} & \cdots & \cdots & m_{2k-1,k} \\ & m_{2k-2,1} & \cdots & \cdots & m_{2k-2,k-1} & \\ & m_{2k-3,1} & & & m_{2k-3,k-1} & \\ & \cdots & \cdots & \cdots & \cdots & \\ & \cdots & \cdots & \cdots & \cdots & \\ & & m_{41} & m_{42} & & \\ & & m_{31} & m_{32} & & \\ & & & m_{21} & & \\ & & & m_{11} & & \end{pmatrix},$$

and the numbers m_{ij} are subject to the conditions

$$\left.\begin{aligned} n_1 \geqslant m_{2k-1} \geqslant n_2 \geqslant m_{2k-1,2} \geqslant \ldots \geqslant n_k \geqslant m_{2k-1,k} \geqslant -n_k, \\ m_{2k-1,1} \geqslant m_{2k-2,1} \geqslant m_{2k-1,2} \geqslant \ldots \geqslant m_{2k-2,k-1} \geqslant |m_{2k-1,k}| \end{aligned}\right\} \quad (6)$$

etc. following inequalities (5). The numbers $n_1, n_2, \ldots, n_k$ are fixed numbers specifying the representation (simultaneously integral or half-integral). With each such array we associate a vector $\xi(\alpha)$ of the space R in which our representation acts. It happens that in this case the vectors $\xi(\alpha)$ constitute a basis of the space R. We shall write down the formulae for the operators $I_{2p+1,\,2p}$ and $I_{2p+2,\,2p+1}$ in the basis $\xi(\alpha)$. We shall denote by $\xi^+(\alpha_r^j)$ the vector corresponding to the array obtained from the array α by changing m_{rj} to $m_{rj}+1$, and by $\xi^-(\alpha_r^j)$ -the vector whose array is obtained from α by changing m_{rj} to $m_{rj}-1$ (it is assumed that in both cases we arrive at arrays which satisfy conditions (5) and (6)).

The formulae for the operators $I_{2p+1,\,p}$ and $I_{2p+2,\,2p+1}$ have the following form

$$I_{2p+1,\,2p}\xi(\alpha) = \sum_{j=1}^{p} A(m_{2p-1,\,j})\,\xi^+(\alpha_{2p-1}^j) - \sum A(m_{2p-1,\,j-1})\,\xi^-(\alpha_{2p-1}^j),$$

$$I_{2p+2,\,2p+1}\xi(\alpha) = \sum_{j=1}^{p} B(m_{2p,\,j})\,\xi^+(\alpha_{2p}^j) - \sum B(m_{2p,\,j-1})\,\xi^-(\alpha_{2p}^j) + iC_{2p}\xi(\alpha).$$

We now write down the expressions for the coefficients A, B and C.

We use the following notation

$$\begin{aligned} m_{2p-1,\,p} &= l_{2p-1,\,p}, & m_{2p,\,p}+1 &= l_{2p,\,p}, \\ m_{2p-1,\,p-1} &= l_{2p-1,\,p-1}, & m_{2p,\,p-1}+2 &= l_{2p,\,p-1}, \\ &\cdots & &\cdots \\ m_{2p-1,\,1}+p-1 &= l_{2p-1}, & m_{2p,\,1}+p &= l_{2p,\,1}. \end{aligned}$$

In which case the numbers defining the representation, namely $n_1, \ldots, n_{k+1}$

for even $n=2k+2$ and $n_1, \ldots, n_k$ for odd $n=2k+1$ are denoted in these equations by $m_{2k+1,1}, m_{2k+1,2}, \ldots, m_{2k+1,k+1}$ and $m_{2k,1}, m_{2k,2}, \ldots, m_{2k,k}$ respectively.

In this notation A, B and C are given by

$$A(m_{2p-1,j}) = \left[\prod_{r=1}^{p-1}(l_{2p-2,r}-l_{2p-1,j}-1)(l_{2p-2,r}+l_{2p-1,j})\right]^{1/2}\times$$

$$\times\left[\prod_{r=1}^{p}(l_{2p,2}-l_{2p-1,j}-1)(l_{2p,r}+l_{2p-1,j})\right]^{1/2}\times$$

$$\times\left\{\prod(l_{2p-1,r}^2-l_{2p-1,j}^2)\left[l_{2p-1,r}^2-(l_{2p-1,j}+1)^2\right]\right\}^{-1/2},$$

$$B(m_{2p,j}) = \left[\frac{\prod_{r=1}^{p}(l_{2p-1,r}^2-l_{2p,j}^2)\prod_{r=1}^{p+1}(l_{2p+1,r}^2-l_{2p,j}^2)}{l_{2p,j}^2(4l_{2p,j}^2-1)\prod_{r\neq j}\left[l_{2p,r}^2-l_{2p,j}^2\right]\left[(l_{2p,r}-1)^2-l_{2p,j}^2\right]}\right]^{-1/2},$$

$$C_{2p} = \frac{\prod_{r=1}^{p} l_{2p-1,r}\prod_{r=1}^{p+1} l_{2p+1,r}}{\prod_{r=1}^{p} l_{2p,r}(l_{2p,2}-1)}.$$

II. Finite Representations of the Group of Non-singular n-th Order Matrices

Here we shall give explicit formulae specifying the infinitesimal operators for all irreducible finite representations of the group of all non-singular linear transformations of a real n-dimensional space. We shall denote this group by A_n. As usual we begin with the construction of the one-parameter subgroups of the group A_n and then find its infinitesimal operators.

We consider the subgroup consisting of matrices of the form

$$a_{ik}(t) = e + e_{ik}t \qquad (i \neq k),$$

where e is the unit matrix, e_{ik} is the matrix with unity at the intersection of the i-th row and the k-th column and with zeros in the remaining places and t is a parameter. It is easily verified that the matrices $a_{ik}(t)$ constitute a subgroup:

$$a_{ik}(t_1)\cdot a_{ik}(t_2) = a_{ik}(t_1+t_2).$$

It is clear that in all there exist $n(n-1)$ such subgroups in the group A_n.

Moreover there are also n subgroups of diagonal matrices

$$a_{ii}(t)=\begin{Vmatrix} 1 & 0 & . & . & . & . & . & . \\ 0 & 1 & . & . & . & . & . & . \\ . & . & . & . & . & . & . & . \\ . & . & . & e^t & 0 & . & . & . \\ . & . & . & 0 & 1 & . & . & . \\ . & . & . & . & . & . & . & . \\ . & . & . & . & . & . & 1 & 0 \\ . & . & . & . & . & . & 0 & 1 \end{Vmatrix}.$$

(e^t occupies the ith place on the principal diagonal).

It may be shown that any element of the group A_n is a product of elements of the one-parameter subgroups $a_{ik}(t)$ $(i,\ k=1,\ 2,\ \ldots,\ n)$.

Thus in order to specify a representation of the group A_n it suffices to specify a representation of the n^2 one-parameter subgroups, or, what comes to the same thing, to specify for each representation the infinitesimal operators I_{ik} corresponding to the subgroups $a_{ik}(t)$.

We write down the commutation relations between the operators I_{ik}. Clearly they coincide with the commutation relations for the infinitesimal operators of the identity representation of the group A_n: $a\rightarrow a$. The matrices e_{ik} serve as the infinitesimal operators of the latter and their commutators have the form

$$\left.\begin{aligned} &[e_{ik}e_{kl}]=e_{il}\quad (i\neq l),\quad [e_{ik}e_{ki}]=e_{ii}-e_{kk},\\ &[e_{i_1k_1}e_{i_2k_2}]=0,\quad \text{if}\quad k_1\neq i_2\ \text{and}\ i_1\neq k_2. \end{aligned}\right\}\tag{1}$$

Analogous relations are also satisfied for the operators I_{ik}.

We now give the formulae specifying the operators I_{ik} for all finite irreducible representations of the group A_n.

We start with particular cases

1. $n=2$. An irreducible finite representation of the group of non-degenerate second-order matrices is specified by two integers m_1 and m_2 $(m_1\geqslant m_2)$. In the space R, where such a representation acts, it is possible to select a basis from the eigenvectors of the operators I_{11} and I_{22}. These vectors may be numbered by the index q which assumes once only every integral value in the range $m_1\geqslant q\geqslant m_2$.

The operators I_{11}, I_{22}, I_{12}, I_{21} act on the basis ξ_q as follows

$$\begin{aligned} &I_{11}\xi_q=q\xi_q,\qquad I_{22}\xi_q=(m_1+m_2-q)\,\xi_q,\\ &I_{12}\xi_q=\sqrt{(m_1-q)(q-m_2+1)}\,\xi_{q+1},\\ &I_{21}\xi_q=\sqrt{(q-m_2)(m_1-q+1)}\,\xi_{q-1}. \end{aligned}$$

2. $n=3$. Each irreducible representation of this group is specified by three integers $m_1 \geqslant m_2 \geqslant m_3$.

We now consider all possible triplets of integers

$$\begin{pmatrix} p_1 & & p_2 \\ & q & \end{pmatrix},$$

which satisfy the conditions

$$m_1 \geqslant p_1 \geqslant m_2 \geqslant p_2 \geqslant m_3, \quad p_1 \geqslant q \geqslant p_2.$$

It happens that it is possible to select a basis for the space where the irreducible representation of the group A_3 acts, from among the eigenvectors of I_{11}, I_{22}, I_{33} and to label each vector of this basis by one of the triplets $\begin{pmatrix} p_1 & & p_2 \\ & q & \end{pmatrix}$. In the following we shall denote these vectors by $\xi\begin{pmatrix} p_1 & & p_2 \\ & q & \end{pmatrix}$.

In terms of the basis $\left\{\xi\begin{pmatrix} p_1 & & p_2 \\ & q & \end{pmatrix}\right\}$, the operators I_{11}, I_{22}, I_{12}, I_{21}, I_{23}, I_{32}, I_{33} are given by

$$I_{11}\xi\begin{pmatrix} p_1 & & p_2 \\ & q & \end{pmatrix} = q\xi\begin{pmatrix} p_1 & & p_2 \\ & q & \end{pmatrix}, \quad I_{22}\xi\begin{pmatrix} p_1 & & p_2 \\ & q & \end{pmatrix} = (p_1+p_2-q)\,\xi\begin{pmatrix} p_1 & & p_2 \\ & q & \end{pmatrix},$$

$$I_{12}\xi\begin{pmatrix} p_1 & & p_2 \\ & q & \end{pmatrix} = \sqrt{(p_1-q)(q-p_2+1)}\,\xi\begin{pmatrix} p_1 & & p_2 \\ & q+1 & \end{pmatrix},$$

$$I_{21}\xi\begin{pmatrix} p_1 & & p_2 \\ & q & \end{pmatrix} = \sqrt{(p-q+1)(q-p_2)}\,\xi\begin{pmatrix} p_1 & & p_2 \\ & q-1 & \end{pmatrix},$$

$$I_{23}\xi\begin{pmatrix} p_1 & & p_2 \\ & q & \end{pmatrix} = \sqrt{\frac{(m_1-p_1)(m_2-p_1-1)(m_3-p_1-2)(p_1-q+1)}{(p_1-p_2+2)(p_1-p_2+1)}}\,\xi\begin{pmatrix} p_1+1 & & p_2 \\ & q & \end{pmatrix} +$$
$$+\sqrt{\frac{(m_1-p_2+1)(m_2-p_2)(m_3-p_2-1)(p_2-q)}{(p_1-p_2+1)(p_1-p_2)}}\,\xi\begin{pmatrix} p_1 & & p_2+1 \\ & q & \end{pmatrix},$$

$$I_{32}\xi\begin{pmatrix} p_1 & & p_2 \\ & q & \end{pmatrix} = \sqrt{\frac{(m_1-p_1+1)(m_2-p_1)(m_3-p_1-1)(p_1-q)}{(p_1-p_2+1)(p_1-p_2)}}\,\xi\begin{pmatrix} p_1-1 & & p_2 \\ & q & \end{pmatrix} +$$
$$+\sqrt{\frac{(m_1-p_2+2)(m_2-p_2+1)(m_3-p_2)(p_2-q)}{(p_1-p_2+2)(p_1-p_2+1)}}\,\xi\begin{pmatrix} p_1 & & p_2-1 \\ & q & \end{pmatrix},$$

$$I_{33}\xi\begin{pmatrix} p_1 & & p_2 \\ & q & \end{pmatrix} = (m_1+m_2+m_3-p_1-p_2)\,\xi\begin{pmatrix} p_1 & & p_2 \\ & q & \end{pmatrix}.$$

3. We now pass on to the general case.

Every finite irreducible representation of the group A_n is specified by n integers $m_1 \geqslant m_2 \geqslant \dots \geqslant m_{n-1} \geqslant m_n$.

We consider all possible arrays of integers, of the form

$$\alpha = \begin{bmatrix} m_{1,\,n-1} & & m_{2,\,n-1} & \ldots\ldots\ldots\ldots\ldots & & m_{n-1,\,n-1} \\ & m_{1,\,n-2} & & m_{2,\,n-2}\ \ldots\ldots\ldots & m_{n-2,\,n-2} & \\ & \ldots\ldots\ldots\ldots\ldots\ldots\ldots\ldots\ldots\ldots & & & & \\ & \ldots\ldots\ldots\ldots\ldots\ldots\ldots\ldots\ldots\ldots & & & & \\ & m_{13} & & m_{23} & m_{33} & \\ & & m_{12} & & m_{22} & \\ & & & m_{11} & & \end{bmatrix},$$

where the numbers m_{pq} satisfy the conditions

$$m_{p,\,q+1} \geqslant m_{pq} \geqslant m_{p+1,\,q+1}.$$

The numbers of the first row vary within the following limits

$$m_1 \geqslant m_{1,\,n-1} \geqslant m_2 \geqslant m_{2,\,n-1} \geqslant \ldots \geqslant m_{n-1,\,n-1} \geqslant m_n$$

(where $m_1, m_2, m_3, \ldots, m_n$ are the numbers specifying the representation) It so happens that to each array α it is possible to assign a vector $\xi(\alpha)$ - an eigenvector for the operators I_{ii} $(i = 1, \ldots, n)$ - of the space R where our irreducible representation acts, so that the vectors $\xi(\alpha)$ constitute a basis of R. We now write down the form of the operators I_{ik} in the basis $\{\xi(\alpha)\}$. It suffices to specify the operators $I_{k-1,\,k}$, I_{kk} and $I_{k,\,k-1}$ since the others are obtained by the commutation of these operators according to formulae (1).

We denote by $\overset{+}{\alpha}{}^i_{k-1,\,k}$ the array which is obtained from the array α by changing $m_{i,\,k-1}$ to $m_{i,\,k-1}+1$, and by $\overset{-}{\alpha}{}^i_{k-1,\,k}$ the array obtained from α by changing $m_{i,\,k-1}$ to $m_{i,\,k-1}-1$. In this notation the operators $I_{k-1,\,k}$, I_{kk}, $I_{k,\,k-1}$ are given by

$$\left.\begin{aligned} I_{k-1,\,k}\xi(\alpha) &= \sum_{i=1}^{n} a^i_{k-1,\,k}\xi\left(\overset{+}{\alpha}{}^i_{k-1,\,k}\right), \\ I_{kk}\xi(\alpha) &= \left(\sum_{i=1}^{k} m_{ik} - \sum_{i=1}^{k-1} m_{i,\,k-1}\right)\xi(\alpha), \\ I_{k,\,k-1}\xi(\alpha) &= \sum_{i=1}^{n} b^i_{k,\,k-1}\xi\left(\overset{-}{\alpha}{}^i_{k,\,k-1}\right), \end{aligned}\right\} \qquad (2)$$

where the numbers $a^i_{k-1,\,k}$ and $b^i_{k,\,k-1}$ are evaluated from the formulae

$$a^j_{k-1,\,k} = \left[(-1)^{k-1}\frac{\prod\limits_{i=1}^{k}(l_{i,\,k}-l_{j,\,k-1})\prod\limits_{i=1}^{k-2}(l_{i,\,k-2}-l_{j,\,k-1}-1)}{\prod\limits_{i\neq j}(l_{i,\,k-1}-l_{j,\,k-1})(l_{i,\,k-1}-l_{j,\,k-1}-1)}\right]^{1/2},$$

$$b^{j}_{k,\,k-1} = \left[(-1)^{k-1} \frac{\prod_{i=1}^{k} (l_{i,\,k} - l_{j,\,k-1} + 1) \prod_{i=1}^{k-2} (l_{i,\,k-2} - l_{j,\,k-1})}{\prod_{i \neq j} (l_{i,\,k-1} - l_{j,\,k-1} + 1)(l_{i,\,k-1} - l_{j,\,k-1})} \right]^{1/2} \qquad (3)$$

The representation is completely defined by formulae (2) and (3).

Nonetheless we shall write down the formulae for all the operators I_{pq} $(p<q)$.

$$I_{pq}\xi(\alpha) = \sum a_{pq}^{i_p \dots i_{q-1}} \xi\left(\alpha_{pq}^{i_p \dots i_{q-1}}\right).$$

Here $\alpha^{i_p \dots i_{q-1}}$ means the array obtained from the array α by increasing each of the indices $m_{i_p p}, \dots, m_{i_{q-1} q-1}$ by 1. The numbers $a_{pq}^{i_p \dots i_{q-1}}$ are defined by formula

$$a_{pq}^{i_p \dots i_{q-1}} = \pm \frac{\prod_{k=p}^{q-1} a^{i_k}_{k,\,k+1}}{\left[\prod_{k=p}^{q-2} (l_{i_k k} - l_{i_{k+1},\,k+1})(l_{i_k k} - l_{i_{k+1},\,k+1})\right]^{1/2}};$$

the sign is determined by the number of inversions in the sequence. $i_p \dots i_{q-1}$. The operators I_{pq} for $p>q$ are given by analogous formulae.

III. An Observation on the Duality between Clebsch-Gordan Coefficients and Jacobi Polynomials

It so happens that there is a surprising analogy between the Clebsch-Gordan coefficients $B^{lm}_{l_1 m_1 l_2 m_2}$ of the group of rotations and the Jacobi polynomials

$$P^{\alpha\beta}_{s}(\mu) = \frac{(-1)^s}{2^s s!}(1-\mu)^{-\alpha}(1+\mu)^{-\beta}\frac{d^s}{d\mu^s}\left[(1-\mu)^{s+\alpha}(1+\mu)^{s+\beta}\right]$$

First of all we introduce some definitions.

1. Let $f(k)$ be a function of an integral argument k. We call the product

$$\{f(k)\}^{[s]}_{k} = \underbrace{f(k)f(k-1)f(k-2)\dots f(k-s+1)}_{s}.$$

the s-th difference power of this function (with respect to the argument k).

Thus for example,

$$\{k\}^{[s]}_{k} = k(k-1)\dots(k-s+1) = \frac{k!}{s!}.$$

2. Suppose, as before, that $f(k)$ is a function of an integral argument.

The first difference $\frac{\Delta}{\Delta k} f(k)$ is defined by

$$\frac{\Delta}{\Delta k} f(k) = f(k) - f(k-1).$$

The second difference is defined similarly by

$$\frac{\Delta^2}{(\Delta k)^2} f(k) = [f(k) - f(k-1)] - [f(k-1) - f(k-2)] =$$
$$= f(k) - 2f(k-1) + f(k-2).$$

It is easily verified that the n-th difference is evaluated from the formula

$$\frac{\Delta^n}{(\Delta k)^n} f(k) = \sum_{s=0}^{n} (-1)^s C_n^s f(k-s).$$

3. We now turn to the Jacobi polynomials. We consider $P_s^{\alpha\beta}\left(\frac{\mu}{k}\right)$:

$$P_s^{\alpha\beta}\left(\frac{\mu}{k}\right) = (-1)^s \left(1 - \frac{\mu}{k}\right)^{-\alpha} \left(1 + \frac{\mu}{k}\right)^{-\beta} \frac{d^s}{d\left(\frac{\mu}{k}\right)^s} \left[\left(1 - \frac{\mu}{k}\right)^{s+\alpha} \left(1 + \frac{\mu}{k}\right)^{s+\beta}\right] =$$
$$= \frac{(-1)^s}{2^s s!} \frac{1}{k^s (k-\mu)^\alpha (k+\mu)^\beta} \frac{d^s}{d\mu^s} \left[(k-\mu)^{s+\alpha} (k+\mu)^{s+\beta}\right].$$

We consider the polynomial

$$T_s^{\alpha\beta}(\mu, k) = 2^s s! k^s (k-\mu)^\alpha (k+\mu)^\beta P_s^{\alpha\beta}\left(\frac{\mu}{k}\right) =$$
$$= (-1)^s \frac{d^s}{d\mu^s} \left[(k-\mu)^{s+\alpha} (k+\mu)^{s+\beta}\right].$$

Let μ and k be integers. In the expression for $T_s^{\alpha\beta}(\mu, k)$ we replace the ordinary power by the corresponding difference power and the differential by a difference, i.e. we take:

$$\tilde{T}_s^{\alpha\beta}(\mu, k) = (-1)^s \frac{\Delta^s}{(\Delta\mu)^s} \left[(k-\mu)_\mu^{[s+\alpha]} (k+\mu)_\mu^{[s+\beta]}\right].$$

We shall call the function $\tilde{T}_s^{\alpha\beta}(\mu, k)$ <u>the difference analogue of the polynomial</u> $T_s^{\alpha\beta}(\mu, k)$.

It turns out that the following remarkable theorem holds: <u>the Clebsch-Gordan coefficients are expressed in terms of the function</u> $\tilde{T}_s^{\alpha\beta}(\mu, k)$ <u>by the formula</u>

$$B_{l_1 m_1 l_2 m_2}^{lm} = \delta_{m, m_1+m_2} \sqrt{\frac{(l+l_1-l_2)!\,(l-l_1+l_2)!\,(l_1+l_2-l)!\,(2l+1)}{(l+l_1+l_2+1)!}} \times$$
$$\times \sqrt{\frac{(l+m)!\,(l-m)!}{(l_1-m_1)!\,(l_1+m_1)!\,(l_2-m_2)!\,(l_2+m_2)!}}\, \tilde{T}_s^{\alpha\beta}(\mu, k),$$

where

$$k=\frac{l_1+l_2+l}{2},\qquad \mu=\frac{l+l_2-l_1}{2}+m_1,$$

$$s=l-l_1+l_2,\qquad \alpha=(l_1-m_1)-(l+m),\qquad \beta=(l_1+m_1)-(l+m).$$

We shall not derive this formula. We merely note that it is obtained by direct expansion if the following expression for the Clebsch-Gordan coefficients is used *:

$$B^{lm}_{l_1m_1l_2m_2}=$$
$$=\delta_{m,\,m_1+m_2}\sqrt{\frac{(l+l_1-l_2)!\,(l-l_1+l_2)!\,(l_1+l_2-l)!\,(2l+1)}{(l+l_1+l_2+1)!}}\times$$
$$\times\sqrt{\frac{(l+m)!\,(l-m)!}{(l_1-m_1)!\,(l_1+m_1)!\,(l_2-m_2)!\,(l_2+m_2)!}}\times$$
$$\times\sum_k\frac{(-1)^{k+l_2+m_2}\,(l+l_2+m_1-k)!\,(l_1-m_1+k)!}{(l-l_1+l_2-k)!\,(l+m-k)!\,k!\,(l_1-l_2-m+k)!}$$

(k assumes all those values for which all the brackets are non-negative). We would remark that the analogy between the Clebsch-Gordan coefficients and the Jacobi polynomials is far reaching: several relations between Jacobi polynomials with distinct indices carry over into analogous relations between the Glebsch-Gordan coefficients, after replacing the power and differentiation symbols by the difference power and difference respectively.

* See, e.g., van der Waerden - "Die Gruppentheoretische Methode in der Quantenmechanik, Springer, Berlin 1932.

BIBLIOGRAPHY

I. GENERAL REPRESENTATION THEORY OF THE ROTATION AND LORENTZ GROUPS

1. H.WEYL. The classical groups, their invariants and representations. Princeton (1946).
2. E.CARTAN. Leçons sur la théorie des spineurs. I. Les spineurs de l'espace à trois dimensions, (Actualités Sci. industr., 643) Paris (1938). II. Les spineurs de l'espace à $m > 3$ dimensions. Les spineurs en géométric Riemannienne. (Actualités sci. industr., 701) Paris (1938).
3. F.D.MURNAGHAN. The theory of group representations, Johns Hopkins Press. Baltimore (1949).
4. M.A.NAIMARK. Linear representations of the Lorentz group, Uspekhi Matem. Nauk, 9:4 (62), 19-93 (1954). English translation A.M.S., 379-458 (1957).
5. L.S.PONTRYAGIN. Continuous groups (Nepreryvnyye gruppy), Gostekhizdat Moscow (1954).
6. Yu.B.RUMER. Spinor analysis (Spinornyi analiz) ONTI, M-L, (1936).

II. APPLICATION OF THE THEORY OF REPRESENTATIONS OF THE ROTATION GROUP

1. V.B.BERESTETSKII. Zh. eks. i. teoret. fiz., No.1, 12-18 (1947).
2. V.B.BERESTETSKII, A.Z.DOLGINOV and K.A.TER-MARTIROSYAN. Zh. eks. i.teoret. fiz, No.6, 527-537 (1950).
3. E.WIGNER. Gruppentheorie und ihre Anwendung auf die Quantenmechanik der Atomspektren, Braunschweig (1931). English translation by Academic Press, New York and London (1959).
4. E.V.CONDON and G.H.SHORTLEY. The theory of atomic spectra. Camb. Univ. Press. (1935)
5. I.B.LEVINSON. The sum of the products of the Wigner coefficients and their graphical representations. Trudy. fiz. tekh. inst. AN Lit. SSR, 2, 17-29 (1956).
6. I.B.LEVINSON. The dependence of the sum of the products of the Wigner coefficients upon the quantum numbers. Trudy. fiz. tekh. inst. AN Lit. SSR, 2, 31-43 (1956).
7. I.B.LEVINSON. Some formulae for the transformation and summation of (j, m)- expressions. Trudy Akad. Nauk Lit. SSR, Series B 4, 3-15, (1957).

8. G.Ya.LYUBARSKII. Group theory and its application to physics (Teoriya grupp i yeye primeneniye v fizike). Gostekhizdat (1957).
9. G.I.PETRASHEN'. Dynamic problems in the case of an isotropic sphere (Dinamicheskiye zadachi v sluchaye izotropnoi sfery). Uch zap. Leningrad gos. univ., No.114, 17 (1949).
10. M.ROUSE. Multipole fields.
11. V.A.FOCK. Elementary quantum mechanics (Nachala kvantovoi mekhaniki). Kubuch (1932).

III. RELATIVISTIC-INVARIANT EQUATIONS

1. A.I.AKHIYEZER and V.B.BERESTETSKII. Quantum electrodynamics. Gostekhizdat, M. (1935).
2. N.J.BHABHA. Relativistic equations for particles with arbitrary spin. Current Science, 14, 89-90 (1945).
3. I.M.GEL'FAND and A.M.YAGLOM. Zh. eks. i. teoret. fiz., 18, No. 8, 703-733 (1948).
4. I.M.GEL'FAND and A.M.YAGLOM. Zh. eks. i. teoret. fiz., 18, No.12, 1096-1104 (1948).
5. I.M.GEL'FAND and A.M.YAGLOM. Zh. eks. i. teoret. fiz., 18, No.12, 1105-1111 (1948).
6. W.PAULI. The relativistic field theories of elementary particles. Rev. Mod. Physics, 13, 203 (1941).

IV. BIBLIOGRAPHY TO THE SUPPLEMENTS

1. I.M.GEL'FAND and M.L.TSEITLIN. Dokl. Akad. Nauk SSSR, 71, No.5, 825-828 (1950).
2. I.M.GEL'FAND and M.L.TSEITLIN. Dokl. Akad. Nauk SSSR, 71, No.6 1017-1020 (1950).